Molecular Spectroscopy

Quantum to Spectrum

Amita Dua
M.Sc., Ph.D.
Department of Chemistry
Dyal Singh College
University of Delhi

Prateek Tyagi
M.Sc., M.Phil., Ph.D.
Department of Chemistry
Zakir Husain Delhi College
University of Delhi

ATLANTIC
PUBLISHERS & DISTRIBUTORS (P) LTD

ATLANTIC
PUBLISHERS & DISTRIBUTORS (P) LTD
7/22, Ansari Road, Darya Ganj, New Delhi
Tel.: +91-11-4077 5252, 2327 3880
E-mail: orders@atlanticbooks.com
Web: www.atlanticbooks.com

Reprint 2025, 2026

Published by Atlantic Publishers & Distributors (P) Ltd.

Printed & bound in India by Atlantic Print Services

Dedicated to my daughter, Sheen

Amita Dua

Dedicated to my mother, Manoj Devi

Prateek Tyagi

Preface

This book has evolved from our constant efforts, real time teaching of Molecular Spectroscopy, and addressing the chemical problems. All major topics of Molecular Spectroscopy have been comprehensively discussed in this book. This work is a very fine blend of quantum and spectroscopy. It has been explicitly designed to be user friendly and easy to read. A series of solved numericals have been inserted after each topic to have a better understanding and absorption of the subject.

This book is need of every science student from undergraduate to postgraduate. It will also be used as a help book by the students appearing in competitive examinations. It will surely prove to be an asset for students and academicians. Constructive suggestions for any further improvement in this work are always welcome.

Amita Dua
Prateek Tyagi

Acknowledgements

We express our heartfelt gratitude towards the guiding light of his Holiness Swamiji Shri Raman.

I, Amita Dua, thank my family for their continuous support and patience that helped me complete this book. It would not have been possible to finalise the book without the cooperation of my daughter, Sheen, who supported me in every manner during my writing days.

I am really thankful to my colleagues, especially Prof. A.K. Bhagi for his valuable suggestions while compiling the book. I would also like to thank our principal, Prof. V.K. Paliwal for providing an academic environment in the college.

I, Prateek Tyagi, express my deepest gratitude to my family for their constant support throughout the writing of this book. I thank my wife, Priyanka, who has stood by me. I would also like to thank Prof. Sangeeta Pandita and Dr. Sulekh Chandra for their vital guidance and valuable suggestions in compiling the book.

For writing this book information has been collected from different sources such as numerous books, articles, and online literature. Thanks to all the contributors.

Last but not the least, our special thanks to Mr. Nitin Sharma and the publishing team of Atlantic Publishers and Distributors (P) Ltd. for their fantastic support.

Amita Dua
Prateek Tyagi

Brief Contents

Detailed Contents

1

Classical and Quantum Mechanics

1.1 Dalton's Atomic Theory

John Dalton in 1808 put forward a theory known as Dalton's atomic theory. The main points of this theory are:

1. Matter is made up of extremely small indivisible particles called atoms.
2. Atoms of the same element are identical in all respects, i.e. size, shape, and mass.
3. Atoms of different elements have different masses, sizes, and chemical properties.
4. Atoms of the same or different elements combine together to form molecules.
5. When atoms combine with one another to form molecules, they combine in simple whole number ratios.
6. Atoms of two elements may combine in different ratios to form more than one compound. Like C combine with O to form CO and CO_2.
7. Atom is the smallest particle that takes part in a chemical reaction.
8. An atom can neither be created nor be destroyed.

Dalton's atomic theory was the first achievement towards the smallest structure of matter called 'atom'. This theory was able to explain the **law of conservation of mass, law of constant composition** and **law of multiple proportion**. However, it failed to explain the results of many other experiments. This leads to the discovery of many different types of sub-atomic particles in the beginning of the twentieth century.

Table 1.1: Discoveries of various sub-atomic particles

Discoveries	Discoverers	Results
Electron/cathode rays	Faraday (1879)	
Charge to mass ratio of electron	J.J. Thomson (1897)	$\frac{e}{m} = 1.76 \times 10^{11}$ C/kg
Charge on electron (oil drop experiment)	R.A. Millikan (1917)	$e = -1.6 \times 10^{-19}$ C $e = 4.8 \times 10^{-10}$ esu

Discoveries	Discoverers	Results
Mass of electron	R.A. Millikan (1917)	$m_e = 9.11 \times 10^{-31}$ kg [It is 1/1837th of H-atom]
Positive rays/Proton rays/ Canal rays/Anode rays (Discharge tube experiment)	Goldstein (1886)	
Charge on proton	Ernest Rutherford (1917)	$+1.6 \times 10^{-19}$ C
Mass of proton	Ernest Rutherford (1917)	$m_p = 1.67 \times 10^{-27}$ kg [same as that of H-atom]
Neutron	Chadwick (1932)	
Charge on neutron	Ernest Rutherford (1920) and confirmed by Chadwick (1932)	Zero
Mass of neutron	Ernest Rutherford (1920) and confirmed by Chadwick (1932)	$m_n = 1.67 \times 10^{-27}$ kg
Radioactivity [phenomena of spontaneous emission of radiations by certain elements like uranium, etc.]	Henri Bequeral and Marie Curie (1896)	After the discovery of electron and proton, it was well-established that atom is divisible and is made of charged particles. This was further confirmed by the phenomena of radioactivity.

1.2 What are Classical Mechanics and Quantum Mechanics?

Classical Mechanics is the Newtonian mechanics, whose foundation was laid by the Newton's laws of motion. The term "Mechanics" means study of motion and Issac Newton gave the laws of motion.

Issac Newton in 1687 gave three laws of motion:

1. **First Law:** An object either remains at rest or continues to move at constant velocity until or unless acted upon by any external force. This law is called as "**Law of Inertia**".
2. **Second Law:** Rate of change of momentum is directly proportional to the force. Mathematically,

$$F \propto \frac{dp}{dt} \qquad (p = \text{momentum} = mv)$$

$$F \propto m \, . \frac{dv}{dt}$$

$$F = m \, . \, a \qquad (a = \text{acceleration})$$

3. **Third Law:** Every action has an equal (in magnitude) and opposite (in direction) reaction.

These are called as **Newton's laws of motion.** Newton's laws explain all the major aspects of a moving body and its motion. They describe the relationship between an object and the forces acting upon it and its motion in response to the said forces.

However, later it was observed that Newton's laws were well-followed by the macroscopic objects (which are visible by naked eye, e.g. falling stone, moving car, recoil of gun etc.) but they failed to explain the motion of microscopic objects (which are not visible by naked eye and we need high resolution microscopes to see them, e.g. electron, atom, molecule, etc.).

This was observed because macroscopic objects essentially possess particle nature, that is why Netwon's laws provide extremely accurate results as long as the area of study is restricted to large objects. But, when the moving object becomes sufficiently small, i.e. microscopic objects then they have essentially a wave nature due to which Newton's laws fail to explain the mechanics of microscopic objects.

The failure of classical mechanics to explain the laws of motion for microscopic objects gave birth to "**Quantum Mechanics**".

Quantum mechanics gives the laws of motion which microscopic objects obey, and they are different from the Newton's laws of motion. Quantum mechanics was introduced in 1900 which was brought to account for the dual nature of microscopic objects, i.e. particle nature and wave nature.

Particle nature	Wave nature
Particle is localized in space i.e., a particle occupies a well-defined position in space. e.g. a cricket ball.	Wave is delocalized in space i.e., a wave is spread out in space. e.g. sound waves.
Two particles cannot occupy the same space simultaneously and do not interfere.	Two or more waves can co-exist in the same region of space and hence waves interfere.

- Actually macroscopic and microscopic objects both have particle nature as well as wave nature. Now, macroscopic objects have essentially particle nature, since the wave attached to a macroscopic object has a small wavelength which is not significant. The classical mechanics treats the moving objects as particle, therefore **macroscopic objects are studied under classical mechanics.** Whereas microscopic objects have essentially wave nature because the size of microscopic particles is too small, thus the wavelength of the wave attached to it is large enough and thus significant. The quantum mechanics treats the moving object as wave, therefore **microscopic objects are studied under quantum mechanics.**
- Earlier (before 1900) when no such quantum mechanics was introduced, all the moving objects (macroscopic or microscopic) were considered to have particle nature and studied under classical mechanics. Thus, the various models of atom, such as Thomson model of atom, Rutherford nuclear model of atom and Bohr model of hydrogen atom are studied under classical mechanics. Since in all these models, the sub-atomic particles (like electron, neutron, proton, atom) are considered to have particle nature, thus all these above models are studied under classical mechanics.

PROBLEM BASED ON CLASSICAL MECHANICS

Q. 1. Write a short note on inadequacy of classical mechanics.

1.3 Reasons for the Failure of Classical Model of Atom or Bohr Model of Atom

The results of Bohr model can be used to explain the motion of macroscopic objects or big objects (like orbiting planets and sun, falling stone, etc.). These macroscopic objects have essentially a particle nature and Newtonian laws of motion can be applied on them. But it fails when applied to explain the motion of microscopic objects or small objects (like electrons, atoms, molecules, etc.). This is because:

1. According to Bohr model, an orbit is a well-defined path and this path can completely be defined only if both the position and velocity of the electron are known exactly at the same time. But this is not possible according to the Heisenberg's uncertainty principle. Thus, **Bohr model contradicts the Heisenberg's uncertainty principle.**
2. According to Bohr model, an electron is regarded as a charged particle moving in well-defined circular orbits concentric about the nucleus. The wave character of the electron is not considered. Thus, **Bohr model ignores the de-Broglie dual nature of matter.**

1.4 Developments Leading to Quantum Mechanical Model of Atom

In view of the shortcomings of Bohr model of atom, efforts were made to develop a new model of atom which could overcome the limitations of Bohr model. The development of the new model was mainly based on the following two concepts that has been put forward:

1. de-Broglie concept of dual nature of matter.
2. Heisenberg uncertainty principle.

1.5 de-Broglie's Dual Nature of Matter

In 1905, Einstein suggested that radiation has dual nature, i.e. wave nature as well as particle nature. Later in 1924, a French physicist, Louis-Victor de-Broglie gave theory of wave particle duality of matter. He suggested that just as radiation possesses both wave and particle nature, all macroscopic as well as microscopic material objects (like electron, proton, atom, molecule, a piece of chalk, a piece of stone, etc.) also possess dual nature, i.e. wave and particle nature. The wave associated with a material particle is called **matter waves** or **de-Broglie waves.**

- **de-Broglie equation.** He calculated the wavelength of the wave associated with a material particle of mass (m) moving with a velocity (v). The de-Brogile relationship is

$$\lambda = \frac{h}{mv} = \frac{h}{p}$$

where, p = momentum of the particle
λ = de-Broglie wavelength
h = Planck's constant.

- **Derivation of de-Broglie relationship.** Consider a photon of light having energy E. If photon is assumed to have particle nature, its energy is given by Planck's quantum theory as

$$E = h\nu \qquad ...(a)$$

If photon is assumed to have wave nature, its energy is given by Einstein equation as

$$E = mc^2 \qquad ...(b)$$

where, h = Planck's constant

c = velocity of light (3×10^8 ms^{-1})

ν = frequency of the wave

m = mass of photon

Equating equations (a) and (b)

$$h\nu = mc^2$$

Now,

$$\nu = \frac{c}{\lambda}$$

$$\frac{hc}{\lambda} = mc^2$$

$$\lambda = \frac{h}{mc}$$

This equation is applicable to any material particle. This is done by replacing the mass of the photon by mass of the material particle and the velocity 'c' of the photon by velocity 'v' of the material particle. Thus, for any material particle like electron, proton, etc. we have

$$\lambda = \frac{h}{mv} \quad \text{or} \quad \lambda = \frac{h}{p} \quad \text{(de-Broglie equation)}$$

Thus, from de-Broglie equation, we conclude that, $\lambda \propto \frac{1}{p}$, i.e. de-Broglie wavelength of the wave associated with material particle is inversely proportional to its momentum. Therefore, de-Broglie equation relates the particle character with wave character of material particles.

	Electromagnetic waves	Matter waves/de-Broglie waves
1.	When electrically charged particles move under acceleration, alternating electric and magnetic fields are produced which are perpendicular to each other. These fields are produced and transmitted in the form of electromagnetic waves.	These waves are not associated with electric and magnetic fields.
2.	They can pass through vacuum.	They cannot pass through vacuum, i.e. they require medium for propagation.

	Electromagnetic waves	Matter waves/de-Broglie waves
3.	They travel with the speed of light.	They do not travel with the same speed as that of light.
4.	The wavelength of the electromagnetic wave is given by $\lambda = \frac{c}{\nu}$.	The wavelength of the de-Broglie wave is given by $\lambda = \frac{h}{mv}$.
5.	They are emitted from a particular source in the form of radiation.	They are not emitted from a source but attached to a moving particle.

- **Significance of de-Broglie equation:** According to de-Broglie, every moving object whether macroscopic or microscopic have a wave character. The wavelength associated with ordinary macroscopic objects (like running car, cricket ball, etc.) are so small (because of their large masses) that their wave properties cannot be detected, i.e. wavelength associated with large objects is not significant. Whereas the wavelength associated with microscopic objects (like electrons and other subatomic particles) is large enough (because of their small masses) and thus can be detected experimentally.

  Since in everyday life, we come across macroscopic objects, therefore de-Broglie relationship has no significance in everyday life.

- **Derivation of Bohr's postulate of angular momentum from de-Broghie equation:** de-Broglie applied his theory of wave particle duality to the Bohr model to explain why only certain orbits are allowed for the electron. He argued that only certain orbits allow the electron to satisfy both its particle and wave properties at the same time because only certain orbits have a circumference ($2\pi r$) that is an integral multiple of the wavelength ($n\lambda$) of the electron.

  In other words, according to de-Broglie concept, electron is not only a particle moving in a circular path but is instead a standing wave* (no energy radiating motion) extending around the nucleus in the circular path. In order that the wave may be completely in phase (Fig. 1.1), the circumference of the orbit must be equal to an integral multiple of wavelength (λ).

  i.e. $$2\pi r = n\lambda \quad \text{...(c)}$$

  where, r is the radius of orbit and, n is an integer

  But, $$\lambda = \frac{h}{mv} \text{ (de-Broglie equation)}$$

  Substituting the value of λ in equation (c)

  $$2\pi r = n\left(\frac{h}{mv}\right)$$

  $$mvr = \frac{nh}{2\pi}$$

  which is Bohr's postulate of angular momentum.

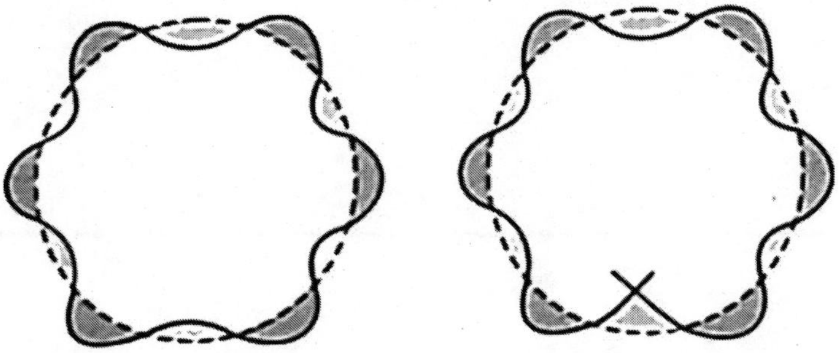

Fig. 1.1: (a) Wave in phase and (b) Wave out of phase

***Stationary or Standing Waves:** If a string is stretched between two fixed points, then a wave is allowed to travel along the string in one direction such that when the wave reaches the other end, it is reflected in the opposite direction with the same wavelength, speed and amplitude. The string is also divided into a number of vibrating segments, separated by points of zero amplitude. Such a vibrating motion is called a stationary or a standing wave. The points of zero amplitude (points a) are called nodes.

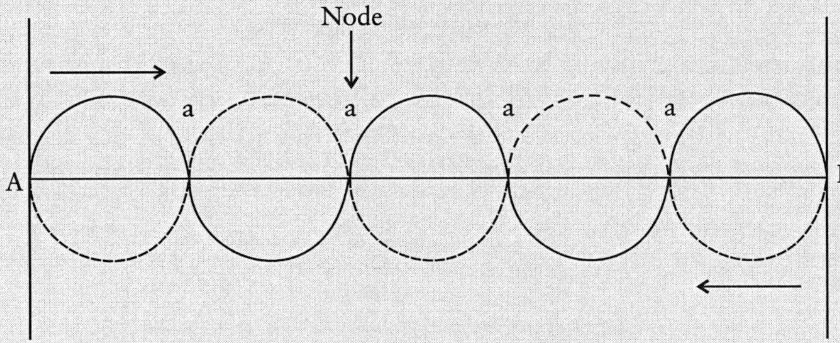

Fig. 1.2: Stationary or standing waves

- **Experimental verification of wave nature of electrons:** In 1927, **C.J. Davisson and Germer** gave experimental verification for the revolutionary postulate given by de-Broglie. It was further verified by **G.P. Thomson.**

1. **C.J. Davisson and Germer's Experiment:** In 1927, Davisson and Germer observed that when a beam of electrons (or neutron, or proton or hydrogen atoms etc.) is allowed to fall on the surface of nickel crystal and the reflected rays are received on a photographic plate, a diffraction pattern (consisting of a number of concentric rings) is obtained (Fig. 1.3). The diffraction rings formed by the electrons were similar to those formed by X-rays.

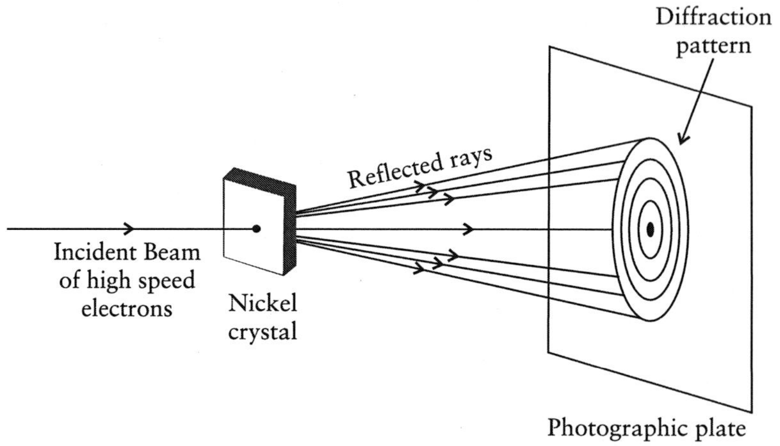

Fig. 1.3: Diffraction of electron beam by nickel crystal

Since, X-rays are electromagnetic waves i.e. they have wave character, therefore, electrons must also have wave character.

Further, the wavelength determined from the diffraction pattern is in good agreement with those calculated from de-Broglie equation.

2. **Thomson's Experiment:** In 1928, G.P. Thomson observed that when an electron beam is allowed to fall on a thin foil of gold in place of nickel crystal and reflected rays are received on a photographic plate, a diffraction pattern (consisting of a number of concentric rings) is obtained (Fig. 1.4). This again confirmed the wave nature of electrons.

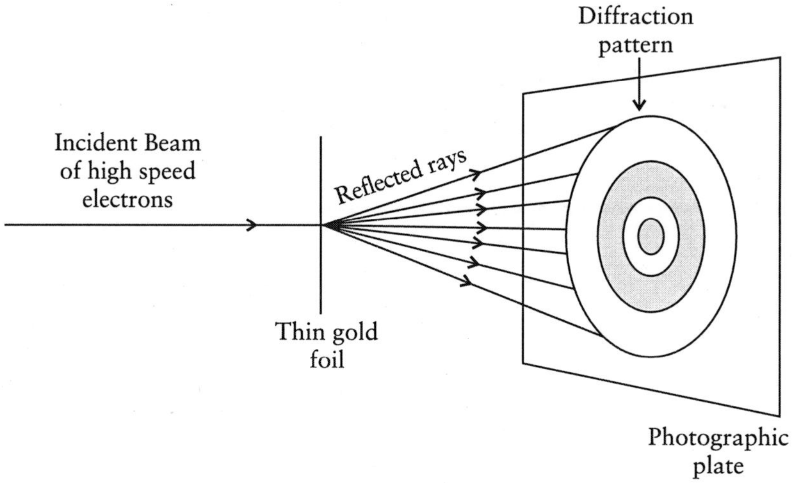

Fig. 1.4: Diffraction of electron beam by thin gold foil

- Thus, de-Broglie's prediction was confirmed experimentally when it was found that an electron beam undergoes diffraction, a phenomena characteristic of waves. This is used in making an electron microscope (used to measure the size of objects) which is based on the wave like behaviour of electrons just as an ordinary microscope utilises the wave nature of light.

- **Experimental verification of particle nature of electrons:** The particle nature of electron can be justified on the basis of following observations.

1. When an electron strikes a screen coated with zinc sulphide, a spot of light known as **scintillation** is produced. One electron produce only one scintillation. Since a scintillation is localised on the screen, thus the striking electron which produces scintillation must also be localised. This suggests that electron possess particle character because localised character is possessed by the particles. Hence, electron has the particle nature.
2. The particle nature of electron can also be shown by **J.J. Thomson's experiment** used for the determination of charge to mass ratio of electron and **Milliken oil drop experiment** used for the determination of charge on electron.

   Thus, we conclude that everything (light and matter) in nature have the properties of both particle as well as wave. In other words, dual character is applicable to everything.

PROBLEMS BASED ON DE-BROGLIE EQUATION

Q. 1. Calculate the wavelength of a body of mass 2 mg moving with a velocity of 20 ms^{-1} ($h = 6.626 \times 10^{-34}$ kgm^2s^{-1}).

Ans.

$$m = 2 \text{ mg} = 2 \times 10^{-3} \text{ g} = 2 \times 10^{-6} \text{ kg}$$

$$v = 20 \text{ ms}^{-1}$$

$$\lambda = \left(\frac{h}{mv}\right) = \frac{6.626 \times 10^{-34} \text{kg m}^2\text{s}^{-1}}{(2 \times 10^{-6})\text{kg} \times (20)\text{ms}^{-1}}$$

$$\lambda = 1.65 \times 10^{-29} \text{ m.}$$

Q. 2. Calculate the momentum of a moving particle which has de-Broglie wavelength of 300 pm.

Ans.

$$\lambda = 300 \text{ pm} = 300 \times 10^{-12} \text{ m.}$$

$$h = 6.626 \times 10^{-34} \text{ kg. m}^2\text{.s}^{-1}.$$

$$\lambda = \frac{h}{p}$$

$$p = \frac{h}{\lambda} = \frac{6.626 \times 10^{-34} \text{kg.m}^2\text{.s}^{-1}}{300 \times 10^{-12} \text{m}}$$

$$p = 2.2 \times 10^{-24} \text{ kg. ms}^{-1}.$$

Q. 3. The mass of an electron is 9.11×10^{-31} kg and its kinetic energy is 5×10^{-25} J. Calculate its de-Broglie wavelength.

Ans.

$$\text{K.E.} = \frac{1}{2}mv^2$$

$$5 \times 10^{-25} = \frac{1}{2} \times (9.11 \times 10^{-31}) \times v^2$$

$$v = \sqrt{\frac{2 \times (5 \times 10^{-25})}{9.11 \times 10^{-31}}}$$

$v = 1.048 \times 10^{3}\ ms^{-1}$

Now $$\lambda = \frac{h}{mv}$$

$$\lambda = \frac{6.626 \times 10^{-34} kg.m^{2}.s^{-1}}{(9.11 \times 10^{-31} kg) \times (1.048 \times 10^{3} ms^{-1})}$$

$\lambda = 6.94 \times 10^{7}$ m.

Q. 4. Calculate the mass of the photon with which a wave is associated of wavelength 2.5Å.

Ans. λ = 2.5 Å = 2.5×10^{-10} m

v = velocity of photon

= velocity of light = $3 \times 10^{8}\ ms^{-1}$

$$\lambda = \frac{h}{mv}$$

$$m = \frac{h}{\lambda v} = \frac{6.626 \times 10^{-34} kg.m^{2}.s^{-1}}{(2.5 \times 10^{-10} m) \times (3 \times 10^{8} ms^{-1})}$$

$m = 0.88 \times 10^{-32}$ kg.

Q. 5. An electron beam is accelerated by a potential difference of 10,000 eV. What is the wavelength of the wave associated with electron beam.

Ans. m = mass of electron = 9.11×10^{-31} kg

K.E. = 1000 eV

K.E. = $10{,}000 \times 1.6 \times 10^{-19}$ J

(Since, 1 eV = 1.6×10^{-19} J)

K.E. = 1.6×10^{-15} J

Now, $$K.E. = \frac{1}{2} mv^{2}$$

$1.6 \times 10^{-15}\ J = 1/2 \times (9.11 \times 10^{-31}\ kg) \times v^{2}$

$$v = \sqrt{\frac{2 \times (1.6 \times 10^{-15} J)}{(9.11 \times 10^{-31} kg)}} = 5.9 \times 10^{7} ms^{-1}$$

Now, $$\lambda = \frac{h}{mv} = \frac{6.626 \times 10^{-34} kg\, m^{2} s^{-1}}{(9.11 \times 10^{-31} kg) \times (5.9 \times 10^{7}\ ms^{-1})}$$

$\lambda = 0.123 \times 10^{-10}$ m

λ = 0.123 Å

Q. 6. What are matter waves? Explain with necessary theory, how Davisson and Germer's experiment established these waves?

Q. 7. Calculate the wavelength of matter waves associated with an electron which has been accelerated, from rest, through a potential difference by 1.25 eV.

Q. 8. Explain de-Broglie hypothesis. Derive an expression for de-Broglie wavelength.

Q. 9. Give an experimental verification of existence of matter waves on the basis of Davisson and Germer experiment.

1.6 Heisenberg's Uncertainty Principle

Werner Heisenberg stated uncertainty principle which is the consequence of dual nature of matter and radiation.

All moving objects that we see in daily life, i.e. macroscopic objects (like moving car, ball thrown in air, etc.) have definite paths or trajectories along which they move. The definite trajectories indicate that their position and velocity can be measured accurately at any moment of time. But, such accurate measurements of position and velocity cannot be done for sub-atomic particles, i.e. microscopic objects (like electron, proton, atom, etc.). There are always some uncertainties in measurement of position and velocity for small moving particles. Thus, for such sub-atomic particles, the existence of definite paths or trajectories is ruled out. Thus in microscopic world, it is found that however refined our instrument there is a fundamental limitation to the accuracy with which the position and velocity of microscopic particle can be known simultaneously.

In 1927, a German physicist, Werner Heisenberg gave a principle about the uncertainties in simultaneous measurements of position and momentum of subatomic particles, which is known as Heisenberg's uncertainty principle. This principle states that: **It is impossible to determine simultaneously the exact position and momentum (or velocity) of small moving particles with accuracy or certainty. If one quantity is known then the determination of other quantity will become impossible**

Mathematically,

$$\Delta x \cdot \Delta p_x \geq \frac{h}{4\pi}$$

$$\Delta x \cdot m\Delta v_x \geq \frac{h}{4\pi}$$

where,
Δx = uncertainty in position
Δp_x = uncertainty in momentum in x-direction
Δv_x = uncertainty in velocity in x-direction
m = mass of the particle
h = Planck's constant

Also, $$\Delta x \propto \frac{1}{\Delta v_x}$$

This implies that the position and velocity of a particle cannot be determined simultaneously with accuracy. If the position of the particle is known with high degree of accuracy (Δx is small), then velocity of particle will be uncertain (Δv_x is large) and vice versa.

The uncertainty principle also applies for the same axis of position and momentum. Like if Δx is along the y-axis, then Δp must also be along the y-axis.

Uncertainty in Time and Energy: Similar to uncertainty in position there is another principle of uncertainty which limits the accuracy in the measurement of time, i.e. if ΔE is the energy uncertainty in time Δt then we have similar expression.

$$\Delta E \times \Delta t \geq h/4\pi$$

- **Justification of Heisenberg Uncertainty Principle:** To understand this principle, let us first know how the position of an object is determined. A light phenomena is used. When a beam of visible light falls on an object, the photons of light gets scattered at the object and the reflected photons enter the retina of our eyes which produces the sensation of sight. If the object is of sufficient size, its position and velocity will not be changed by the energy of the photons of light falling on it. As such it will be possible to know both the position and the velocity of the object at the same time.

  But in case of microscopic objects (such as an electron), an entirely different situation exists due to very small size of the particle to be located. According to principle of optics, the accuracy with which the position of a particle can be located depends upon the wavelength (λ) of light used and the position cannot be located more accurately than $\pm \lambda$. Therefore, due to the very small size of the electron its position can be determined with sufficient accuracy only by choosing light of very small wavelength. But shorter wavelength of light means that its photons would possess higher energy. When such a high energy photon falls on the electron, part of the energy of the photon falls on the electron and part of the energy of the photon would be transferred to the electron. As a result, the electron would undergo considerable change in its velocity (therefore, momentum) (Fig. 1.5). Thus we find that if a photon of smaller wavelength is used there will be large inaccuracy in the determination of momentum of the electron. On the other hand, if a photon of larger wavelength is used, there will be large uncertainty in the location of its position.

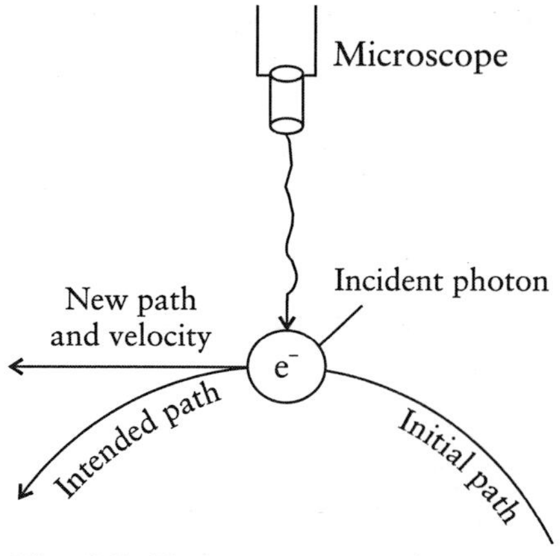

Fig. 1.5: The momentum of electron changes on collision with photon

Significance of Heisenberg Uncertainty Principle

1. Although Heisenberg's uncertainty principle holds good for all the objects but its effect is significant only for motion of microscopic objects and is negligible for that of macroscopic objects. Since in everyday life, we deal with macroscopic objects, thus Heisenberg uncertainty principle has no significance in everyday life.

The reason why the uncertainty principle is of no importance in our daily life is that Planck's constant 'h' is so small (6.625×10^{-34} joule-seconds) that the uncertainties in position and momentum of even quiet small (not microscopic objects) objects are far too small to be experimentally observed. For microscopic phenomena such as atomic processes, the displacements and momentum are such that the uncertainty relation is critically applicable.

2. It rules out the existence of definite path or trajectories of small microscopic particles. This is because position and velocity of a moving particle fix its trajectory. Since for microscopic object, it is impossible to determine simultaneously the position and velocity at any given instant with accuracy. Thus, it is not possible to talk of the trajectory of the small moving microscopic particles.

3. The principle describes the incompleteness of Bohr's atomic theory. This is the reason that Bohr's concept of fixed circular path with definite position and momentum of the electron (subatomic particle) have been replaced by stating that the electron has the **probability** of having a given position and momentum. This later on, forms the basis of quantum or wave mechanical model of atom.

PROBLEMS BASED ON HEISENBERG UNCERTAINTY PRINCIPLE

Q. 1. State and explain Heisenberg Uncertainty Principle.

Q. 2. According to Heisenberg's principle of uncertainty, why it is impossible to measure simultaneously the exact position and momentum of fast moving electron?

1.7 Necessity of Quantum Mechanics

In the 17th century, it was thought that motion of macroscopic and microscopic objects could be well-expressed by using the laws of classical mechanics. But towards the end of 19th century, experimental studies showed that classical mechanics failed when applied to microscopic objects. This was mainly because classical mechanics did not take into account the concept of de-Broglie's dual nature of matter and Heisenberg's uncertainty principle. Also, classical mechanics does not put any restrictions on the values of dynamical properties (like position, momentum, energy, etc.) calculated for microscopic object. According to classical mechanics, microscopic objects can have **any value** of dynamical properties (**exact, large** or **small** is permissible). Also **any number** of properties can be calculated **simultaneously** for a system.

However, experimental measurements for microscopic objects show that there exists a well-defined set of energy levels. Like, an electron moving around the nucleus of an atom have a discrete set of energy levels known as **quantised energy levels.** So to understand such behaviour of microscopic objects, it was necessary to postulate a new system of mechanics called **Quantum mechanics.** Quantum mechanics deals with the study of motion of matter at microscopic level. It is a new branch of science which takes into account the dual nature of matter.

According to classical mechanics, it is possible to write down equations which tells exact values of position and momentum of a system at any time. Thus, the state of a given system (containing a number of particles) can be completely explained by knowing the position and momentum of every particle in the system at any particular time. Whereas the state of a given system (containing a number of particles) in quantum mechanics is not given merely by its position and momentum but by something called "**wavefunction**". By knowing the state of a system, i.e. wavefunction does not enable to predict the results of measurements with certainty, but rather gives a set of probabilities for the possible outcomes of any measurement. **Thus, the word exactness in classical mechanics has been replaced by the word probability in quantum mechanics.**

When quantum mechanics is applied to macroscopic objects, the results are the same as those from the classical mechanics.

1.8 Schrodinger Wave Equation

The two important principles laid the foundation of Quantum mechanics, the de-Broglie dual nature of matter and Heisenberg uncertainty principle. Quantum mechanics, developed by Erwin Schrodinger in 1926, is based on the wave motion associated with the small moving objects, i.e. microscopic objects.

For the wave motion of the electron in the three dimensional space around the nucleus, Schrodinger put forward an equation, known after his name as **Schrodinger wave equation,** which is considered as the heart of quantum mechanics. This equation takes into account the dual nature of matter and is given as:

$$\frac{\partial^2\psi}{\partial x^2}+\frac{\partial^2\psi}{\partial y^2}+\frac{\partial^2\psi}{\partial z^2}+\frac{8\pi^2 m}{h^2}(E-V)\psi=0$$

where, ψ = amplitude of the electron wave

x, y, z = coordinates of the electron

E = total energy of the electron

V = potential energy of the electron

m = mass of the electron

h = Planck's constant

$\frac{\partial^2\psi}{\partial x^2}$ = second derivative of ψ w.r.t. x.

The solution of Schrodinger wave equation for an electron give the values of E and ψ. The values of E represent the possible values of energy which the electron in the atom can occupy. The corresponding values of ψ are called **wavefunctions.** The wavefunction corresponding to an energy state contains all the information about an electron which is present in that state.

Classical Mechanics. Deals with the study of behaviour of large objects under the action of forces.

Quantum Mechanics. Deals with the study of behaviour of small objects under the action of forces.

Quantum Chemistry. The quantum mechanics when applied to chemistry.

Wave Mechanics. Deals with the mathematical treatment of the behaviour of microscopic object (like electron).

Statistical Mechanics. Deals with the mathematical treatment of the behaviour of macroscopic object.

1.9 Derivation of Time Independent Schrodinger Wave Equation

Schrodinger equation was proposed by the Austrian physicist Erwin Schrodinger in 1926.

Schrodinger developed a mathematical wave equation to describe the wave motion of the electron in a atom. He considered that the nucleus is surrounded by a vibrating electron wave which may be mathematically compared with stationary or standing wave formed by a vibrating string fixed between two points as shown in Fig. 1.6. The electron in an atom is under constraint imposed by the attraction of the nucleus. Therefore, there must be a wave equation which describe the wave motion of electron analogous to that used to describe a "standing or stationary" wave system (wave appearing in stretched string) whose solution in one dimension is given by the sine wave equation (1.1)

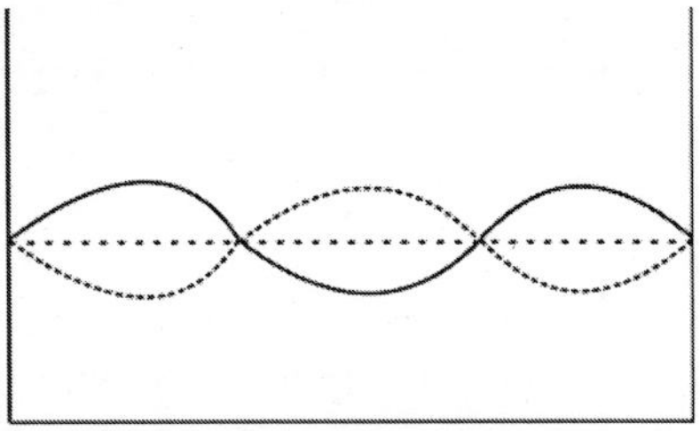

Fig. 1.6: A standing wave

$$\psi = A\sin\frac{2\pi x}{\lambda} \quad ...(1.1)$$

where, ψ is the amplitude of the wave of wavelength λ at a distance x from the origin and A is a constant.

On double differentiation with respect to x, we get

$$\frac{\partial\psi}{\partial x} = A\cdot\frac{2\pi}{\lambda}\cdot\cos\frac{2\pi x}{\lambda} \qquad \left[\frac{d}{dx}(\sin x) = \cos x\right]$$

$$\frac{\partial^2\psi}{\partial x^2} = A\cdot\frac{2\pi}{\lambda}\cdot\frac{2\pi}{\lambda}\left(\frac{-\sin 2\pi x}{\lambda}\right) \qquad \left[\frac{d}{dx}(\cos x) = -\sin x\right]$$

$$\frac{\partial^2\psi}{\partial x^2} = \frac{-4\pi^2}{\lambda^2}\left(A\sin\frac{2\pi x}{\lambda}\right) \quad ...(1.2)$$

Substituting $\psi = A\sin\frac{2\pi x}{\lambda}$ from equation (1.1) into equation (1.2), we get

$$\frac{\partial^2\psi}{\partial x^2} = -\frac{4\pi^2}{\lambda^2}(\psi)$$

or,

$$\frac{\partial^2\psi}{\partial x^2} + \frac{4\pi^2}{\lambda^2}(\psi) = 0$$

This is the equation for one dimensional standing wave which can be extended to describe motion of an electron in three dimensions by a second order partial differential equation (1.3)

$$\frac{\partial^2\psi}{\partial x^2} + \frac{\partial^2\psi}{\partial y^2} + \frac{\partial^2\psi}{\partial z^2} + \frac{4\pi^2}{\lambda^2}\psi = 0 \quad ...(1.3)$$

where, ψ is a function of cartesian coordinates x, y, z. This may be written more precisely as

$$\nabla^2\psi + \frac{4\pi^2}{\lambda^2}\psi = 0 \quad ...(1.4)$$

where, $\nabla^2 = \left(\frac{\partial^2}{\partial x^2} + \frac{\partial^2}{\partial y^2} + \frac{\partial^2}{\partial z^2}\right)$ is the **Laplacian operator** (discussed later in the topic operators).

Incorporating de-Broglie relationship into equation (1.4) for an electron of wavelength λ,

$$\lambda = \frac{h}{mv}$$

we get $$\nabla^2\psi + \frac{4\pi^2 m^2 v^2}{h^2}\psi = 0 \qquad ...(1.5)$$

Considering the electron as a particle, the total energy (E) of the system is sum of kinetic ($1/2mv^2$) and potential (V) energies, i.e.

$$E = \frac{1}{2}mv^2 + V$$

$$v^2 = \frac{2(E - V)}{m} \qquad ...(1.6)$$

In order to calculate values for the energy states, v^2 from equation (1.6) is substituted into equation (1.5) to obtain a fundamental form of Schrodinger wave equation for a single particle in three dimensions.

$$\nabla^2\psi + \frac{8\pi^2 m}{h^2}(E - V)\psi = 0 \qquad ...(1.7)$$

or $$\frac{\partial^2\psi}{\partial x^2} + \frac{\partial^2\psi}{\partial y^2} + \frac{\partial^2\psi}{\partial z^2} + \frac{8\pi^2 m}{h^2}(E - V)\psi = 0 \qquad ...(1.8)$$

If the electron is at rest at an infinite distance from the nucleus, its potential energy with respect to the nucleus at a distance r is given by

$$V = \frac{-Ze^2}{r}$$

where, Z = atomic number

$+Ze$ = charge on the nucleus

$-e$ = charge on the electron.

Consequently, the equation (1.8) for one electron atomic system in three dimensions may be written as,

$$\frac{\partial^2\psi}{\partial x^2} + \frac{\partial^2\psi}{\partial y^2} + \frac{\partial^2\psi}{\partial z^2} + \frac{8\pi^2 m}{h^2}\left(E + \frac{Ze^2}{r}\right)\psi = 0$$

Equation (1.7) is known as **Schrodinger's fundamental wave equation** with respect to space. The Schrodinger equation is valid for a particle executing simple harmonic motion (SHM). The permitted solutions of Schrodinger equation are known as **wavefunction** (ψ) which denotes the definite energy state.

It is a second order differential equation. When it is solved, it gives infinite solutions of ψ, represented as ψ_1, ψ_2, ψ_3, ψ_4, ... etc. with corresponding energies E_1, E_2, E_3, E_4, ... etc. But all the solutions are not acceptable. Only those solutions are acceptable which have some special properties, i.e. only well behaved ψ's are acceptable.

The equation (1.7) may also be expressed as

$$\nabla^2\psi + \frac{8\pi^2 mE}{h^2}\psi - \frac{8\pi^2 mV}{h^2}\psi = 0$$

$$-\nabla^2\psi + \frac{8\pi^2 mV}{h^2}\psi = \frac{8\pi^2 mE}{h^2}\psi$$

Dividing throughout by $\frac{8\pi^2 m}{h^2}$ we get:

$$-\frac{h^2}{8\pi^2 m}\nabla^2\psi + V\psi = E\psi$$

$$\left[-\frac{h^2}{8\pi^2 m}\nabla^2 + V\right]\psi = E\psi$$

$$\hat{H}\psi = E\psi \qquad ...(1.9)$$

where, $\hat{H} = -\frac{h^2}{8\pi^2 m}\nabla^2 + V$ is the Hamiltonian operator (discussed later in the topic operators)

Total energy of electron = K.E. + P.E.

$$\hat{H} = \hat{T} + \hat{V}$$

$$\left(\hat{T} + \hat{V}\right) = E\psi$$

PROBLEM BASED ON SCHRODINGER WAVE EQUATION

Q. 1. Write down Schrodinger wave equation and explain the terms involved in it.

Ans. $$\frac{\partial^2\psi}{\partial x^2} + \frac{\partial^2\psi}{\partial y^2} + \frac{\partial^2\psi}{\partial z^2} + \frac{8\pi^2 m}{h^2}(E - V)\psi = 0$$

where, ψ = amplitude of the wave associated with the electron

x, y, z = cartesian coordinates of the electron

m = mass of the electron

h = Planck's constant

E = total energy of the electron

V = potential energy of the electron.

1.10 Physical Significance of Wavefunction (ψ) and Probability Density (ψ^2)

The wavefunction ψ in the Schrodinger wave equation is the amplitude of the electron wave or amplitude of vibration at any point surrounding the nucleus. The wavefunction for an electron in an atom has no physical significance by itself. It can neither be measured nor be observed. It has only mathematical significance. It is a mathematical function whose value depends upon the coordinates of the electron in an atom, $\psi(x, y, z)$. A complete description of the state of the system can be described by wavefunction ψ.

In the classical theory of light or sound, the square of the amplitude of the wave at any point gives the intensity of the sound or light at that point. Similarly, very similar concept was suggested by Max Born in quantum mechanics, according to which the square of the amplitude of the electron wave, i.e. ψ^2 at any point gives the intensity of the electron wave at that point. Thus, ψ^2 at any point gives the probability of finding the electron.

According to **Max Born**, a German physicit, probability concept of the uncertainty principle should be utilized to get an idea of distribution of the electron in the atom in terms of the wavefunction ψ. When the amplitude of the wave is large, the probability of the event is large and when the amplitude is small, the probability of the event is small. But the wavefunction itself cannot represent probability because probability of finding an electron at a given point in space is always real and positive. Whereas wavefunction, ψ may be positive, negative and in some cases a complex quantity, i.e. contains imaginary number $\left(i = \sqrt{-1}\right)$. Thus, the square of wavefunction, i.e. ψ^2 is proposed. By taking square of wavefunction ($\psi^2 = \psi \cdot \psi^*$), the imaginary or negative value vanishes ($\therefore$ ψ^* is conjugate of ψ). Thus, the ψ^2 obtained has a real and positive value.

If ψ has imaginary value, then it can be expressed as

$$\psi = a + ib$$

where a and b are real numbers and i (known as iota) $= \sqrt{-1}$. In such a case, ψ is made real by multiplying it by its own complex conjugate (ψ^*) which is expressed as

$$\psi^* = a - ib$$

The product obtained by multiplying ψ with ψ^* is

$$\begin{aligned} \psi\psi^* &= (a + ib)(a - ib) = a^2 - (i^2b^2) \\ &= a^2 - (-1 \times b^2) \qquad (\because i = \sqrt{-1} \text{ and } i^2 = -1) \\ &= (a^2 + b^2) \end{aligned}$$

However, if ψ is real, then $\psi = \psi^*$, then $\psi\psi^* = \psi^2$.

The square of the wavefunction ψ^2 has physical significance and it represents intensity. Greater the intensity of wavefunction at a particular point, greater is the probability of locating the electron at that point. Thus, ψ^2 is interpreted being proportional to electron density. When ψ^2 is high, electron density is high, that is, the probability of finding an electron is high. If ψ^2 is low, probability of finding an electron is low.

If ψ extends over a finite volume in space $dx \cdot dy \cdot dz$ ($= d\tau$), then the electron would be found in that volume. Thus, the probability of finding electron in the small volume element $dx \cdot dy \cdot dz$ ($= d\tau$) situated at a point in space is given as:

$$\text{Probability density } \left(\psi^2\right) = \int_{-\infty}^{+\infty} \psi\psi^* \, dx\, dy\, dz = \int_{-\infty}^{+\infty} \psi\psi^* \, d\tau = \int_{-\infty}^{+\infty} \psi\psi^* \, dV.$$

The probability of finding the electron is finite which means $\psi\psi^* d\tau$ must have real value.

1.11 Concept of Atomic Orbital

Taking into consideration the wave nature of electron and the concept of probability, the movement of the electron around the nucleus is like that of bee around the hive. Sometimes, it may come close to the nucleus and sometimes it may move away from it. Further, it does not move in one plane (as wrongly postulated by Bohr) but moves in all directions and in all planes around the nucleus. Thus, in the light of uncertainty principle, Bohr model of atom postulating that electrons move along definite circular paths and hence their position and momentum could be determined with absolute accuracy became unacceptable. Now, we speak in terms of 'probability' of finding the electron at any particular position about the nucleus at any instant of time.

The region of space around the nucleus which describes the probability of finding an electron of given energy is called an electron cloud.

Also the probability of finding an electron is never zero. Even at large distance from the nucleus, there is a finite, though small, probability of finding an electron of a given energy. This means that electron clouds do not have sharp boundaries (Fig. 1.7).

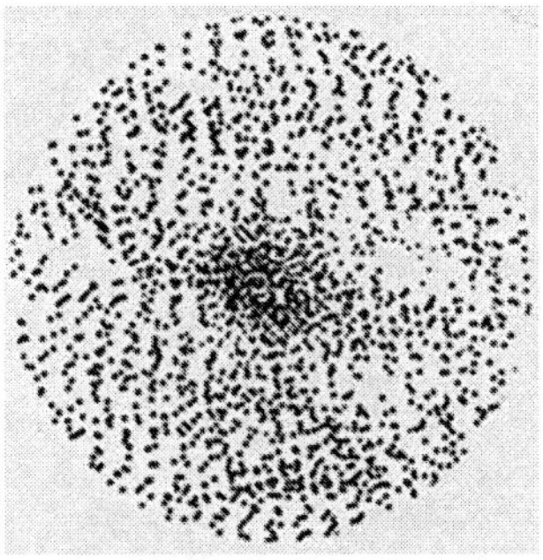

Fig. 1.7: Electron cloud for the only electron of the hydrogen atom.

However, for the sake of pictorial clarity and for convenience of representation, a **boundary surface** (Fig. 1.8) may be drawn which connects points of equal probability and encloses a certain volume of the space around nucleus within which the probability of finding an electron of given energy is maximum (say upto 90%). This is called **atomic orbital.**

Thus, atomic orbital may be defined as the three dimensional space around the nucleus within which the probability of finding an electron of given energy is maximum (say upto 90%).

By finding ψ^2 at different points around the nucleus in an atom, we can predict the region of space around the nucleus within which the probability of finding the electron with a definite value of energy is maximum. This space around the nucleus which represents the electron density at different points is called an atomic orbital. That is why, the wavefunction for an electron in an atom is called **orbital wavefunction** or **atomic orbital.**

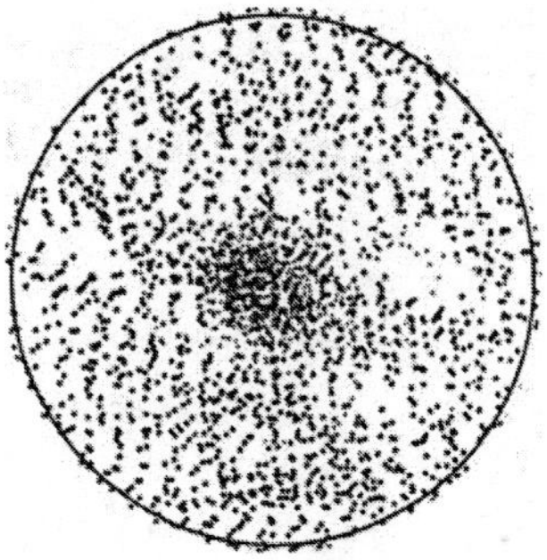

Fig. 1.8: Boundary surface for maximum probability density.

	Orbit	Orbital
1.	An orbit (proposed by Bohr) is a circular path around the nucleus in which an electron moves.	An orbital is a quantum mechanical concept and it refers to the three dimensional space around the nucleus within which the probability of finding an electron is maximum (upto 90%).
2.	It represents the planar motion of electron around the nucleus.	It represents the three dimensional motion of an electron around the nucleus.
3.	The concept of orbit is not in accordance with the wave nature of electrons and uncertainty principle. Thus, a precise description of this path of the electron is impossible according to Heisenberg uncertainty principle.	The concept of an orbital is in accordance with the wave character of electrons and uncertainty principle.
4.	Bohr orbits have no real meaning and their existence can never be demonstrated experimentally.	It is a quantum mechanical concept and refers to the one electron wavefunction ψ in an atom. The value of ψ depends on the coordinates of the electron. Also, ψ itself has no physical significance but ψ^2 has a physical significance.
5.	The maximum number of electrons in an orbit is given by $2n^2$, where, n is the number of the orbit.	The maximum number of electrons present in any orbital is two.

1.12 Quantum Mechanical Model of Atom

Just as Bohr model of atom was developed on the basis of classical mechanics, the quantum mechanical model of atom has been developed on the basis of quantum mechanics, i.e. Schrodinger wave equation. This quantum mechanical model of atom takes into account the two concepts, i.e. de-Broglie dual nature of matter and Heisenberg uncertainty principle. Thus, quantum mechanical model takes wave nature of electron into consideration. The Bohr concept of fixed circular path with definite position and momentum of the electron have been replaced in quantum mechanical model by stating that the electron has the probability of having a given position and momentum. According to this model, an electron does not move in one plane but moves in all directions and in all planes around the nucleus. There is a finite, though small, probability of finding an electron of a given energy around the nucleus. The probability of finding an electron is never zero. The electrons in an atom have only quantised values of energy. These quantised values of energy are obtained from the solution of Schrodinger wave equation. The corresponding values of the wavefunction ψ are also obtained from the solution of Schrodinger wave equation. This wavefunction ψ is simply a function of coordinates of the electron and has no physical significance as such. However, ψ^2 gives the probability of finding the electron at that point, i.e. electron density at that point. The concept of probability is justified in view of Heisenberg's uncertainty principle. By finding ψ^2 at different points around the nucleus in an atom, we can predict the region of space around the nucleus within which the probability of finding the electron with a definite value of energy is maximum. This space around the nucleus is called orbital. This wavefunction ψ is called **orbital wavefunction** or **atomic orbital.** Since an electron can have many wavefunctions, therefore, there are many atomic orbitals in an atom. The wavefunction ψ contains all the information about an electron in an atom.

PROBLEM BASED ON QUANTUM MECHANICAL MODEL OF ATOM

Q. 1. Orbital picture of atom accommodate Heisenberg's uncertainty principle. Explain.

1.13 Eigen Value and Eigen Wavefunction

Schrodinger wave equation is a second order differential equation. When it is solved, it gives infinite solution of ψ represented as ψ_1, ψ_2, ψ_3, ... etc. with corresponding energies E_1, E_2, E_3, ... etc. But all the solutions are not acceptable. Only those values of ψ are acceptable which correspond to some definite values of the total energy E. The acceptable values of the wavefunction ψ are called **eigen wavefunction** or **acceptable wavefunction.** The value of E given by a particular eigen wavefunction is called **eigen value** or **proper value** or **characteristic value.** Thus, every value of wavefunction is not an eigen wavefunction.

The properties of an eigen function are: **S, C, F, B N**

1. **ψ must be single valued** (S)
   ψ must be single valued at any given point.

Like:

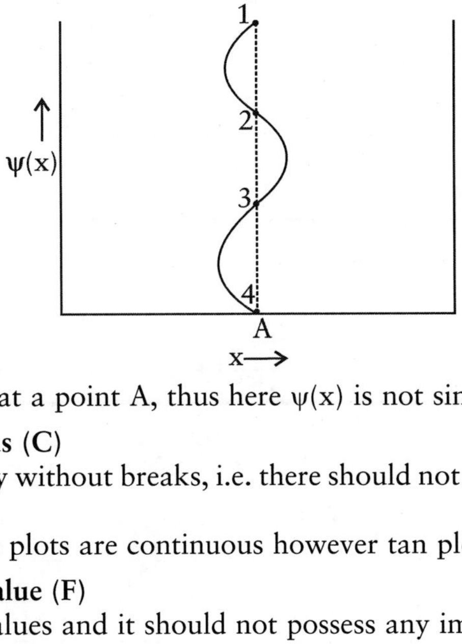

Four values are permitted at a point A, thus here ψ(x) is not single valued.

2. **ψ must be continuous** (C)
   ψ must vary smoothly without breaks, i.e. there should not be any gap or discontinuity in the function.
   Like: sine and cosine plots are continuous however tan plot is not continuous.
3. **ψ must have finite value** (F)
   ψ must have finite values and it should not possess any imaginary values
   Like: iota $\left(i = \sqrt{-1}\right)$ values.
4. **ψ must be bounded in the space** (B)
   ψ must have boundary conditions in the space.

5. **ψ must be normalised (N)**
ψ being normalised means that the probability of finding the particle inside the bounded space is unity (i.e., sure success in probability).

$$\int_{-\infty}^{+\infty} \psi\psi^* \, d\tau = 1$$

The function which possess all the above properties is called well-behaved wavefunction or acceptable wavefunction.

PROBLEMS BASED ON EIGEN FUNCTION

Q. 1. Which of the following functions are acceptable in quantum mechanics?

Ans.

$\psi = \sin x$; for $0 \le x \le \frac{\pi}{2}$	acceptable when x lies between 0 and $\frac{\pi}{2}$
$\psi = \cos x + \sin x$; for $0 \le x \le \frac{\pi}{2}$	acceptable when x lies between 0 and $\frac{\pi}{2}$
$\psi = \tan x$; for $0 \le x \le \frac{\pi}{2}$	not acceptable because $\psi \to \infty$ as $x \to \frac{\pi}{2}$
$\psi = \text{cosec } x$; for $0 \le x \le \frac{\pi}{2}$	not acceptable because $\psi \to \infty$ as $x \to 0$
$\psi = e^{-ax}$; for $x \ge 0$	acceptable
$\psi = xe^{-ax}$; for $x \ge 0$	acceptable because $\psi \to 0$ as $x \to 0$
$\psi = e^{-ax}$; for $x \le 0$	acceptable
$\psi = e^{-ax^2}$; for $x \le 0$	not acceptable
$\psi = x$	not acceptable because $\psi \to \infty$ as $x \to \infty$
$\psi = x^2$	not acceptable because $\psi \to \infty$ as $x \to \infty$
$\psi = e^x$	not acceptable because $\psi \to \infty$ as $x \to \infty$
$\psi = e^{-x}$	acceptable because $\psi \to 0$ as $x \to \pm\infty$
$\psi = e^{-x^2}$	acceptable because $\psi \to 0$ as $x \to \pm\infty$

Q. 2. What are the properties of an acceptable wavefunction?

or

What are conditions that ψ must obey as a wavefunction?

or

What are the conditions under which a wavefunction is said to be a well behaved one?

or

What is meant by well-behaved wavefunction and give it properties?

Ans. Properties or Conditions of a wavefunction to be acceptable are:

S = Single valued, ψ must be single valued at any given point

C = Continuous, ψ must vary smoothly without breaks

F = Finite, ψ must have finite values (imaginary values not allowed)

B = Boundary conditions, ψ must be bounded in the space

N = Normalised, i.e. probability of finding the particle in the bounded space must be equal to unity.

Q. 3. What is the significance of ψ and ψ^2?

Ans. The wavefunction ψ in the Schrodinger equation is the amplitude of the electron wave. It has no physical significance by itself but has only mathematical significance. It is a mathematical function whose value depends upon the coordinates of the electron in an atom. The wavefunction can be positive, negative or a complex quantity.

Whereas the square of the wavefunction ψ^2, i.e. square of the amplitude of the electron wave, at any point give the intensity of electron wave at that point. It gives the probability of finding the electron at a given point in space. The ψ^2 have always real and positive values.

Q. 4. Explain as to why 'ψ' should be finite, single valued and continuous.

Ans. When Schrodinger equation is solved, it gives infinite solutions of ψ with corresponding energies. But all the solutions are not acceptable. Only those values of ψ are acceptable which fulfills certain properties like it must be finite, single valued and continuous.

Q. 5. What do you understand by normalisation of wavefunction?

Q. 6. Write four properties of a function to make it acceptable as a solution of Schrodinger equation. Support graphically by plotting f(x) vs x in each case.

1.14 Normalised, Orthogonal and Orthonormal Wavefunction

1. Normalised Wavefunction:

- The square of wavefunction ($\psi\psi^*$) is proportional rather than equal to the probability of finding the electron in a given volume element $dx \cdot dy \cdot dz$ (= $d\tau$ or = dV). For most of the purposes, it is taken as equal to the probability which implies that the electron is in a given volume element $d\tau$. In such a case, the integration of $\psi\psi^*\, d\tau$ over whole of the configuration space, which gives the total probability in a given volume element must be equal to unity

$$\int_{-\infty}^{+\infty} \psi\psi^* \, d\tau = 1 \quad \text{or} \quad \int_{-\infty}^{+\infty} \psi^2 \, d\tau = 1$$

The wavefunction which satisfies the above relation is said to be normalised.

- But very often, ψ is not normalised. Thus, ψ is multiplied by a constant N to give a new wavefunction, $N\psi$, which is also a solution of the wave equation. The value of N used to make the new wavefunction a normalised wavefunction can be calculated as follows:

$$\int_{-\infty}^{+\infty} (N\psi)(N\psi^*) d\tau = 1$$

$$N^2 \int_{-\infty}^{+\infty} \psi\psi^* \, d\tau = 1$$

$$N = \left(\frac{1}{\int_{-\infty}^{+\infty} \psi\psi^* \, d\tau} \right)^{1/2}$$

The factor N thus introduced is called **Normalisation Constant** and is independent of the coordinates (x, y, z). The function $N\psi$ is called **Normalised Wavefunction.**

- The solution of Schrodinger equation may provide more than one wavefunction, i.e. each wavefunction satisfies the Schrodinger equation with its own characteristic eigenvalue. If ψ_i and ψ_j are the two different eigen or acceptable wavefunctions of a given system, then these functions will be normalised if they meet the requirement.

$$\int_{-\infty}^{+\infty} \psi_i \psi_i^* \, d\tau = 1 \quad \text{and} \quad \int_{-\infty}^{+\infty} \psi_j \psi_j^* \, d\tau = 1$$

2. **Orthogonal Wavefunction:**

Let ψ_i and ψ_j are the two independent eigen wavefunctions obtained on solving the Schrodinger equation of a given system and each wavefunction satisfies the Schrodinger equation with its own characteristic eigen values. If two such wavefunctions satisfy the expression

$$\int_{-\infty}^{+\infty} \psi_i \psi_j^* \, d\tau = 0$$

then the two functions are said to be orthogonal to each other. Orthogonality implies the independence of energy states.

3. **Orthonormal Wavefunction (Orthonormal = Normalised + Orthogonal):**

If the solutions of Schrodinger equation satisfy the expression

$$\int_{-\infty}^{+\infty} \psi_i \psi_j^* \, d\tau = \delta_{ij} \qquad \begin{bmatrix} \delta_{ij} = 0, \text{ if } i \neq j \\ \delta_{ij} = 1, \text{ if } i = j \end{bmatrix}$$

then the two solutions (ψ_i and ψ_j) form orthonormal set of wavefunction. The symbol δ_{ij} is called **Kronecker's delta.**

1.15 Operators

An operator is a symbol for a mathematical calculation/procedure/command which changes one function into another.

Operator (Function) = New Function

An operator written alone has no significance. The left hand side of the above equation does not mean that the function is multiplied with the operator. The function on which the operation is carried out is called an **operand.**

Let, $(x^2 + a)$ is a function, then

Operation	Operator	Result
(i) Multiply by C	$C \times (x^2 + a)$	$Cx^2 + Ca$
(ii) Take the square	$(x^2 + a)^2$	$x^4 + a^2 + 2ax^2$
(iii) Differential	$\frac{d}{dx}(x^2 + a)$	$2x$
(iv) Integration	$\int (x^2 + a)dx$	$\frac{x^3}{3} + ax$
(v) Addition of x	$x + (x^2 + a)$	$x + x^2 + a$

An operator is used with symbol (^) over it, i.e. a CAROT is put over the symbol of the operator. Thus, P operator is represented as $\hat{P}$ and Q operator is represented as $\hat{Q}$.

- **Algebra of Operators**

1. **Addition and Subtraction of Operators:** If $\hat{A}$ and $\hat{B}$ are two different operators and f(x) is the function, then

$$\left(\hat{A} + \hat{B}\right)f(x) = \hat{A}f(x) + \hat{B}f(x)$$

$$\left(\hat{A} - \hat{B}\right)f(x) = \hat{A}f(x) - \hat{B}f(x)$$

2. **Multiplication of Operators:** If $\hat{A}$ and $\hat{B}$ are two different operators and f(x) is the function, then order of the operation is very important.

$$\hat{A}\hat{B}\left[f(x)\right] = \hat{A}\left[\hat{B}f(x)\right] = \hat{A}\left[f'(x)\right] = f''(x)$$

$$\left(\text{where } \hat{B}f(x) = f'(x)\right)$$

This above expression implies that first operator $\hat{B}$ operates on function f(x) to get the result, say $f'(x)$ and then operator $\hat{A}$ operates on function $f'(x)$ to get the final result say $f''(x)$.

When we are dealing with operators, the order of operation should be maintained. Usually, the order of operation is from RHS to LHS.

Like: $\hat{A} = x, \hat{B} = \frac{d}{dx}$ and $f(x) = x^2$

$$\hat{A}\hat{B}f[(x)] = \hat{A}[Bf(x)] = x \cdot \left[\frac{d}{dx}(x^2)\right] = x(2x) = 2x^2$$

Also, if the same operation is to be done a number of times in succession, it is shown by a power of operator.

$$\hat{A}\hat{A}f(x) = \hat{A}^2 f(x).$$

Like: $\hat{A} = \frac{d}{dx}$ and $f(x) = \sin x$

$$\hat{A}^2 f(x) = \left(\frac{d}{dx}\right)^2 \sin x = \frac{d^2}{dx^2}(\sin x)$$

$$= \frac{d}{dx}\left[\frac{d}{dx}(\sin x)\right] = \frac{d}{dx}[\cos x] = -\sin x.$$

3. **Commutative Property:** If two operators are such that the result of their successive applications is the same irrespective of the order of operations, then the two operators are said to be commutative.

   If, $\hat{A}\hat{B}f[(x)] = \hat{B}\hat{A}f[(x)]$, then the two operators $\hat{A}$ and $\hat{B}$ commute with each other, i.e. the normal product of two operators is equal to their product in reverse order.

   The expression $\left[\hat{A}\hat{B} - \hat{B}\hat{A}\right]$ is called **commutator** and is represented as $\left[\hat{A} \cdot \hat{B}\right]$.

   Commutator $\left[\hat{A} \cdot \hat{B}\right] = \hat{A}\hat{B}[f(x)] - \hat{B}\hat{A}[f(x)] = 0.$

   If the two operators commute with one another, then the value of commutator is zero, i.e.

   $$\left[\hat{A} \cdot \hat{B}\right] = \hat{A}\hat{B}[f(x)] - \hat{B}\hat{A}[f(x)] = 0.$$

   Like: $\hat{A} = \frac{d}{dx}, \hat{B} = \frac{d^2}{dx^2}$ and $f(x) = \sin x$

   $$\hat{A}\hat{B}[f(x)] = \hat{A}\left[\hat{B}f(x)\right] = \frac{d}{dx}\left(\frac{d^2}{dx^2}\sin x\right) = -\cos x$$

   $$\hat{B}\hat{A}[f(x)] = \hat{B}\left[\hat{A}f(x)\right] = \frac{d^2}{dx^2}\left(\frac{d}{dx}(\sin x)\right) = -\cos x$$

   $$\therefore \quad \left[\hat{A} \cdot \hat{B}\right] = \hat{A}\hat{B}[f(x)] - \hat{B}\hat{A}[f(x)] = (-\cos x) - (-\cos x) = 0$$

   Thus, $\hat{A}$ and $\hat{B}$ commute with each other because value of commutator is zero.

On the other hand, the two operators are said to be non-commutative if $\hat{A}\hat{B}[f(x)] \neq \hat{B}\hat{A}[f(x)]$. Also if the two operators do not commute with one another, then the commutator has some value, i.e. non-zero.

$$\text{Commutator} = \left[\hat{A}\cdot\hat{B}\right] = \hat{A}\hat{B}[f(x)] - \hat{B}\hat{A}[f(x)] \neq 0$$

Like: $\hat{A} = x,\ \hat{B} = \dfrac{d}{dx}$ and $f(x) = x^2$

$$\hat{A}\hat{B}[f(x)] = \hat{A}\left[\hat{B}f(x)\right] = x\cdot\left[\frac{d}{dx}(x^2)\right] = x(2x) = 2x^2$$

$$\hat{B}\hat{A}[f(x)] = \hat{B}\left[\hat{A}f(x)\right] = \frac{d}{dx}\left[x.(x^2)\right] = \frac{d}{dx}(x^3) = 3x^2$$

$$\left[\hat{A}\cdot\hat{B}\right] = \hat{A}\hat{B}[f(x)] - \hat{B}\hat{A}[f(x)] = 2x^2 - 3x^2 = -x^2 \neq 0$$

$\therefore$ Thus, $\hat{A}$ and $\hat{B}$ do not commute with each other because value of commutator is non-zero.

- **Significance of Commutation:** If the two operators commute i.e.,

$$\left[\hat{A}\cdot\hat{B}\right] = \hat{A}\hat{B}[f(x)] - \hat{B}\hat{A}[f(x)] = 0$$

then, the two observable properties represented by operators $\hat{A}$ and $\hat{B}$ can be simultaneously and exactly determined. However, if the operators do not commute, i.e.

$$\left[\hat{A}\cdot\hat{B}\right] = \hat{A}\hat{B}[f(x)] - \hat{B}\hat{A}[f(x)] \neq 0$$

then, the two observable properties represented by operators $\hat{A}$ and $\hat{B}$ cannot be simultaneously and exactly determined.

- **Eigen Function, Eigen Value and Eigen Value Equation:** If an operator $\hat{A}$ operates on a well-behaved function f(x), and give the same function multiplied by a constant, then the function f(x) is called the **eigen function** and the constant is called the **eigen value** of the operator. The equation formed is called as **Eigen value equation,** i.e. equation (1.10).

$$\underset{\text{eigen function}}{\underset{\downarrow}{\hat{A}f(x)}} = \underset{\text{eigen value}}{\underset{\downarrow}{af(x)}} \qquad \text{...(1.10)}$$

Like: (i) $$\frac{d}{dx}\underset{\text{eigen function}}{\underset{\downarrow}{\left(e^{ax}\right)}} = \underset{\text{eigen value}}{\underset{\downarrow}{ae^{ax}}}$$

(ii) Schrodinger equation is an important eigen value equation, $\hat{H}\psi = E\psi$.

- **Some of the common operators used in quantum mechanics:**

1. **Laplacian operator** (∇^2)**:** It is an important differential operator used in quantum mechanics.

$$\vec{\nabla} = \hat{i}\frac{\partial}{\partial x} + \hat{j}\frac{\partial}{\partial y} + \hat{k}\frac{\partial}{\partial z}$$

where $\hat{i}$, $\hat{j}$ and $\hat{k}$ are unit vectors along, x, y and z axes respectively

$$\nabla^2 = \frac{\partial^2}{\partial x^2} + \frac{\partial^2}{\partial y^2} + \frac{\partial^2}{\partial z^2} \qquad \left(\hat{i}\cdot\hat{i} = \hat{j}\cdot\hat{j} = \hat{k}\cdot\hat{k} = 1\right)$$

2. **Momentum Operator** $(\hat{p})$ **:** For an electron wave, the wavefunction may be represented by the function

$$\psi = A\exp\left(\frac{2\pi ix}{\lambda}\right)$$

Differentiating w.r.t. x

$$\frac{\partial\psi}{\partial x} = \frac{2\pi i}{\lambda}\left[A\exp\left(\frac{2\pi ix}{\lambda}\right)\right] = \frac{2\pi i}{\lambda}\psi \qquad \text{...(1.11)}$$

Using de-Broglie relationship

$$\lambda = \frac{h}{p_x}$$

$$\frac{1}{\lambda} = \frac{p_x}{h} \qquad \text{...(1.12)}$$

Putting (1.12) in (1.11), we have

$$\frac{\partial\psi}{\partial x} = 2\pi i\left(\frac{p_x}{h}\right)\psi$$

On rearranging, we have:

or, $$\hat{p}_x\psi = \frac{h}{2\pi i}\frac{\partial\psi}{\partial x}$$

But ψ is a function which on being removed from the equation reduces it to a differential operator,

$$\hat{p}_x = \frac{h}{2\pi i}\frac{\partial}{\partial x} = \frac{\hbar}{i}\frac{\partial}{\partial x} = -i\hbar\frac{\partial}{\partial x} \qquad \left(\text{where, } \hbar = \frac{h}{2\pi}\right)$$

Similarly, $$\hat{p}_y = \frac{h}{2\pi i}\frac{\partial}{\partial y} = \frac{\hbar}{i}\frac{\partial}{\partial y} = -i\hbar\frac{\partial}{\partial y}$$

$$\hat{p}_z = \frac{h}{2\pi i}\frac{\partial}{\partial z} = \frac{\hbar}{i}\frac{\partial}{\partial z} = -i\hbar\frac{\partial}{\partial z}$$

Thus, the momentum operator $\hat{p}$, is given as

$$\hat{p} = \hat{p}_x + \hat{p}_y + \hat{p}_z = -i\hbar\left(\frac{\partial}{\partial x} + \frac{\partial}{\partial y} + \frac{\partial}{\partial z}\right)$$

$$\hat{p} = -i\hbar\nabla$$

3. **Square of Momentum Operator $(\hat{p}^2)$:**

$$\hat{p}_x^2 = \left(\frac{\hbar}{i}\frac{\partial}{\partial x}\right)\left(\frac{\hbar}{i}\frac{\partial}{\partial x}\right) = -\hbar^2\frac{\partial^2}{\partial x^2} \qquad (i \times i = i^2 = -1)$$

$$\hat{p}_y^2 = \left(\frac{\hbar}{i}\frac{\partial}{\partial y}\right)\left(\frac{\hbar}{i}\frac{\partial}{\partial y}\right) = -\hbar^2\frac{\partial^2}{\partial y^2}$$

$$\hat{p}_z^2 = \left(\frac{\hbar}{i}\frac{\partial}{\partial z}\right)\left(\frac{\hbar}{i}\frac{\partial}{\partial z}\right) = -\hbar^2\frac{\partial^2}{\partial z^2}$$

$$\hat{p}^2 = \hat{p}_x^2 + \hat{p}_y^2 + \hat{p}_z^2 = -\hbar^2\left(\frac{\partial^2}{\partial x^2} + \frac{\partial^2}{\partial y^2} + \frac{\partial^2}{\partial z^2}\right)$$

$$\hat{p}^2 = -\hbar^2\nabla^2$$

4. **Angular Momentum Operator ($\hat{L}$):** Angular momentum is given by the vector product of position ($\vec{r}$) and linear momentum ($\vec{p}$)

$$\vec{r} = x\hat{i} + y\hat{j} + z\hat{k}$$

$$\vec{p} = p_x\hat{i} + p_y\hat{j} + p_z\hat{k}$$

where, $\hat{i}$, $\hat{j}$ and $\hat{k}$ are unit vectors along x, y, z axes respectively.

Now,

$$\vec{L} = \vec{r} \times \vec{p}$$

$$\vec{L} = (x\hat{i} + y\hat{j} + z\hat{k}) \times (p_x\hat{i} + p_y\hat{j} + p_z\hat{k})$$

$$= \begin{vmatrix} \hat{i} & \hat{j} & \hat{k} \\ x & y & z \\ p_x & p_y & p_z \end{vmatrix}$$

$$\vec{L} = \hat{i}(yp_z - zp_y) - \hat{j}(xp_z - zp_x) + \hat{k}(xp_y - yp_x)$$

$$\vec{L} = \hat{i}(yp_z - zp_y) + \hat{j}(zp_x - xp_z) + \hat{k}(xp_y - yp_x) \qquad \text{...(1.13)}$$

Also, by definition:

By definition, $\vec{L} = L_x\hat{i} + L_y\hat{j} + L_z\hat{k}$ ...(1.14)

where L_x, L_y and L_z are the three components of L.

Comparing equations (1.13) and (1.14)

$$\hat{L}_x = y\hat{p}_z - z\hat{p}_y = y\frac{\hbar}{i}\frac{\partial}{\partial z} - z\frac{\hbar}{i}\frac{\partial}{\partial y} = \frac{\hbar}{i}\left(y\frac{\partial}{\partial z} - z\frac{\partial}{\partial y}\right)$$

$$\hat{L}_y = z\hat{p}_x - x\hat{p}_z = z\frac{\hbar}{i}\frac{\partial}{\partial x} - x\frac{\hbar}{i}\frac{\partial}{\partial z} = \frac{\hbar}{i}\left(z\frac{\partial}{\partial x} - x\frac{\partial}{\partial z}\right)$$

$$\hat{L}_z = x\hat{p}_y - y\hat{p}_x = x\frac{\hbar}{i}\frac{\partial}{\partial y} - y\frac{\hbar}{i}\frac{\partial}{\partial x} = \frac{\hbar}{i}\left(x\frac{\partial}{\partial y} - y\frac{\partial}{\partial x}\right)$$

5. **Kinetic Energy Operator ($\hat{T}$):** Classically, kinetic energy is given as

$$T = \frac{1}{2}mv^2 \qquad \text{...(1.15)}$$

Now, $p = mv$

$$p^2 = m^2v^2$$

Dividing throughout by 2m:

$$\frac{p^2}{2m} = \frac{m^{\not 2}v^2}{2\not m}$$

$$\frac{p^2}{2m} = \frac{1}{2}mv^2 \qquad \text{...(1.16)}$$

On comparing equations (1.15) and (1.16), we have

$$\therefore\ T = \frac{p^2}{2m}$$

$$\hat{T} = \frac{1}{2m}\left(\hat{p}_x^2 + \hat{p}_y^2 + \hat{p}_z^2\right)$$

$$\hat{T} = \frac{1}{2m}\left[\left(-\hbar^2\frac{\partial^2}{\partial x^2}\right) + \left(-\hbar^2\frac{\partial^2}{\partial y^2}\right) + \left(-\hbar^2\frac{\partial^2}{\partial z^2}\right)\right]$$

$$\hat{T} = -\frac{\hbar^2}{2m}\left[\frac{\partial^2}{\partial x^2} + \frac{\partial^2}{\partial y^2} + \frac{\partial^2}{\partial z^2}\right]$$

$$\hat{T} = -\frac{\hbar^2}{2m}\nabla^2$$

6. **Hamiltonian Operator** ($\hat{H}$): The total energy of a system is the sum of kinetic and potential energies. The operator corresponding to the total energy is called Hamiltonian operator ($\hat{H}$).

The total energy of a system containing a particle of mass 'm' is given as

$$E = K.E. + P.E.$$

$$\hat{H} = \hat{T} + \hat{V}$$

$$\hat{H} = -\frac{\hbar^2}{2m}\nabla^2 + V(x, y, z)$$

Suppose, there is a system containing two particles, each of mass m_1 and m_2. The total energy of the system is given as

$$\hat{H} = -\frac{\hbar^2}{2m_1}\nabla_1^2 - \frac{\hbar^2}{2m_2}\nabla_2^2 + V\left(x_1, y_1, z_1; x_2, y_2, z_2\right)$$

In general for a large number of particles say i

$$\hat{H} = -\frac{\hbar^2}{2}\sum\frac{1}{m_i}\nabla_i^2 + V(x, y, z)$$

- In quantum mechanics, we deal with only **Linear Hermitian operators**

1. **Linear Operators:** By linear operator, we mean that if it is applied on the sum of two functions, the result is equal to the sum of the operations on the two functions separately.

Let operator $\hat{A}$ operates on two function of f and g, then

$$\hat{A}[f + g] = \hat{A}f + \hat{A}g \qquad ...(1.17)$$

Like: Differential and Integral are linear operators.

$$\frac{d}{dx}(f + g) = \frac{d}{dx}(f) + \frac{d}{dx}(g)$$

$$\int[C_1f_1(x) + C_2f_2(x)]dx = C_1\int f_1(x)dx + C_2\int f_2(x)dx$$

But Square and Square root are non-linear operators.

$$[C_1f_1(x) + C_2f_2(x)]^2 \neq C_1^2[f_1(x)]^2 + C_2^2[f_2(x)]^2 + 2C_1C_2f_1(x)f_2(x)$$

$$\sqrt{C_1f_1(x) + C_2f_2(x)} \neq \sqrt{C_1f_1(x)} + \sqrt{C_2f_2(x)}$$

2. **Hermitian Operators:** Suppose ψ and ϕ are the two eigen functions of the operator $\hat{A}$, and if

$$\int_{-\infty}^{+\infty}\psi^*\left(\hat{A}\phi\right)d\tau = \int_{-\infty}^{+\infty}\phi\left(\hat{A}\psi\right)^* d\tau \qquad ...(1.18)$$

then the operator Â is called **Hermitian operator.**

Like: Show that $\frac{d^2}{dx^2}$ is a Hermitian operator.

Let $\psi = e^{ix}$ and $\phi = \sin x$ are the two eigen functions.

If $\psi = e^{ix}$, then $\psi^* = e^{-ix}$.

$$\int \psi^*(A\phi)dx = \int e^{-ix}\left(\frac{d^2}{dx^2}(\sin x)\right)dx = -\int e^{-ix}\sin x\, dx$$

$$\int \phi(\hat{A}\psi)^* dx = \int \sin x\left(\frac{d^2}{dx^2}(e^{-ix})\right)dx$$

$$= \int \sin x(-e^{-ix})dx = -\int \sin x \cdot e^{-ix}\, dx$$

Thus, the two integrals are same; therefore $\frac{d^2}{dx^2}$ is Hermitian operator.

- **Two Properties of Hermitian Operator:**

(i) Eigen Values of a Hermitian Operator are Real (Positive or Negative):

Let Â operator be a Hermitian operator and ψ its eigen function, then 'a' would be the real eigen value

$\hat{A}\psi = a\psi$ → Â: Hermitian operator; ψ: Eigen function; a: Real eigen value

Proof: Let Â be a Hermitian operator, ψ its eigen function, and 'a' the eigen value.

$$\hat{A}\psi = a\psi \qquad \text{...(1.19)}$$

and

$$(\hat{A}\psi)^* = a^*\psi^* \qquad \text{...(1.20)}$$

Multiply equation (1.19) by ψ* and on integration

$$\int \psi^*(A\psi)d\tau = \int \psi^*(a\psi)d\tau$$

$$\int \psi^*(\hat{A}\psi)d\tau = a\int \psi^*\psi\, d\tau \qquad \text{...(1.21)}$$

Multiply equation (1.20) by ψ and on integration

$$\int \psi(\hat{A}\psi)^* d\tau = \int \psi(a^*\psi^*)d\tau$$

$$\int \psi(\hat{A}\psi)^* d\tau = a^*\int \psi\psi^* d\tau \qquad \text{...(1.22)}$$

LHS of equation (1.21) and equation (1.22) is equal, since Â is Hermitian,

i.e. $$\int \psi^*(\hat{A}\psi)d\tau = \int \psi(\hat{A}\psi)^* d\tau$$

Therefore, RHS of equation (1.21) and equation (1.22) is also equal

i.e. $$a\int \psi^* \psi \, d\tau = a^* \int \psi\psi^* \, d\tau$$

Therefore, $a = a^*$

i.e. a is real.

(ii) Eigen Functions of a Hermitian Operator Corresponding to Different Eigen Values are Orthogonal:

Let ψ_1 and ψ_2 are two eigen functions of a Hermitian operator $\hat{A}$ corresponding to two eigen values a_1 and a_2 respectively.

The condition of orthogonality is

$$\int \psi_1 \psi_2^* \, d\tau = 0$$

or, $$\int \psi_1^* \psi_2 \, d\tau = 0$$

Proof: The eigen value equations are

$$\hat{A}\psi_1 = a_1\psi_1 \qquad ...(1.23)$$

$$\hat{A}\psi_2 = a_2\psi_2 \qquad ...(1.24)$$

Multiply equation (1.23) by ψ_2^* and on integration

$$\int \psi_2^* \left(\hat{A}\psi_1\right) d\tau = \int \psi_2^* \left(a_1\psi_1\right) d\tau$$

$$\int \psi_2^* \left(\hat{A}\psi_1\right) d\tau = a_1 \int \psi_2^* \psi_1 \, d\tau \qquad ...(1.25)$$

Since $\hat{A}$ is Hermitian, LHS of equation (1.25) becomes

$$\int \psi_2^* \left(\hat{A}\psi_1\right) d\tau = \int \psi_1 \left(\hat{A}\psi_2\right)^* d\tau$$

$$= \int \psi_1 \left(a_2\psi_2\right)^* d\tau \qquad \text{[using equation (1.24)]}$$

$$= a_2^* \int \psi_1 \psi_2^* \, d\tau$$

$$\int \psi_2^* \left(\hat{A}\psi_1\right) d\tau = a_2 \int \psi_1 \psi_2^* \, d\tau \qquad ...(1.26)$$

[$a_2^* = a_2$ because Hermitian operators have real value]

Equating RHS of equation (1.25) and equation (1.26), we have:

$$a_1 \int \psi_2^* \psi_1 \, d\tau = a_2 \int \psi_1 \psi_2^* \, d\tau$$

$$(a_1 - a_2) \int \psi_1 \psi_2^* \, d\tau = 0$$

But $a_1 \neq a_2$ then, $\int \psi_1 \psi_2^* \, d\tau = 0$

i.e. ψ_1 and ψ_2 are orthogonal if $a_1 \neq a_2$

However, if $a_1 = a_2$ then, $\int \psi_1 \psi_2^* \, d\tau \neq 0$

$\therefore$ ψ_1 and ψ_2 may not be orthogonal if $a_1 = a_2$.

1.16 Postulates of Quantum Mechanics

Quantum mechanics is based on few principles or postulates.

1. **Postulate I: Wavefunction or state function is an important property in quantum mechanics. To every state of a physical system, there is a function ψ which defines the state of the system.** Knowledge of this function permits a complete description of the state of the system.

   - The wavefunction $\underline{\overline{\psi}}$ (r, t) depends on the space co-ordinates of the particles, r = (x_1, y_1, z_1, x_2, y_2, z_2...) and on time, t. Suppose there is a dynamic system with 'n' particles (1, 2, 3, ...n) and their space co-ordinates are (x_1, y_1, z_1), (x_2, y_2, z_2), etc. Then the state of these particles can be explained by a wavefunction, $\underline{\overline{\psi}}$.

   $$\underline{\overline{\psi}}(r \cdot t) = \underline{\overline{\psi}}(x_1, y_1, z_1, x_2, y_2, z_2, \ldots x_n, y_n, z_n, t).$$

   Generally, the wavefunction is a complex function, which means we have to do with its complex conjugate $\underline{\overline{\psi}}$ *, as well, obtained by changing the imaginary number $i = \sqrt{-1}$ into –i. [Basically, this imaginary number, 'i', is not present in classical mechanics but is an inherent property which is present in quantum mechanics.]
   Thus, we calculate

   $$\int_{-\infty}^{+\infty} \underline{\overline{\psi}}(r, t) \cdot \underline{\overline{\psi}}^*(r, t) d\tau = \int_{-\infty}^{+\infty} \underline{\overline{\psi}}^2 d\tau$$

   By doing this, the imaginary number 'i' vanishes. This $\underline{\overline{\psi}}^2$ is the probability of finding the particle in a particular volume element $d\tau$ = dx_1, dy_1, dz_1, dx_2, dy_2, dz_2, ..., located at distance r and time, t.

   - If the wavefunction depends on time, t, then called **time dependent wavefunction,** $\underline{\overline{\psi}}$(r, t). But, if it is independent of time, t, then called **time independent wavefunctions,** ψ(r).

   $$\underline{\overline{\psi}}(r, t) = \psi(r) \cdot \phi(t)$$

   $$\underline{\overline{\psi}}(x, y, z, t) = \psi(x, y, z) \cdot \phi(t)$$

   - Actually, when we speak about evolution, the time dependent wavefunction, $\underline{\overline{\psi}}$ (r, t) is taken into account. But when we speak about the state of a physical system, then time independent wavefunction, ψ(r) is taken into account.
   - Thus, according to this postulate, state of a physical system is described by wavefunction ψ and all the information about a state is contained inside ψ. It is actually the storehouse of information.

2. **Postulate II: The function ψ must be single valued, finite, smoothly varying and continuous** between the boundaries at which it vanishes, but at the boundary the

wavefunction has to vanish. Thus, ψ vanishes at infinity, i.e. probability of finding the particle at infinity is zero. The function which possess all these properties is a well-behaved wavefunction.

Dynamic variable: It is any property of the system. Like energy, position, velocity, momentum, etc.

Observable: It is that property which can be experimentally observed.

Usually, dynamic variable is an observable. But spin component and angle 'x' are dynamic variable and not observable.

3. **Postulate III: For every observable of a system in classical mechanics (like position, velocity, momentum, energy, etc.), there corresponds a mathematical operator in quantum mechanics.** The form of the operator can be obtained from the classical expressions for the observables according to the definite rules. The physical properties of the observable can be inferred from the mathematical properties of its associated operator.

Table 1.2: Classical mechanical observable and their corresponding quantum mechanical operators
[Refer Sec. 1.15: Operators]

Observable		Operator	
Name	*Symbol*	*Symbol*	*Operation*
Position	x	$\hat{x}$	multiplication by x
x-component of momentum	p_x	$\hat{p}_x$	$\frac{\hbar}{i}\frac{\partial}{\partial x}$
Square of x-component of momentum	p_x^2	$\hat{p}_x^2$	$-\hbar^2\frac{\partial^2}{\partial x^2}$
Linear momentum	p	$\hat{p}^2$	$\frac{\hbar}{i}\nabla$
Square of linear momentum	p^2	$\hat{p}^2$	$-\hbar^2\nabla^2$
x-component of angular momentum	L_x	$\hat{L}_x$	$\frac{\hbar}{i}\left(y\frac{\partial}{\partial z}-z\frac{\partial}{\partial y}\right)$
Kinetic energy	$T=\frac{p^2}{2m}$	$\hat{T}=\frac{\hat{p}^2}{2m}$	$\frac{-\hbar^2}{2m}\nabla^2$
Potential energy	V	$\hat{V}(x,y,z)$	Multiplication by V
Total energy	$E = T + V$	$\hat{H}=\hat{T}+\hat{V}$	$\frac{-\hbar^2}{2m}\nabla^2+V(x,y,z)$

In quantum mechanics, we deal with **Linear Hermitian operators.**

[Refer sec. 1.15: Operators]

A **Linear operator** $\hat{A}$ has the property, i.e. equation (1.17)

$$\hat{A}[f+g]=\hat{A}(f)+\hat{A}(g)$$

A **Hermitian operator** $\hat{A}$ has the property, i.e. equation (1.18)

$$\int_{-\infty}^{+\infty}\psi^*\left(\hat{A}\phi\right)d\tau=\int_{-\infty}^{+\infty}\phi\left(\hat{A}\psi\right)^*d\tau$$

The Hermitian property of the operator ensures that the value of any observable obtained will be real.

4. **Postulate IV: For any measurement of the observable associated with the operator $\hat{A}$, the only value that will ever be observed is the eigen value 'a' which can be obtained from eigen value equation,** i.e., equation (1.10).

$$\hat{A}\psi = a\psi$$

An eigen value is just a real number and when an observable has a definite value, the value of the observable is the eigen value.

An eigen value equation is that equation in which an operator $\hat{A}$ when operates upon the wavefunction ψ, it gives a constant called eigen value 'a' and wavefunction ψ back. While the eigen function and the operator may be real or complex, but the eigen value must be real because it represents observable physical quantities.

Moreover, quantum mechanical operators of interest (like momentum, energy, etc.) are Hermitian operators which have real eigen values. The Hermitian property of the operator ensures that value of any observable obtained will be real.

5. **Postulate V:** If ψ is an eigen function of operator $\hat{A}$, then $\hat{A}\psi = a\psi$. But it has to be noted that ψ is not necessarily an eigen function of $\hat{A}$. **When ψ is not an eigen function of $\hat{A}$, then we talk of average value in quantum mechanics.**

The average (or expectation) value of an observable of a system whose state function is ψ, is given by **mean value theorem:**

$$\underset{\text{average value}}{\bar{a}} = \underset{\text{expectation value}}{\langle a\rangle} = \frac{\int_{-\infty}^{+\infty}\psi^*\hat{A}\psi\,d\tau}{\int_{-\infty}^{+\infty}\psi^*\psi\,d\tau}$$

Also, if ψ is an eigen function of operator $\hat{A}$, the expectation value is equal to the eigen value 'a'.

$$\bar{a}=\langle a\rangle=\frac{\int_{-\infty}^{+\infty}\psi^*\left(\hat{A}\psi\right)d\tau}{\int_{-\infty}^{+\infty}\psi^*\psi\,d\tau}=\frac{a\int_{-\infty}^{+\infty}\psi^*(a\psi)d\tau}{\int_{-\infty}^{+\infty}\psi^*\psi\,d\tau}=\frac{a\int_{-\infty}^{+\infty}\psi^*\psi\,d\tau}{\int_{-\infty}^{+\infty}\psi^*\psi\,d\tau}=a$$

The integrals are overall permitted value of the coordinate of the system. The expectation value is defined as the prediction of the mean of a large number of measurements of the mechanical quantity.

2
Spectroscopy

2.1 Introduction

Quantum mechanics suggested that the energy levels of the system are quantized and are represented by the appropriate quantum numbers. These energy levels are obtained by the solution of Schrodinger wave equation.

Spectroscopy is defined as the branch of science which deals with the study of interaction of electromagnetic radiation with matter. It deals with absorption and emission of electromagnetic radiation by an atom, ion or molecule. The energy is emitted out as radiation when an electron jumps from higher energy level to lower energy level. Whereas the energy is absorbed when an electron jumps from lower to higher energy level.

The interaction of electromagnetic radiations with matter produces **spectra**. A spectra can be used for both qualitative and quantitative analysis. It is possible to correlate the experimentally observed data on absorption and emission with the nature of atoms or molecules. The detailed information regarding molecular structure (molecular symmetry, bond distances and bond angle) and chemical properties (electronic distribution, bond strength and intra-and inter-molecular processes) can be obtained from the spectra.

Thus, spectroscopy is a powerful tool used in study of the atomic and molecular structure and analysis of wide range of compounds in chemistry.

PROBLEMS BASED ON SPECTROSCOPY

Q. 1. What is spectroscopy and what is the importance of spectroscopy?

Q. 2. What is the role of quantum mechanics in spectroscopy?

2.2 Dual Nature of Electromagnetic Radiation

Since the study of spectroscopy involves the interactions of electromagnetic radiation, so it becomes essential to know about the behaviour of electromagnetic radiation.

A. Classical theory of radiation or Electromagnetic wave theory (radiation has wave nature)

B. Quantum theory of radiation or Planck's quantum theory (radiation has particle nature)

A. Classical Theory of Radiation or Electromagnetic Wave Theory

James Clark Maxwell in 1864 put forward the electromagnetic wave theory. The main points of this theory are:

1. When electrically charged particles move under acceleration, alternating electric field and magnetic field are produced and transmitted. These fields are transmitted in the form of waves called **Electromagnetic waves (EM waves)** or **Electromagnetic radiations.**
2. The oscillating electric and magnetic fields produced by oscillating charged particles are perpendicular to each other and both are perpendicular to the direction of electromagnetic radiations.
3. The radiations possess wave nature and travel with the velocity of light ($c = 3 \times 10^8$ ms^{-1}).
4. Unlike sound waves or water waves, these electromagnetic radiations do not require any medium for propagation and can move in vacuum.

Some important characteristics of a wave:

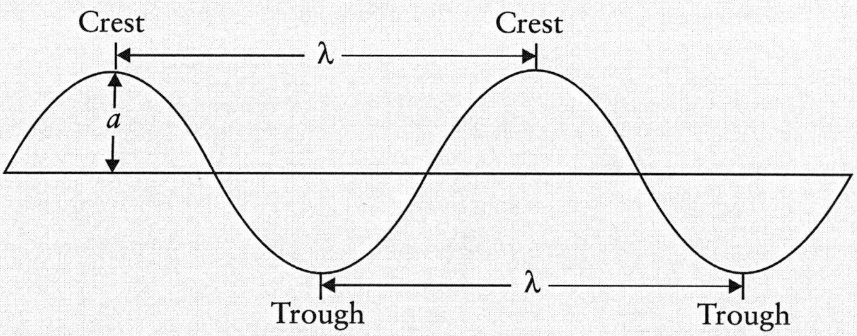

1. **Wavelength (λ):** It is defined as the distance between any two consecutive crests or troughs.
   Unit: Å or m or cm or nm or pm.
   1 Å = 10^{-10} m; 1 nm = 10^{-9} m; 1 pm = 10^{-12} m; 1 cm = 10^{-2} m.
2. **Frequency (ν):** It is defined as the number of waves passing through a point in one second.
   Unit: Hertz (Hz) or cycles/sec (cps) or sec^{-1}
   $$1 \text{ Hz} = 1 \text{ cps} = 1 \text{ sec}^{-1}$$
3. **Velocity (c):** It is defined as the distance travelled by the wave in one second.
   Unit: cms^{-1} or ms^{-1}
   In vacuum, all types of electromagnetic waves, regardless of wavelength, travel at same speed, i.e. 3×10^8 ms^{-1}. This is called speed of light (c).
   $$c = 3 \times 10^8 \text{ ms}^{-1}$$
4. **Amplitude (a):** It is defined as the height of the crest or depth of the trough.
   Unit: Å or m or cm or nm or pm.
5. **Wavenumber $(\bar{v})$:** It is defined as the number of waves present per unit length. It is equal to the reciprocal of the wavelength.
   $$\bar{v} = \frac{1}{\lambda}$$
   Unit: cm^{-1} or m^{-1}.

There are many types of electromagnetic radiations which differ from one another in wavelength and hence in frequencies. When these electromagnetic radiations are arranged in order of their increasing wavelengths or decreasing frequencies, the complete arrangement obtained is called **electromagnetic spectrum** (as shown in Fig. 2.1).

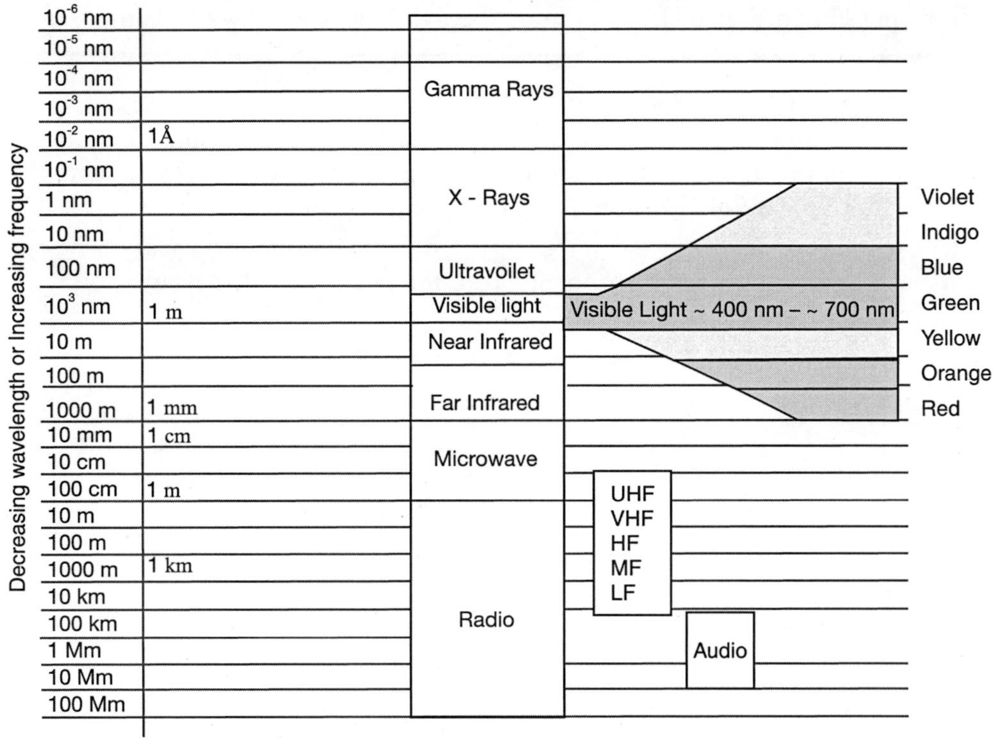

Fig. 2.1: The Electromagnetic Spectrum

Limitations of Classical theory of radiation or Electromagnetic wave theory: Some of the experimental phenomena such as reflection, refraction, diffraction, interference, polarization, etc. can be explained by the wave nature of electromagnetic radiations, but it could not explain the following observations:

1. **Black body radiation** (the nature of emission of radiation from hot bodies).
2. **Photoelectric effect** (ejection of electrons from metal surface when radiation strikes it).
3. **Compton effect** (Scattering of X-rays).
4. **Variation of heat capacity of solids** as a function of temperature.
5. **Line spectra of atoms** with special reference to hydrogen.

These phenomena could be explained only if electromagnetic radiations are supposed to have particle nature instead of wave nature. Thus to explain the above facts, Quantum theory of radiation was given by Max Planck.

B. Planck's Quantum Theory or Quantum Theory of Radiation or Corpuscular Theory of Radiation

To explain the phenomena such as black body radiation, photoelectric effect, compton effect, variation of heat capacity of solids and line spectra of atoms, Max Planck in 1900 put forward Planck's quantum theory. This theory was further extended by Einstein in 1905.

The main points of this theory are:

1. The radiant energy is emitted or absorbed not continuously but discontinuously in the form of small discrete packets of energy. Each such packet of energy is called **quanta.** In case of light, the packet of energy is called **photon.**
2. Each wave packet is associated with a definite amount of energy. The energy of each quanta is directly proportional to the frequency of the radiation

$$E \propto \nu$$

$$E = h\nu$$

as, $$c = \nu\lambda$$

$$E = \frac{hc}{\lambda}$$

where, h is a proportionality constant called **Planck's constant.**

$$h = 6.626 \times 10^{-27} \text{ erg sec} = 6.626 \times 10^{-34} \text{ J sec.}$$

3. A body can emit or absorb energy in terms of the integral multiples of quanta

$$E = Nh\nu$$

or $$E = \frac{Nhc}{\lambda}$$

where, N is an integral number.

- **Dual Nature of Radiation:** Some properties of radiation like reflection, refraction, diffraction, interference, polarization, etc. can be explained only if we consider radiation to have wave nature, whereas some properties of radiation like black body radiation, compton effect, photoelectric effect, etc. can be explained if we consider radiation to have particle nature. Thus, Einstein in 1905 suggested that radiation has dual nature, i.e. wave nature as well as particle nature.

PROBLEM BASED ON DUAL NATURE OF RADIATION

Q. 1. Write a short note on wave-particle duality.

PROBLEMS BASED ON PLANCK'S QUANTUM THEORY

Q. 1. Calculate the wavenumber and frequency of radiation having wavelength 3×10^6 nm.

Ans. $\lambda = 3 \times 10^6 \text{ nm} = (3 \times 10^6) \times 10^{-9} \text{ m} = 3 \times 10^{-3} \text{ m}$

(i) Wavenumber $= \bar{v} = \frac{1}{\lambda} = \frac{1}{3\times10^{-3}} = 0.33\times10^{3}\text{m}$

(ii) Frequency $= \nu = \dfrac{c}{\lambda} = \dfrac{3\times10^{8}\,ms^{-1}}{3\times10^{-3}\,m} = 10^{11}\,s^{-1}$

Q. 2. Calculate the wavelength of the electromagnetic radiation having frequency of 1367 kHz (kilohertz). Which part of the electromagnetic spectrum does it belong to?

Ans. $\nu = 1367$ kHz

$= 1367 \times 10^{3}$ Hz $= 1367 \times 10^{3}\,s^{-1}$ (1 Hz = 1 s^{-1})

$c = 3 \times 10^{8}\,ms^{-1}$,

$$\lambda = \frac{c}{\nu} = \frac{3\times10^{8}\,ms^{-1}}{1367\times10^{3}\,s^{-1}} = 219.5\,m$$

It belongs to the radiowave region of the electromagnetic spectrum.

Q. 3. Calculate the frequency and energy of a photon of light having wavelength 3000 Å.

Ans. $\lambda = 3000$ Å $= 3000 \times 10^{-10}$ m

$c = 3 \times 10^{8}\,ms^{-1}$

$h = 6.626 \times 10^{-34}$ Js

(i) $\nu = \dfrac{c}{\lambda} = \dfrac{3\times10^{8}\,ms^{-1}}{3000\times10^{-10}\,m} = 10^{-15}\,s^{-1}$

(ii) Energy of a photon = E = hν

$E = (6.626 \times 10^{-34}\,Js) \times (10^{-15}\,s^{-1})$

$E = 6.6 \times 10^{-19}$ Js

Q. 4. Calculate the energy of one mole of photons of light having frequency $3 \times 10^{14}\,s^{-1}$.

Ans. $\nu = 3 \times 10^{14}\,s^{-1}$

$h = 6.626 \times 10^{-34}$ Js

Energy of one photon = E = hν

$= (6.626 \times 10^{-34}) \times (3 \times 10^{14})$

$= 19.8 \times 10^{-20}$ J

Number of atoms in one mole $= N_A = 6.023 \times 10^{23}$

Energy of one mole of photons $= N_A h\nu$

$E = (6.023 \times 10^{-23}) \times (6.626 \times 10^{-34}) \times (3 \times 10^{14})$

$E = 119.25 \times 10^{3}$ J

Q. 5. A 100 Watt bulb emits light of wavelength 6000 Å. Calculate the number of photons emitted by the bulb in one minute.

Ans. Power (P) = 100 W

Time (t) = 1 min = 60 sec

Energy = Power × Time

E = 100 × 60 = 6000 J

λ = 6000 Å $= 6000 \times 10^{-10}$ m

$$E = Nh\nu = \frac{Nhc}{\lambda}$$

$$N = \frac{E \times \lambda}{h \times c} = \frac{(6000) \times (6000 \times 10^{-10})}{(6.626 \times 10^{-34}) \times (3 \times 10^{8})}$$

$$N = 1.81 \times 10^{22} \text{ photons}$$

Q. 6. 3×10^{18} photons of a certain light are found to produce 1.5 J of energy. Calculate the wavelength of the light.

Ans.

$$N = 3 \times 10^{18}$$

$$E = 1.5 \text{ J}$$

$$c = 3 \times 10^{8} \text{ ms}^{-1}$$

$$h = 6.626 \times 10^{-34} \text{ Js}$$

$$E = \frac{Nhc}{\lambda}$$

$$\lambda = \frac{Nhc}{E}$$

$$\lambda = \frac{(3 \times 10^{18}) \times (6.626 \times 10^{-34}) \times (3 \times 10^{8})}{1.5}$$

$$\lambda = 39.78 \times 10^{-8} \text{ m}$$

Q. 7. Calculate and compare the energies of two radiations, one with a wavelength of 4000 Å and other with wavelength 8000 Å.

Ans. Case I: $\lambda_1 = 4000$ Å $= 4000 \times 10^{-10}$ m

$$E_1 = \frac{hc}{\lambda_1} = \frac{(6.626 \times 10^{-34}) \times (3 \times 10^{8})}{(4000 \times 10^{-10})}$$

$$E_1 = 4.95 \times 10^{-19} \text{ J}$$

Case II: $\lambda_2 = 8000$ Å $= 8000 \times 10^{-10}$ m

$$E_2 = \frac{hc}{\lambda_2} = \frac{(6.626 \times 10^{-34}) \times (3 \times 10^{8})}{(8000 \times 10^{-10})}$$

$$E_2 = 2.48 \times 10^{-19} \text{ J}$$

On comparing the two value of energy, i.e. taking ratio, we have

$$\frac{E_1}{E_2} = \frac{4.95 \times 10^{-19}}{2.48 \times 10^{-19}}$$

$$E_1 = 2E_2$$

Q. 8. The bond energy of the carbonyl bond in acetone is 720 kJ/mole. What will be the wavelength of light absorbed by this bond? Will a 254 nm Hg source be effective in breaking this bond?

Ans.

$$E = 728 \text{ kJ/mole} = 728000 \text{ J/mole}$$

$$E = \frac{hc}{\lambda}$$

$$\lambda = \frac{hc}{E} = \frac{(6.6\times10^{-34}\,\text{J sec})\times(3\times10^{8}\,\text{ms}^{-1})}{(728000\,\text{J/mole})}$$

Q. 9. Express 3000 cm^{-1} wavenumber in

(i) the frequency in MHz

(ii) energy in joules.

Q. 10. The bond energy of carbonyl bond in acetone is 728 kJ/mole. What will be the wavelength of light absorbed by this bond? Will a 264 nm mercury source be effective in breaking this bond?

2.3 Emission and Absorption Spectra

Interaction of electromagnetic radiation with matter produces **spectra.**

An instrument used to separate the radiations of different wavelengths (or frequencies) is called **spectroscope.** A spectroscope consists of a prism or a diffraction grating for the dispersion of radiations and a telescope to examine the emergent radiation with the human eye.

- The spectra are broadly classified into:

  A. Emission spectra

  B. Absorption spectra

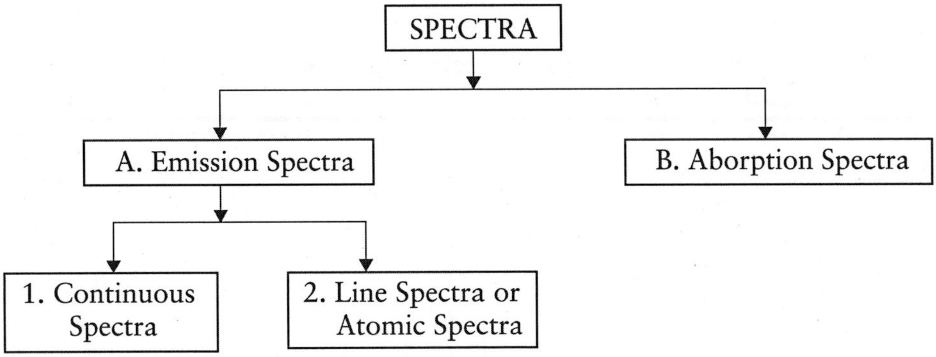

A. Emission Spectra: When electromagnetic radiation interacts with matter (like by passing electric discharge through a gas at low pressure or by heating some substance at high temperature, etc.), then atoms and molecules may absorb energy and reach to a higher energy state. With higher energy, these are in an unstable state. For returning to their normal energy state, the atoms and molecules emit radiations. When these emitted radiations are passed directly through the prism and then received on the photographic plate, the spectra obtained is called '**Emission Spectra**'.

Depending upon the source of radiation, the emission spectra are mainly of two types:

1. **Continuous Spectra:** When radiations from any source such as sun, a bulb or any hot glowing body are passed directly through a prism, it splits up into seven different colours from violet to red (like rainbow). These colours are so continuous that each of them merges into the next. The spectra obtained is called **Continuous Spectra.**

2. **Line Spectra or Atomic Spectra:** When some substance (like volatile salt) is heated at high temperature or an electric discharge is passed through a gas at low pressure, the radiations are emitted. When these radiations are passed directly through a prism, it is found that no continuous spectra is obtained but some isolated coloured lines are obtained on the photographic plate separated from each other by dark spaces. The spectra obtained is called **Line Spectra.** Each line in the spectra corresponds to a particular wavelength.

   Each element gives its own characteristic line spectra, differing from those of all other elements. Hence, spectra of the elements are regarded as their finger prints differing from each other like the finger prints of the human beings. Since spectra of most of the elements consists of a number of bright lines separated by dark lines, that is why **line spectra** is also called **atomic spectra.**

B. Absorption Spectra: When white light from any source is passed through the solution or vapours of a chemical substance, then some wavelengths are absorbed, depending upon the nature of the element. After absorption of certain wavelengths, the transmitted light is analysed by the spectroscope, then it is observed that some dark lines are obtained in the continuous spectra of the white light which is irradiated on the source. Further, it is observed that the dark lines are at the same place where coloured lines are obtained in the emission spectra of the same substance. The spectra obtained is called **absorption spectra.** Absorption spectra is always discontinuous spectra consisting of dark lines.

- **Line Emission Spectra of Hydrogen**

  When an electric discharge is passed through gaseous hydrogen at low pressure, hydrogen molecules dissociate and the energetically excited hydrogen atoms produced emit radiations of discrete frequencies. The emitted radiations are examined with a spectroscope, and the spectra obtained is called the emission spectra of hydrogen (as shown in Fig. 2.2). The hydrogen spectra consists of a large number of lines which are grouped into different series, named after the discoverers.

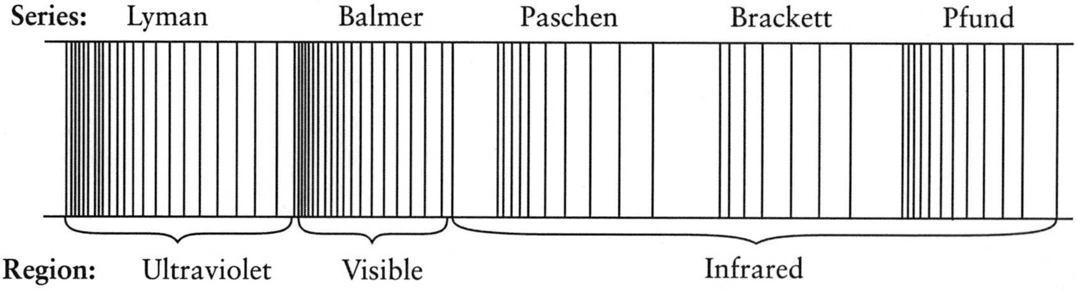

Fig. 2.2: Line emission spectra of hydrogen

Line spectra of hydrogen atom is simplest because hydrogen being the lightest element, the number of lines produced in the spectra is minimum. As the molecular mass of the gas increases, the spectra becomes more and more complex.

2.4 Atomic and Molecular Spectroscopy

The study of spectroscopy can be carried out under the following heads—Atomic spectroscopy and Molecular spectroscopy.

(a) **Atomic Spectroscopy:** It deals with the interaction of electromagnetic radiation with atoms which are most commonly in their lowest energy state called the **ground state.**

Monoatomic substances normally exist in the gaseous state and are able to absorb electromagnetic radiation. When atoms absorb electromagnetic radiation, this results in an increase in their electronic energy, resulting in transition of electrons from one electronic energy level to another. The absorption can only occur if the attacking photon of electromagnetic radiation possesses energy which is equal to the energy difference between two quantized energy levels of the absorbing atom, i.e.

$$\Delta E = h\nu$$

where, ΔE is the energy difference between two quantum levels and ν is the frequency of photon which can result in the electronic excitation.

When the electron is present in any orbit other than the orbit of least energy for it, the electron as well as the atom as whole, are said to be in the excited state. When the excited atom reverts to the ground state, it loses energy in one straight jump or through a number of steps in the form of a single quantum or photon of electromagnetic radiation. When it happens for a large number of atoms, the total effect is seen in the **atomic spectra** as **spectral lines.**

The atomic spectra obtained from atoms after absorbing radiation consists of a number of bright lines separated by dark lines. These spectral lines are of discrete wavelengths extending throughout the spectra and are generally obtained from the light sources like mercury, sodium, neon discharge tube, etc. Some salient features of the atomic spectra are:

(i) The spectral lines are regularly spaced.

(ii) The wavelength of the lines emitted are the characteristics of the element under consideration.

(iii) When the spectral lines are observed by a high resolving power instrument, the different lines are regularly spaced but they differ in their intensities.

(b) **Molecular Spectroscopy:** It deals with the interaction of electromagnetic radiation with molecule. The spectra of molecules are much more complicated than those of atoms because it involves transition between rotational and vibrational energy levels in addition to electronic transitions.

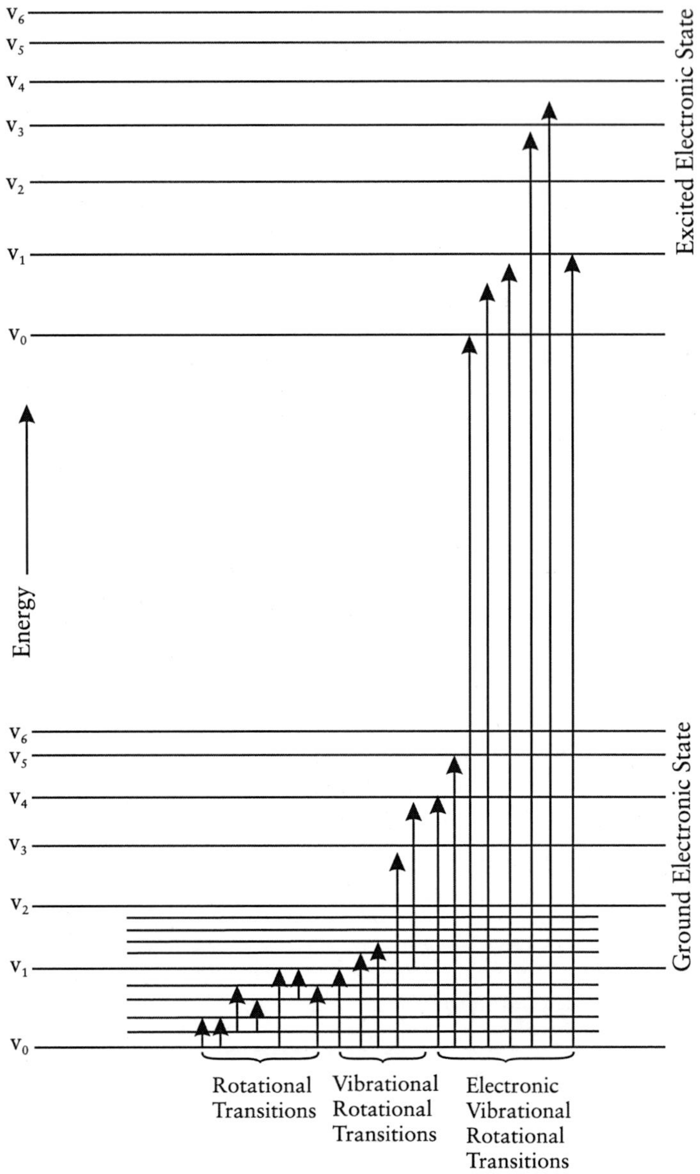

Fig. 2.3: Two electronic levels

In case of atoms when the energy is absorbed, the electrons jumps to some outer orbits and then they start jumping back with the emission of energy equal to the difference between the energy of the two energy levels, and the result is an **atomic spectra**. However, in case of molecules when the energy is absorbed it may result into rotation, vibration or electronic transition depending upon the amount of energy absorbed. Just as electronic energy is quantized, i.e. there are discrete electronic energy levels in an atom or a molecule, the rotational and vibrational energies are also quantized. The discrete rotational, vibrational and electronic energy levels of a molecule are collectively called **molecular energy levels**. The transitions of energies can take place only between these energy levels. This results in a **molecular spectra**.

The energy absorbed by a molecule is so large that an electron can jump from one electronic energy level to another (say n = 1 to n = 2) but as the energy required for transition from one vibrational energy level to another is much less and for one rotational level to another is still less. Hence, the electronic transition is also accompanied by the transition between the vibrational levels as well as rotational levels. The complete picture of the various molecular energy levels for the first two electronic levels is shown in Fig. 2.3.

It is evident that each electronic level consists of a number of vibrational sub-levels represented by v = 0, 1, 2, 3, etc. and further each vibrational levels consists of a number of rotational sub-levels represented by J = 1, 2, 3, etc. The v and J are called **vibrational and rotational quantum number** respectively.

Because of a large number of energy levels involved for transition, the molecular spectra are much more complex than atomic spectra. However, just as atomic spectra help in the study of the structure of the atoms, the study of molecular spectra gives even more information about the structure and properties of molecules such as bond lengths, bond angles, bond strength, shapes, dipole moments, etc.

2.5 Types of Molecular Energies

A molecule may possess internal energy which can be sub-divided into four classes: **translational energy, rotational energy, vibrational energy and electronic energy**. Thus, the total molecular energy (E) can be separated into different contributions.

$$E = E_{trans} + E_{rot} + E_{vib} + E_{elec}$$

These four contributions to the energy are not completely independent. For example, rotation of a molecule introduces an effective centrifugal force which stretches the bonds and therefore affects the vibrational energy. However, treating these energy terms as separate is good for understanding.

(a) **Translational energy:** This energy is associated with the uniform motion (velocity) of a molecule as a whole. This motion is generally described with respect to the centre of mass of the molecule (Fig. 2.4). The classical energy due to translational motion is given by

$$E_{trans} = \frac{1}{2}mv^2$$

where E_{trans} is the translational energy of molecular mass 'm' moving with velocity 'v' with respect to the centre of mass.

Due to translational motion, the molecule is free to move in the three perpendicular directions x, y, z. It means that it has three degrees of freedom. But the translational energy of a molecule of gas by the classical theorem of the equipartition of energy along any axis is $\frac{1}{2}$ kT or $\frac{1}{2}$ RT per mole.

Therefore, the translational energy of molecule of gas along three mutually perpendicular axes will be $3\times\frac{1}{2}$ kT or $\frac{3}{2}$ kT or $\frac{3}{2}$ RT per mole. In solids, the translational freedom is restricted.

As the number of translational energy levels is exceedingly large and the difference between them is negligible, it means that continuous range of translational energies are available. The concept of translational energy is not of much use in molecular spectroscopy.

(b) **Rotational Energy:** This type of energy is associated with the overall rotation of the molecule with atoms considered as fixed point masses. According to the classical theory, the value of rotational energy is given by

$$E_{rot} = \frac{1}{2}I\omega^2$$

where I is the moment of inertia and ω the angular velocity of rotating molecule. However, the quantum mechanical formula for the rotational energy of a simple linear molecule is given by

$$E_{rot} = J(J+1)\frac{h^2}{8\pi^2 I} \qquad ...(2.1)$$

where I is the moment of inertia of a simple linear molecule and J is zero or a positive integer called the **rotational quantum number**. Each rotation level has a (2J + 1) fold degeneracy.

It is important to remark here that only the polar molecules can rotate in an electromagnetic field, i.e. having permanent dipole moment.

A monoatomic molecule possesses only one rotational degree of freedom. But a nonlinear tri-or-poly-atomic molecule may rotate about the three perpendicular axes passing through the centre of gravity (Fig. 2.4). It is therefore said to have three rotational degrees of freedom.

A linear molecule cannot rotate about the axis passing through the nuclei of the bonded atoms. It means that one of its rotational degree of freedom is restricted. Hence, a linear molecule possesses only two rotational degrees of freedom.

(c) **Vibrational Energy:** This type of energy is associated with oscillation of atoms (of the molecule) which are considered as point masses about equilibrium positions. This type of energy can be treated on a quantum mechanical basis. The vibrational energy of a molecule may be given as

$$E_{vib} = h\nu\left(v + \frac{1}{2}\right) \qquad ...(2.2)$$

where ν is the vibrational frequency and v is zero or a positive integer called the **vibrational quantum number.**

It is to be noted that the form of equation (2.2) does not fulfil the convention that the energy shall be zero for the ground state. The ground state corresponds to v = 0 with an energy $h\nu/2$, this is generally known as **zero point energy.**

The vibrational degree of freedom for non-linear molecules such as H_2O, NO_2 and CH_4 are 3N–6. But there are (3N–5) vibrational degrees of freedom for linear molecules, e.g. CO_2, C_2H_2 and I_3^-.

When a molecule absorbs energy, the increase in vibration motion of the molecule is usually accompanied by increased rotation of the same molecule. This combination provides the basis of IR absorption spectroscopy.

(d) **Electronic Energy:** This type of energy is associated with the 'motion' of electrons while considering the nuclei of atoms (of a molecule) as fixed points. The increase in the electronic energy of a molecule occurs due to an increase in the kinetic energy required to move the electron from the ground state to an excited state and varies with the type of molecular bond in which the electron exists. Further, the vibrational and rotational energies of the molecules are added to the electronic energies.

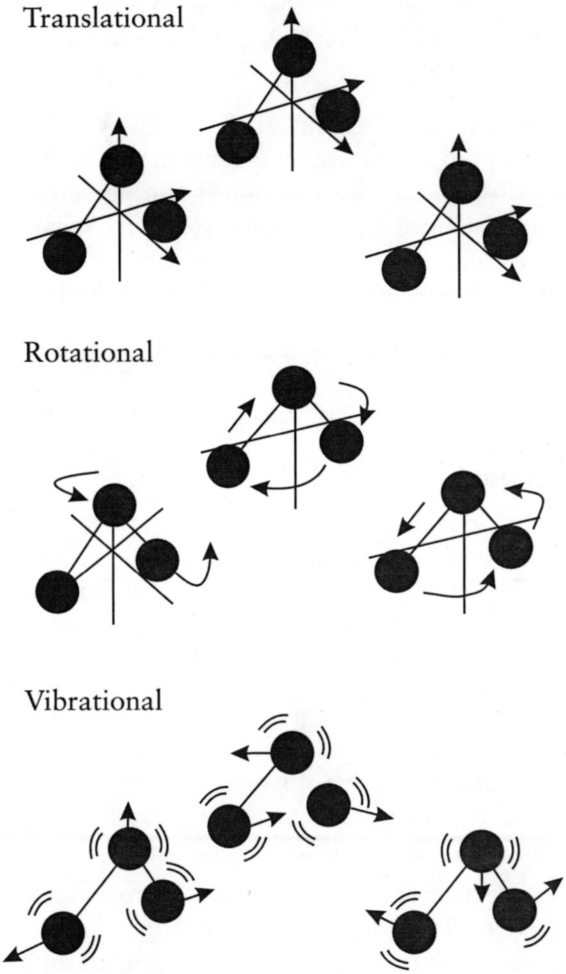

Fig. 2.4: Three different ways in which molecules can store energy as a result of motion of atoms in a non-linear triatomic molecule

PROBLEM BASED ON MOLECULAR ENERGIES

Q. 1. Explain the different molecular energies (Translation, Vibration, Rotation and Electronic) of a diatomic molecule.

2.6 Born-Oppenheimer Approximation

The exact solution of the Schrodinger wave equation for any molecule is not possible because the Hamiltonian operator contains many terms

$$\hat{H} = \hat{T}_n + \hat{T}_e + \hat{V}_{ee} + \hat{V}_{en} + \hat{V}_{nn}$$

where, $\hat{T}_n$ stands for the kinetic energy operator of the nuclei, $\hat{T}_e$ stands for the kinetic energy for electron, $\hat{V}_{ee}$ is the electron repulsion term, $\hat{V}_{en}$ is the electron-nuclear attraction term and $\hat{V}_{nn}$ is the nuclear-nuclear repulsion terms.

Since nuclei are much heavier than electrons, their velocities are much smaller. The electrons carry out many cycles of motion in the time the nucleus moves a short distance. Therefore, we consider the nuclei to be fixed while the electron moves through entire volume of the molecule. To a good approximation, the Schrodinger equation can be separated into two parts—one part depending upon the electronic motion while the other on the fixed nuclear position. This approximation is known as the **Born-Oppenheimer approximation.**

PROBLEM BASED ON BORN-OPPENHEIMER APPROXIMATION

Q. 1. Explain the Born-Oppenheimer approximation and its relevance.

2.7 Types of Molecular Spectra

A molecule has different types of quantized energy levels, i.e. translational, rotational, vibrational and electronic. The molecular spectra arise due to the transitions taking place between these energy levels. The energy absorbed for any transition is equal to the difference in the energies of two levels involved. It is found that these energies for transition are in the order.

$$E_t << E_r << E_v << E_e$$

The difference between the successive translational energy levels is so small ($\simeq 10^{-60}$ J mole^{-1}) that it cannot be observed experimentally. For this reason, for practical purposes, **translational energy is considered as continuous and we do not observe any translational spectrum.** The different types of molecular spectra are generally observed, depending upon the energies absorbed and the regions of the electromagnetic spectrum.

Rotational (Microwave) Spectra: These spectra arise due to transitions between the rotational energy levels of a **gaseous** molecule on the absorption of radiations in the microwave region. These spectra are shown by molecules which **possess a permanent dipole moment,** e.g. HCl, CO, H_2O vapour, NO, etc. Homonuclear diatomic molecules like H_2, Cl_2, etc. and linear polyatomic molecules such as CO_2, which do not have a dipole moment, are microwave inactive. Microwave spectra occur in the spectral range of 1–100 cm^{-1}.

Vibrational (Infrared) Spectra: These spectra arise due to transitions induced between the vibrational energy levels of a molecule on the absorption of radiation belonging to the infrared region. IR spectra are shown by molecules when vibrational motion is accompanied by a **change in the dipole moment of the molecule.** These spectra occur in the spectral range of 500-400 cm^{-1}.

Raman Spectra: Raman spectra is related to vibrational and/or rotational transitions in molecules but in a different manner. In this case, what we measure is the scattering and not the absorption of radiation. An intense beam of monochromatic radiation in the visible region is allowed to fall on a sample and the intensity of scattered light is observed at right angles to the incident beam. Most of the scattered light has the same frequency as the incident beam (this is called Rayleigh scattering). However, a small amount of light has different frequencies than the incident beam. This is called Raman Scattering. The energy differences between these weak lines and the main Rayleigh line correspond to vibrational and/or rotational transitions in the molecule under investigation. Raman spectra are observed in the visible region, viz. 12,500-25,000 cm^{-1}.

Electronic Spectra: Electronic spectra arise due to electronic transitions in a molecule by absorption of radiations falling in the visible and ultraviolet regions. While electronic spectra in the visible region span 12,500-25,000 cm^{-1}, those in the ultraviolet region span 25,000-70,000 cm^{-1}. As electronic transitions in a molecule are invariably accompanied by vibrational and rotational transitions, the electronic spectra of molecules are highly complex.

2.8 Selection Rules

Atomic spectra as well as molecular spectra are obtained due to transition taking place between energy levels. However, such transitions can take place only between definite energy levels and not between any two energy levels. **The restrictions thus applied on the transition are governed by certain rules which are called the selection rules.** If these rules are followed, the transition can take place and it is called an **allowed transition.** If these rules are not followed, the transition cannot take place and it is called a **forbidden transition.** The selection rules to be followed depend upon the type of transition (in case of molecular spectra).

The **selection rules** are generally expressed in terms of changes in quantum numbers for the allowed transitions. For example, for the pure rotational transition, the selection rule is $\Delta J = \pm 1$ where J represents rotational quantum number ($\Delta J = +1$ corresponds to absorption and $\Delta J = -1$ corresponds to emission). Similarly, for pure vibrational transition, the selection rule is $\Delta v = \pm 1$ where v represents the vibrational quantum number.

PROBLEMS BASED ON MOLECULAR TRANSITION AND SELECTION RULE

Q. 1. Why do transitions takes place when electromagnetic radiations interact with a molecule?

Q. 2. State the conditions for a molecule to undergo transitions.

2.9 Signal to Noise Ratio (S/N)

Modern spectrophotometer use some form of electronic amplification device and these are used to magnify the signals which are produced by detector. The recorded spectrum has a background of random fluctuations caused by electronic signals of source or generated by the amplifying equipment. These fluctuations are known as '**noise**'. In order to distinguish the real spectral peak from noise one, it must be three or four times more intense to that noise fluctuations (S/N, 3 ≃ or 4). The background noise imposes a limitation on the sensitivity of spectrometer.

There are different ways to improve the signal to noise ratio (S/N) but all require the expenditure of time.

2.10 Resolving Power

The performance of a spectrometer is generally expressed in terms of its **resolving power.**

The molecular absorption never takes place at a single frequency, but always spread over a range of frequencies. This is the reason, that the spectra obtained have broadened line spectra.

To resolve lines whose wavelengths are close together (that is, to make the lines distinguishable), the line should be as narrow as possible.

2.11 Widths and Intensities of the Spectral Lines

In any type of molecular spectroscopy the two important characteristics of any spectral line are:

(i) How sharp (or broad) is the time?
This is decided by the **width** of the spectral line.

(ii) How strong (or weak) is the line?
This is decided by the **intensity** of the spectral line.

(i) Width of the Spectral Line: If the spectral line is sharp, it will have no width as shown in Fig. 2.5(a). On the other hand, if the line is not sharp, it will have a certain width as shown in Fig. 2.5(b), i.e. it spreads over a range of frequencies (or wavenumbers).

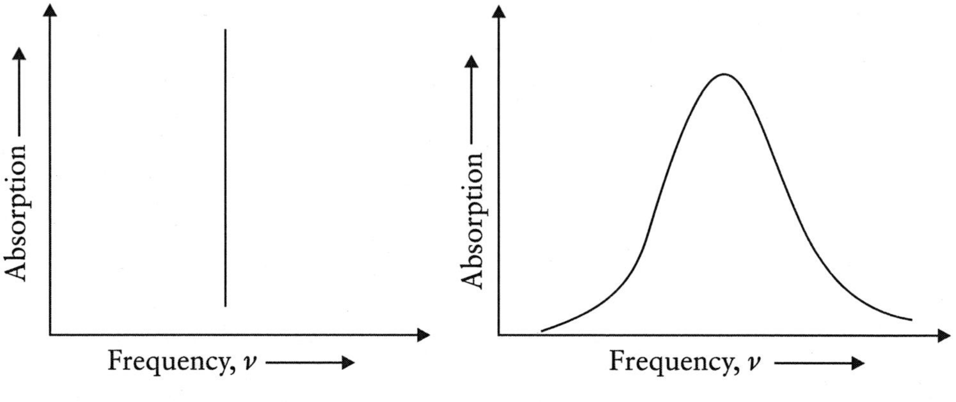

Fig. 2.5 (a): A sharp spectral line (b) A spectral line having width

Three factors contribute to broadening of spectral line.

(a) **Collision Broadening:** Matter consists of solid, liquid and gaseous phases. Solid phase shows sharp line spectra because in case of solids, particles are bounded by a definite space so they are not free to move and motion of particles is very limited, so spectra obtained is sharp. Whereas when molecules in the gaseous or liquid phase collide with one another, they deform the charge clouds of the outer electrons, thereby slightly perturbing the energy levels of these electrons. Hence, the spectral transitions between these perturbed energy levels are broadened.

(b) **Pressure Broadening:** As the pressure increases, there is an increase in the collisions, which results in deformation of electron charge cloud, which further leads to perturbance of energy levels. As a result, the transitions between these perturbed (disturbed) energy levels get broadened.

(c) **Doppler Broadening:** It arises when the molecule under investigation has a velocity higher as compare to the observing instrument. This is generally the case with gaseous samples where the molecules are undergoing random motion according to the postulates of the kinetic theory of gases.

(ii) Intensity of the Spectral Line: The strength or intensity of a spectral line depends upon the following factors:

(a) **Strength of the incident radiation**

(b) **Boltzmann population of energy level:** According to Boltzmann, if, at temperature T, N_0 is the number of molecules in the ground state and N is the number of molecules in the excited state, then as per Boltzmann distribution the relative population at equilibrium is given as

$$\frac{N}{N_0} = e^{-\Delta E/kT}$$

$$N = N_0 e^{-\Delta E/kT}$$

where, ΔE is the energy difference between ground and excited states and k is the Boltzmann constant.

Evidently, if ΔE is large, N/N_0 is small, i.e. the number of molecules in the excited state is less than that in the ground state. In fact, at room temperature, most of the molecules are in the ground state. Hence, the spectral lines originating from transitions from the ground state to a higher, say, third excited state would be more intense than those originating from transitions from the first excited state to the third excited state.

(c) **Transition probability:** It means the probability that a transition will take place between the energy levels. A molecule on absorption of a photon does not just go anywhere it wants, it ends up in an energy level determined by the selection rules for that particular transition. The selection rules, in turn, determine the allowed transition and the forbidden transitions. If the transition is allowed, a strong (or intense) spectral line will be obtained. If it is forbidden, no spectral line (or weak line) will be obtained.

PROBLEMS BASED ON WIDTH AND INTENSITY OF SPECTRAL LINES

Q. 1. State the factors that control intensity of spectral lines/band.

Q. 2. What are Doppler broadening and Collision broadening that contribute to the width of spectral lines?

Q. 3. What cause broadening of spectral band?

Q. 4. What are the factors that affect the intensity of a spectral line? Explain.

Q. 5. Calculate the energy difference, ΔE, and the relative populations of the upper and lower energy levels for transitions giving rise to absorption of the following at 298 K.

(a) an IR photon with wavenumber 2000 cm^{-1};

(b) a microwave photon with frequency 20 GHz;

(c) visible light with wavelength 500 nm;

(d) an X-ray with wavelength 4 nm;

(e) a radio wave with frequency 10 MHz.

[Assume the degeneracy of all levels is g = 1]

Ans. (a) A wavenumber of 2000 cm^{-1} is equivalent to an energy

$$\Delta E = hc\bar{v} = 6.626 \times 10^{-34} \text{ Js} \times 3 \times 10^{8} \text{ ms}^{-1} \times 2000 \times 10^{2} \text{ m}^{-1} = 3.97 \times 10^{-20} \text{ J}$$

Thus, the relative populations between non-degenerated levels of separation is

$$\frac{\eta_{upper}}{\eta_{lower}} = e^{-[3.97\times10^{-20}\,J/1.381\times10^{-23}\,JK^{-1}\times298\,K]} = 6.4\times10^{-5}$$

The population of the upper level is, essentially, negligible in comparison with that of the lower level.

(b) A frequency of 20 GHz = 20 × 10^9 Hz = 20 × 10^9 s^{-1} corresponds to an energy of

$$\Delta E = h\nu = 6.626 \times 10^{-34} \text{ Js} \times 20 \times 10^{9} \text{ s}^{-1} = 1.33 \times 10^{-23} \text{ J}$$

and so $$\frac{\eta_{upper}}{\eta_{lower}} = e^{-[1.33\times10^{-23}\,J/1.38\times10^{-23}\,JK^{-1}\times298\,K]}$$

(c) The energy of a photon of wavelength 500 nm = 500 × 10^{-9} m

$$\Delta E = \frac{hc}{\lambda} = 6.626\times10^{-34}\text{ Js}\times3\times10^{8}\text{ ms}^{-1}/500\times10^{-9}\text{ m} = 3.97\times10^{-19}\text{J}$$

and so $$\frac{\eta_{upper}}{\eta_{lower}} = e^{-[3.97\times10^{-19}\,J/1.381\times10^{-23}\,JK^{-1}\times298\,K]} = 1.3\times10^{-42}$$

(d) Similarly, an X-ray with wavelength 4 nm = 4 × 10^{-9} m has an energy

$$\Delta E = \frac{hc}{\lambda} = 6.626\times10^{-34}\text{ Js}\times3\times10^{8}\text{ ms}^{-1}/4\times10^{-9}\text{ m} = 4.97\times10^{-17}\text{J}$$

and so $$\frac{\eta_{upper}}{\eta_{lower}} = e^{-[4.97\times10^{-17}\,J/1.38\times10^{-23}\,JK^{-1}\times298\,K]}$$

(e) A radiowave with a frequency of 10 MHz = 10 × 10^6 Hz = 10 ×10^6 s^{-1} corresponds to an energy of

$$\Delta E = h\nu = 6.626 \times 10^{-34} \text{ Js} \times 10 \times 10^{6} \text{ s}^{-1} = 6.63 \times 10^{-27} \text{ J}$$

and so $$\frac{\eta_{upper}}{\eta_{lower}} = e^{-[6.63\times10^{-27}\,J/1.38\times10^{-23}\,JK^{-1}\times298\,K]}$$

Q. 6. What wavelength of radiation is sufficient to break a C–H bond in methane, where the bond dissociation energy is +439 kJ mol^{-1}?

$$CH_4 \rightarrow \dot{C}H_3 + \dot{H}$$

Ans. The molar energy of photons of wavelength λ is

$$E = \frac{N_A hc}{\lambda}$$

On rearranging

$$\lambda = \frac{N_A hc}{E}$$

$$= \frac{6.022\times10^{23}\,mol^{-1}\times6.626\times10^{-34}\,Js\times3\times10^{8}\,ms^{-1}}{439\times10^{3}\,J\,mol^{-1}}$$

$$= 272 \times 10^{-9}\ m = 272\ nm$$

3

Rotational Motion and Microwave Spectroscopy

3.1 Introduction

Rigid rotator is the simplest case of a diatomic molecule with a rigid bond length i.e., whose bond length does not change during rotation. It is an important system as its study makes us understand the rotational spectra of diatomic molecules in spectroscopy.

The rotational spectrum arises due to the transitions between the rotational energy levels of gaseous molecules on absorption of microwaves (1-100 cm^{-1}). The molecules having permanent dipole moment give a rotational spectrum. The molecules without permanent dipoles do not exhibit spectra in such microwave region. It is observed that a molecule cannot have any amount of rotational energy, but rotational energy is quantized depending on the shape and size of the molecule concerned.

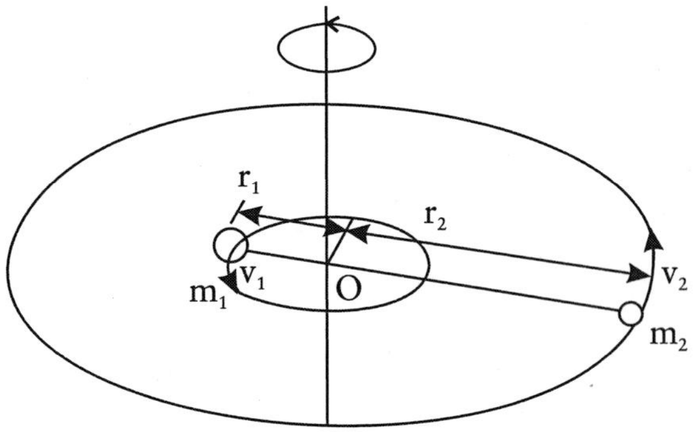

Fig. 3.1: A rigid rotator, rotating perpendicular to its own axis

3.2 Classical Treatment of Rigid Rotator

We consider a diatomic which consists of two atoms of mass 'm_1' and 'm_2' connected through a rigid bond of length 'r' (equal to the distance between the two atoms of the molecule). Assume that the centre of mass of the system is fixed at point O.

Let the distance of m_1 from point O (point of centre of mass) be r_1 and distance of m_2 from point O be r_2. Let this rigid rotator be rotated around an axis perpendicular to its own axis and passing through point O. The interatomic distance does not change during rotation, i.e. bond length is fixed.

For a microscopic diatomic system, we consider reduced mass (μ) instead of individual masses m_1 and m_2.

Total mass = M = $m_1 + m_2$

The reduced mass is given by the expression

$$\frac{1}{\mu}=\frac{1}{m_1}+\frac{1}{m_2}$$

$$\frac{1}{\mu}=\frac{m_1+m_2}{m_1 m_2}$$

$$\mu=\frac{m_1 m_2}{m_1+m_2}=\frac{m_1 m_2}{M}$$

Reduced mass is always less than either of the two masses, m_1 and m_2 and the lighter mass of the two will always have a greater contribution to the reduced mass.

If $m_1 << m_2$, we neglect m_1 in the denominator

$$\mu=\frac{m_1 \cancel{m_2}}{\cancel{m_2}}=m_1$$

3.2.1 Kinetic Energy of Rigid Rotator

A rigid rotator rotating around a fixed axis (axis of centre of mass) has a rotational frequency, ν_{rot} and rotational velocity, v_{rot} given as

Rotational velocity = Distance travelled in a rotation × Rotational frequency

$$v_{rot} = 2\pi r \times \nu_{rot}$$

$$v_{rot} = r(2\pi\nu_{rot})$$

$$v_{rot} = r\omega \quad \text{...(3.1)}$$

where, $\omega = 2\pi\nu_{rot}$ = angular velocity.

Let v_1 and v_2 be the respective rotational velocities of the two atoms with which they are rotating around the axis. The kinetic energy of the rotator is

$$\text{K.E.} = T = \frac{1}{2}m_1 v_1^2 + \frac{1}{2}m_2 v_2^2 \quad \text{...(3.2)}$$

For the two atoms of rigid rotator, using equation (3.1) we have

$$\left.\begin{aligned} v_1 &= r_1\omega \\ v_2 &= r_2\omega \end{aligned}\right] \quad \text{...(3.3)}$$

Put equation (3.3) in equation (3.2):

$$T=\frac{1}{2}m_1(r_1\omega)^2+\frac{1}{2}m_2(r_2\omega)^2$$

$$T=\frac{1}{2}\left[m_1 r_1^2+m_2 r_2^2\right]\omega^2$$

$$T = \frac{1}{2} I\omega^2 \qquad \text{...(3.4)}$$

where, $I = \sum_{i=1}^{m} \mu r_i^2$ = moment of inertia of the system.

Moment of Inertia

From the location of centre of mass (Fig. 3.1) we have

$$m_1 r_1 = m_2 r_2 \qquad \text{...(1)}$$

Also, $$r = r_1 + r_2 \qquad \text{...(2)}$$

Using (1) and (2), we have

(i) $$r = r_1 + r_2$$

$$r_2 = r - r_1$$

Now, $$m_1 r_1 = m_2 r_2$$

$$m_1 r_1 = m_2 (r - r_1)$$

$$m_1 r_1 = m_2 r - m_2 r_1$$

$$m_2 r = m_1 r_1 + m_2 r_1$$

$$m_2 r = (m_1 + m_2) r_1$$

$$r_1 = \left(\frac{m_2}{m_1 + m_2}\right) r \qquad \text{...(3)}$$

(ii) $$r = r_1 + r_2$$

$$r_1 = r - r_2$$

Now, $$m_1 r_1 = m_2 r_2$$

$$m_1 (r - r_2) = m_2 r_2$$

$$m_1 r - m_1 r_2 = m_2 r_2$$

$$m_1 r = m_1 r_2 + m_2 r_2$$

$$m_1 r = (m_1 + m_2) r_2$$

$$r_2 = \left(\frac{m_1}{m_1 + m_2}\right) r \qquad \text{...(4)}$$

Now, expression for the moment of inertia for a diatomic system is given as

$$I = \sum_{i=1}^{2} m_i r_i^2$$

$$I = m_1 r_1^2 + m_2 r_2^2$$

Using equation (3) and equation (4):

$$I = m_1\left(\frac{m_2 r}{m_1 + m_2}\right)^2 + m_2\left(\frac{m_1 r}{m_1 + m_2}\right)^2$$

$$I = \frac{m_1 m_2^2 r^2}{(m_1 + m_2)^2} + \frac{m_2 m_1^2 r^2}{(m_1 + m_2)^2}$$

$$I = \frac{m_1 m_2 \cancel{(m_1 + m_2)} r^2}{(m_1 + m_2)^{\cancel{2}}}$$

$$I = \left(\frac{m_1 m_2}{(m_1 + m_2)}\right) r^2$$

$$I = \mu r^2$$

where, μ is the reduced mass of the system and $\mu = \frac{m_1 m_2}{m_1 + m_2}$

For rotating models all the linear properties get converted to angular properties.

Linear properties	Angular properties
Linear momentum (p)	Angular momentum (L)
Linear velocity (v)	Angular velocity (ω)
Mass (M)	Moment of inertia (I)

Therefore, the classical kinetic energy expression for linear systems [equation (3.4)] can be written for rotating models (i.e. in terms of angular momentum) as:

$$T = \frac{1}{2} I\omega^2 = \frac{I\omega^2}{2} \times \frac{I}{I}$$

$$= \frac{I^2\omega^2}{2I} = \frac{(I\omega)^2}{2I}$$

$$T = \frac{L^2}{2I} \qquad \text{...(3.5)}$$

where, $L = I\omega$ = angular momentum.

Thus, according to classical mechanics, the kinetic energy of the rotator can have any value. However, quantum mechanically, it was proved that energy of rigid rotator is quantised and thus cannot posses any value.

3.3 Quantum Mechanical Treatment: Schrodinger Wave Equation for Rigid Rotator

The rigid rotator can be solved for energy using Schrodinger wave equation. The Schrodinger equation for rigid rotator can be derived by two ways:

(i) One in which angular momentum of kinetic energy operator is reduced in polar coordinates to obtain rigid rotator wave equation.

(ii) Other in which Laplacian of kinetic energy operator is reduced in polar coordinates to obtain rigid rotator wave equation.

3.3.1 Deriving Schrodinger Equation by Reducing Angular Momentum of Kinetic Energy Operator in Polar Coordinates

Total energy of the system is the sum of kinetic energy and potential energy, i.e. $\hat{H} = \hat{T} + \hat{V}$.

In this system, we assume potential energy to zero (i.e. V = 0) because the distance between the two atoms is fixed during the rotation. Thus, the total energy is solved by taking kinetic energy only.

From the kinetic energy expression (K.E. = $L^2/2I$), it is seen that, kinetic energy depends only on the square of angular momentum (L^2) which is in the numerator. Since moment of inertia (I) in the denominator is specific to system and thus has a constant value.

We have deduced from equation (3.5) that

$$\text{K.E.} = T = \frac{L^2}{2I}$$

Now, $$\vec{L} = \vec{r} \times \vec{p}$$

where, $$\vec{r} = \hat{i}x + \hat{j}y + \hat{k}z$$

$$\vec{p} = \hat{i}p_x + \hat{j}p_y + \hat{k}p_z$$

The solution for the cross product of $\vec{r}$ and $\vec{p}$ may be written in the form of determinant

$$\vec{L} = \begin{vmatrix} \hat{i} & \hat{j} & \hat{k} \\ x & y & z \\ p_x & p_y & p_z \end{vmatrix}$$

$$\vec{L} = \hat{i}\left(yp_z - zp_y\right) - \hat{j}\left(xp_z - zp_x\right) + \hat{k}\left(xp_y - yp_x\right)$$

$$\vec{L} = \hat{i}\left(yp_z - zp_y\right) + \hat{j}\left(zp_x - xp_z\right) + \hat{k}\left(xp_y - yp_x\right)$$

Also, $$\vec{L} = \hat{i}\vec{L}_x + \hat{j}\vec{L}_y + \hat{k}\vec{L}_z$$

Thus, on comparing both the expression of $\vec{L}$, we get

$$\left.\begin{aligned}\vec{L}_x &= yp_z - zp_y = -i\hbar\left[y\frac{\partial}{\partial z} - z\frac{\partial}{\partial y}\right]\\ \vec{L}_y &= zp_x - xp_z = -i\hbar\left[z\frac{\partial}{\partial x} - x\frac{\partial}{\partial z}\right]\\ \vec{L}_z &= xp_y - yp_x = -i\hbar\left[x\frac{\partial}{\partial y} - y\frac{\partial}{\partial x}\right]\end{aligned}\right\}\text{ Refer operators (chapter 1)}$$

Now, $\hat{L}^2 = \vec{L}\cdot\vec{L}$

$$\hat{L}^2 = \vec{L}_x^2 + \vec{L}_y^2 + \vec{L}_z^2$$

as $\hat{i}\cdot\hat{i} = \hat{j}\cdot\hat{j} = \hat{k}\cdot\hat{k} = 1$

$$\hat{i}\times\hat{i} = \hat{j}\times\hat{j} = \hat{k}\times\hat{k} = 0$$

This L^2 operator which is in cartesian coordinates is converted into polar coordinates.

i.e. $\hat{L}^2 = L_x^2 + L_y^2 + L_z^2$ (Cartesian coordinates)

$\downarrow$ Transformation of $\hat{L}^2$ to polar coordinates (not discussed here)

$$\hat{L}^2 = -\hbar^2\left[\frac{1}{\sin\theta}\frac{\partial}{\partial\theta}\left(\sin\theta\frac{\partial}{\partial\theta}\right) + \frac{1}{\sin^2\theta}\frac{\partial^2}{\partial\phi^2}\right] \quad \text{(Polar coordinates)}$$

Thus, the kinetic energy operator in terms of polar coordinates for rigid rotator becomes:

$$\therefore \quad \text{K.E.} = \hat{T} = \frac{\hat{L}^2}{2I} = \frac{\hat{L}^2}{2\mu r^2} \qquad (\because I = \mu r^2)$$

Put the expression of $\hat{L}^2$ in terms of polar coordinates into the expression of $\hat{T}$.

$$\hat{T} = -\frac{\hbar^2}{2\mu r^2}\left[\frac{1}{\sin\theta}\frac{\partial}{\partial\theta}\left(\sin\theta\frac{\partial}{\partial\theta}\right) + \frac{1}{\sin^2\theta}\frac{\partial^2}{\partial\phi^2}\right]$$

$$\hat{T} = -\frac{h^2}{8\pi^2\mu}\left[\frac{1}{r^2\sin\theta}\frac{\partial}{\partial\theta}\left(\sin\theta\frac{\partial}{\partial\theta}\right) + \frac{1}{r^2}\frac{1}{\sin^2\theta}\frac{\partial^2}{\partial\phi^2}\right] \qquad \text{...(3.6)}$$

For a rotator rotating freely, the potential energy is zero (V = 0) and hence its total energy is equal to its kinetic energy $(\hat{H} = \hat{T})$.

Therefore, the Schrodinger equation becomes

$$\hat{H}\psi = E\psi$$

$$\hat{T}\psi = E\psi$$

$$\frac{\hat{L}^2}{2I}\psi = E\psi$$

Using equation (3.6), we have:

$$-\frac{h^2}{8\pi^2\mu}\left[\frac{1}{r^2\sin\theta}\frac{\partial}{\partial\theta}\left(\sin\theta\frac{\partial}{\partial\theta}\right)+\frac{1}{r^2\sin^2\theta}\frac{\partial^2}{\partial\phi^2}\right]\psi = E\psi \qquad ...(3.7)$$

3.3.2 Deriving Schrodinger Education by Reducing Laplacian of Kinetic Energy Operator in Polar Coordinates

The kinetic energy operator in terms of cartesian coordinates is

$$\hat{T} = -\frac{h^2}{8\pi^2 m}\nabla^2 = -\frac{h^2}{8\pi^2 m}\left(\frac{\partial^2}{\partial x^2}+\frac{\partial^2}{\partial y^2}+\frac{\partial^2}{\partial z^2}\right) \qquad \text{(Cartesian coordinates)}$$

↓ Transformation of Laplacian operator (∇^2) to polar coordinates (not discussed here).

$$\hat{T} = -\frac{h^2}{8\pi^2 m}\left[\frac{1}{r^2}\frac{\partial}{\partial r}\left(r^2\frac{\partial}{\partial r}\right)+\frac{1}{r^2\sin\theta}\frac{\partial}{\partial\theta}\left(\sin\theta\frac{\partial}{\partial\theta}\right)+\frac{1}{r^2\sin^2\theta}\frac{\partial^2}{\partial\phi^2}\right] \qquad \text{(Polar coordinates)}$$

Since the distance 'r' (as shown in Fig. 3.1), between the two masses of the rigid rotator remains constant, the derivative with respect to r will vanish in the kinetic energy operator of rigid rotator in terms of polar coordinates. Thus, the above equation for kinetic energy operator for rigid rotator, gets reduced to:

$$\hat{T} = -\frac{h^2}{8\pi^2 m}\left[\frac{1}{r^2\sin\theta}\frac{\partial}{\partial\theta}\left(\sin\theta\frac{\partial}{\partial\theta}\right)+\frac{1}{r^2\sin^2\theta}\frac{\partial^2}{\partial\phi^2}\right]$$

For a microscopic diatomic rigid rotator, we consider reduced mass (μ) instead of individual masses m_1 and m_2. Thus, the above kinetic energy operator for rigid rotator becomes:

$$\hat{T} = -\frac{h^2}{8\pi^2\mu}\left[\frac{1}{r^2\sin\theta}\frac{\partial}{\partial\theta}\left(\sin\theta\frac{\partial}{\partial\theta}\right)+\frac{1}{r^2\sin^2\theta}\frac{\partial^2}{\partial\phi^2}\right]$$

For a rotator rotating freely, the potential energy (V = 0) is zero and hence its total energy is equal to its kinetic energy

$$\hat{H} = \hat{T} + \hat{V}$$

$$\hat{H} = \hat{T} \qquad (\because\ V = 0)$$

Therefore, the Schrodinger equation for rigid rotator becomes

$$\hat{H}\psi = E\psi$$

$$-\frac{h^2}{8\pi^2\mu}\left[\frac{1}{r^2\sin\theta}\frac{\partial}{\partial\theta}\left(\sin\theta\frac{\partial}{\partial\theta}\right)+\frac{1}{r^2\sin^2\theta}\frac{\partial^2}{\partial\phi^2}\right]\psi = E\psi \qquad ...(3.8)$$

This equation (3.8) is same as equation (3.7) derived earlier in section 3.3.1.

3.4 Wavefunction of Rigid Rotator

We have derived the Schrodinger equation by two methods in sec. 3.3, both of them generating the same equation, i.e. (3.7 and 3.8)

$$-\frac{h^2}{8\pi^2\mu}\left[\frac{1}{r^2\sin\theta}\frac{\partial}{\partial\theta}\left(\sin\theta\frac{\partial}{\partial\theta}\right)+\frac{1}{r^2\sin^2\theta}\frac{\partial^2}{\partial\phi^2}\right]\psi = E\psi$$

Since the distance 'r' between the two masses of the rotator remains constant, the derivative w.r.t. r will not appear in this equation. Also, the wavefunction of rigid rotator depends only upon the angular orientation part (θ and ϕ) and is independent of the radial part (r). Thus, the wavefunction ψ(r, θ, ϕ) for rigid rotator has been replaced by a function Y(θ, ϕ), known as **Spherical Harmonics** which depends only on the coordinates θ and ϕ.

$$\text{Spherical harmonics} = Y(\theta, \phi) = \Theta(\theta)\Phi(\phi)$$

Thus, the Schrodinger equation becomes:

$$-\frac{h^2}{8\pi^2(\mu r^2)}\left[\frac{1}{\sin\theta}\frac{\partial}{\partial\theta}\left(\sin\theta\frac{\partial}{\partial\theta}\right)+\frac{1}{\sin^2\theta}\frac{\partial^2}{\partial\phi^2}\right]Y(\theta,\phi) = EY(\theta,\phi)$$

$$-\frac{h^2}{8\pi^2 I}\left[\frac{1}{\sin\theta}\frac{\partial}{\partial\theta}\left(\sin\theta\frac{\partial}{\partial\theta}\right)+\frac{1}{\sin^2\theta}\frac{\partial^2}{\partial\phi^2}\right]Y(\theta,\phi) = EY(\theta,\phi)\ \left[\because\ I = \mu r^2\right]$$

Multiply both sides by $(-8\pi^2 I/h^2)$:

$$\left[\frac{1}{\sin\theta}\frac{\partial}{\partial\theta}\left(\sin\theta\frac{\partial}{\partial\theta}\right)+\frac{1}{\sin^2\theta}\frac{\partial^2}{\partial\phi^2}\right]Y(\theta,\phi) = -\frac{8\pi^2 IE}{h^2}Y(\theta,\phi)$$

$$\frac{1}{\sin\theta}\frac{\partial}{\partial\theta}\left(\sin\theta\frac{\partial Y}{\partial\theta}(\theta,\phi)\right)+\frac{1}{\sin^2\theta}\frac{\partial^2 Y}{\partial\phi^2}(\theta,\phi) = -\frac{8\pi^2 IE}{h^2}Y(\theta,\phi)$$

Put $Y(\theta, \phi) = \Theta \cdot \Phi$

$$\frac{\Phi}{\sin\theta}\frac{\partial}{\partial\theta}\left(\sin\theta\frac{\partial\Theta}{\partial\theta}\right)+\frac{\Theta}{\sin^2\theta}\frac{\partial^2\Phi}{\partial\phi^2} = -\frac{8\pi^2 IE}{h^2}\Theta\Phi$$

Multiply by $\dfrac{\sin^2\theta}{\Theta\cdot\Phi}$:

$$\frac{\sin\theta}{\Theta}\frac{\partial}{\partial\theta}\left(\sin\theta\frac{\partial\Theta}{\partial\theta}\right)+\frac{1}{\Phi}\frac{\partial^2\Phi}{\partial\phi^2}+\frac{8\pi^2 IE}{h^2}\sin^2\theta = 0$$

$$\frac{\sin\theta}{\Theta}\frac{\partial}{\partial\theta}\left(\sin\theta\frac{\partial\Theta}{\partial\theta}\right)+\frac{8\pi^2 IE}{h^2}\sin^2\theta = -\frac{1}{\Phi}\frac{\partial^2\Phi}{\partial\phi^2}$$

Setting both the sides of the equation equal to m^2, we get a pair of differential equations each in one variable θ and ϕ. Thus, the Schrodinger equation for rigid rotator can be solved by using the technique of **Separation of Variables** method.

Thus, the two differential equations are:

$$-\frac{1}{\Phi}\frac{\partial^2\Phi}{\partial\phi^2} = m^2$$

$$\frac{1}{\Phi}\frac{\partial^2\Phi}{\partial\phi^2} = -m^2$$

$$\frac{\partial^2\Phi}{\partial\phi^2} + m^2\Phi = 0 \quad \text{...(3.9)}$$

The equation (3.9) is Φ part of wavefunction.

and $$\frac{\sin\theta}{\Theta}\frac{\partial}{\partial\theta}\left(\sin\theta\frac{\partial\Theta}{\partial\theta}\right) + \frac{8\pi^2 IE}{h^2}\sin^2\theta = m^2$$

Put, $$\beta = \frac{8\pi^2 IE}{h^2}$$

$$\frac{\sin\theta}{\Theta}\frac{\partial}{\partial\theta}\left(\sin\theta\frac{\partial\Theta}{\partial\theta}\right) + \beta\sin^2\theta - m^2 = 0 \quad \text{...(3.10)}$$

The equation (3.10) is Θ part of wavefunction.

Thus for solving the Schrodinger equation, we have to first solve the Φ and Θ part of the wavefunction.

3.4.1 Solving the Φ Part of Wavefunction, i.e. Solving equation (3.9)

$$\frac{\partial^2\Phi}{\partial\phi^2} + m^2\Phi = 0 \quad \text{...(3.9)}$$

The above equation is a differential equation whose solution is given as

$$\Phi_m(\phi) = C\cdot e^{\pm im\phi} \quad \text{...(3.11)}$$

$\downarrow$ Normalised solution of this equation

$$\Phi_m(\phi) = \frac{1}{\sqrt{2\pi}}\cdot e^{\pm im\phi} \quad \text{...(3.12)}$$

The function Φ should be finite, and consistent in order to be an acceptable wavefunction. Also in a rigid rotator the angle ϕ varies from 0 to 2π (Fig. 3.2). Thus, for function Φ to be continuous, the value of Φ at ϕ should be equal to the value of function at $(\phi + 2\pi)$. Thus, when the angle ϕ is replaced by $(\phi + 2\pi)$, the same set of points in space is obtained.

$$\Phi(\phi) = \Phi(\phi + 2\pi)$$

(or) $$\Phi(0) = \Phi(2\pi)$$

Multiply both sides by 'im' and taking exponential, we get

$$\exp(im\phi) = \exp[im(\phi + 2\pi)]$$

$$\cancel{e^{im\phi}} = \cancel{e^{im\phi}} \cdot e^{im(2\pi)}$$

$$e^{im(2\pi)} = 1$$

Also, $\cos 2\pi m + i \sin 2\pi m = 1$

[From **De Moivre's Theorem**)

$e^{im(2\pi)} = \cos 2\pi m + i \sin 2\pi m = 1$

and this is possible only if $m = 0, \pm 1, \pm 2, \ldots$

Hence 'm' is the quantum number satisfying the above values according to the solution. The result is same as that of magnetic quantum number of Bohr Sommerfield model.

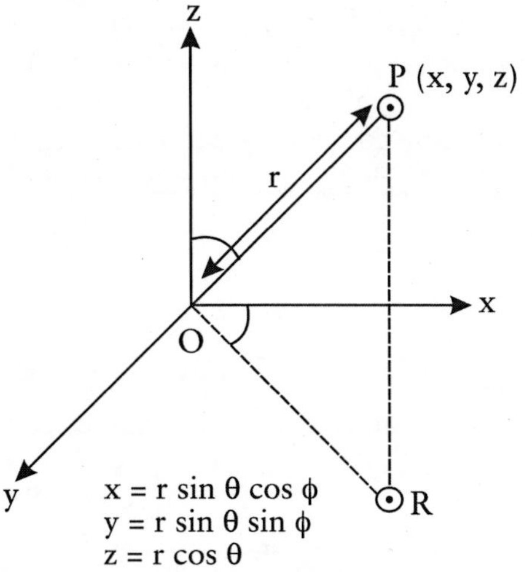

Fig. 3.2: Conversion of cartesian coordinates (x, y, z) to polar coordinates (r, θ, ϕ)

3.4.2 Solving the Θ Part of Wavefunction, i.e. Solving equation (3.10)

$$\frac{\sin\theta}{\Theta}\frac{\partial}{\partial\theta}\left(\sin\theta\frac{\partial\Theta}{\partial\theta}\right) + \beta\sin^2\theta - m^2 = 0 \qquad \ldots(3.10)$$

↓ Transformation to Legendre equation

$$\left(1-x^2\right)\frac{\partial^2 P(x)}{\partial x^2} - 2x\frac{\partial P(x)}{\partial x} - \left[\ell(\ell+1) - \frac{m^2}{\left(1-x^2\right)}\right]P(x) = 0 \text{ : Legendre Equation}$$

↓ Normalised solution of this equation

$$\Theta_{\ell,m}(\theta) = N_{\ell m} \cdot P_\ell^{|m|}(x)$$

or

$$\Theta_{\ell,m}(\theta) = \sqrt{\frac{(2\ell+1)}{2}\frac{(\ell-|m|)!}{(\ell+|m|)!}}P_\ell^{|m|}(x) \qquad \ldots(3.13)$$

where,

$$N_{\ell m} = \sqrt{\frac{(2\ell+1)}{2}\frac{(\ell-|m|)!}{(\ell+|m|)!}} \text{ : Normalisation constant}$$

$$P_\ell^{|m|}(x) = \left(1-x^2\right)^{|m|/2}\frac{\partial^{|m|}}{\partial x^{|m|}} \cdot P_\ell(x) \text{: Associated Legendre polynominal}$$

$$P_\ell(x) = \frac{1}{2^\ell \cdot \ell!}\frac{\partial^\ell}{\partial x^\ell}\left(x^2-1\right)^\ell \text{ : Legendre polynominal}$$

and $x = \cos\theta$

$P_\ell^{|m|}$ is the Associated Legendre polynomial of degree ℓ and order |m|. The above equation has introduced another coefficient 'ℓ', whose allowed values are 0, 1, 2, 3, ... This is same as azimuthal quantum number in Bohr-Sommerfeld model. The above equation also impose restrictions on allowed value of m. In above equation, a term $(\ell - |m|)!$ appears, that means if |m| is allowed to be greater than ℓ, the factorial of a negative number would result, which is undefined. **Hence, m can have values, 0, ± 1, ± 2, upto ± ℓ, i.e. m cannot be greater than ℓ.**

Thus, the function Θ depends on the value of two quantum numbers, i.e. ℓ, and |m|. These solutions are obtained, provided the following conditions are satisfied.

$$\ell = 0, 1, 2, 3, \ldots$$

$$m = 0, \pm 1, \pm 2, \pm 3, \ldots, \pm \ell$$

m can have $(2\ell + 1)$ values for a given values of ℓ.

PROBLEMS BASED ON RIGID ROTATOR

Q. 1. What is the relation between polar coordinates and cartesian coordinates?

Ans. Cartesian coordinates (x, y, z) → Polar coordinates (r, θ, ϕ)

One of the atoms is kept at origin 'O' and the other in the three dimensional space (x, y, z) at point 'P' for the diatomic rigid rotator, held through a fixed bond length, which is the distance from the origin (OP) signified 'r'. This bond length (r) does not vary and remain constant during rotation. However, the angular orientations are defined by θ and ϕ in polar coordinates. The angle θ with the *z*-axis (POZ) may vary from 0 to π during rotation and ϕ is the angle made with the projection on the *x*-axis (XOR) which varies from 0 to 2π during rotation.

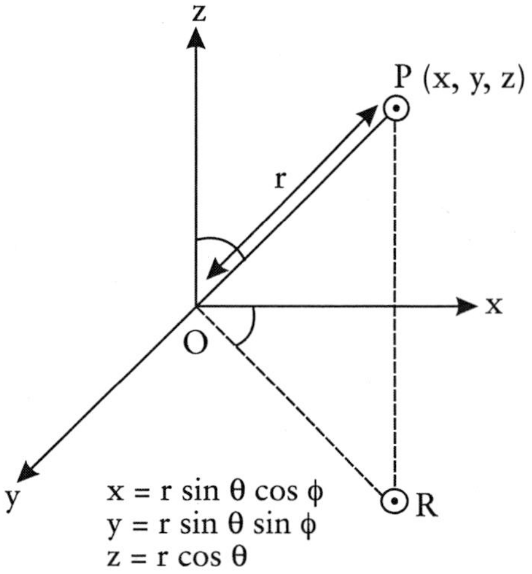

Q. 2. What are spherical harmonics?

Ans. Wavefunction for a system depends on three coordinates (r, θ and ϕ) in terms of polar coordinates.

$$\psi (r, \theta, \phi) = R(r) \cdot \Theta(\theta) \cdot \Phi(\phi).$$

The wavefunction of rigid rotator depends only upon the angular orientation part (θ and ϕ) and is independent of the radial part (r). Thus, the wavefunction ψ(r, θ, ϕ) for rigid rotator has been replaced by a function Y(θ · ϕ) known as **spherical harmonics.**

Spherical harmonics depends only on the coordinates θ and ϕ.

Spherical Harmonics = $Y(\theta, \phi) = \Theta(\theta) \,.\, \Phi(\phi)$.

Q. 3. What is a rigid rotator?

Ans. Rigid rotator is a simplest case of a diatomic molecule with a rigid bond length, i.e. whose bond length does not change during rotation. It is considered to be a diatomic molecule which consists of two atoms of mass m_1 and m_2 connected through a rigid bond of length 'r' (equal to the distance between the two atoms of the molecule).

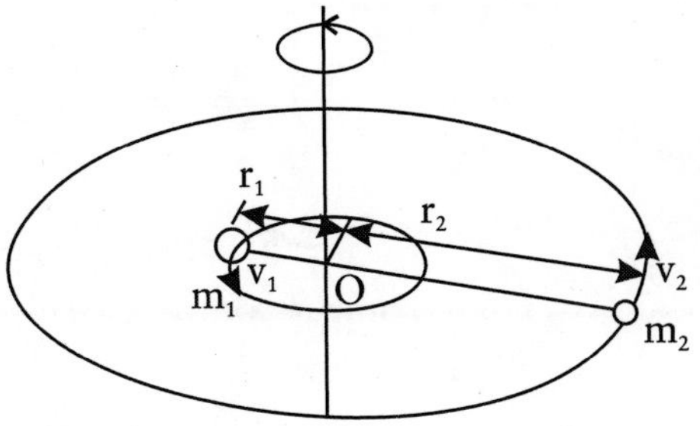

Rigid rotator rotating along centre of mass axis

Q. 4. Utilizing the classical mechanics, show that for a rigid rotator.

(a) $E = \frac{1}{2}I\omega^2$

Ans. A rigid rotator rotating around a fixed axis (axis of centre of mass) has a rotational frequency, ν_{rot} and its rotational velocity, v_{rot} is given as:

rotational velocity = distance travelled in rotation × rotational frequency

$$v_{rot} = 2\pi r \times \nu_{rot}$$

$$v_{rot} = r(2\pi\, \nu_{rot})$$

$$v_{rot} = r\omega \qquad ...(1)$$

[$\omega = 2\pi\nu_{rot}$ is the angular velocity]

Let v_1 and v_2 be the respective rotational velocities of the two atoms with which they are rotating around the axis. The kinetic energy of the rotator is

$$K.E. = T = \frac{1}{2}m_1v_1^2 + \frac{1}{2}m_2v_2^2 \qquad ...(2)$$

Using equation (1)

$$\left.\begin{aligned} v_1 &= r_1\omega \\ v_2 &= r_2\omega \end{aligned}\right] \qquad ...(3)$$

Putting equation (3) in equation (2):

$$K.E. = T = \frac{1}{2}m_1(r_1\,\omega)^2 + \frac{1}{2}m_2(r_2\,\omega)^2$$

$$T = \frac{1}{2}m_1\,r_1^2\,\omega^2 + \frac{1}{2}m_2\,r_2^2\,\omega^2$$

For microscopic subatomic rigid rotator, we consider reduced mass (μ) instead of individual masses (m_1 or m_2)

$$T = \frac{1}{2}\mu\,r_1^2\,\omega^2 + \frac{1}{2}\mu r_2^2\,\omega^2$$

$$T = \frac{1}{2}\left(\mu\,r_1^2 + \mu r_2^2\right)\omega^2$$

$$T = \frac{1}{2} I\omega^2$$

where, $I = \sum_{i=1}^{n} \mu r_i^2$ = moment of inertia of the system.

For a rotator rotating freely, the potential energy (V = 0) is zero and hence its total energy is equal to its kinetic energy

$$E = T + V$$

$$E = \frac{1}{2} I\omega^2 + V$$

$$E = \frac{1}{2} I\omega^2 \qquad [\text{as } V = 0]$$

(b) $I = \mu r^2$

Ans. From the location of centre of mass (Fig. 3.1) we have,

$$m_1 r_1 = m_2 r_2 \qquad ...(1)$$

Also, $$r = r_1 + r_2 \qquad ...(2)$$

Using (1) and (2), we have

(i) $$r = r_1 + r_2$$

$$r_2 = r - r_1$$

Now, $$m_1 r_1 = m_2 r_2$$

$$m_1 r_1 = m_2 (r - r_1)$$

$$m_1 r_1 = m_2 r - m_2 r_1$$

$$m_2 r = m_1 r_1 + m_2 r_1$$

$$m_2 r = (m_1 + m_2) r_1$$

$$r_1 = \left(\frac{m_2}{m_1 + m_2}\right) r \qquad ...(3)$$

(ii) $$r = r_1 + r_2$$

$$r_1 = r - r_2$$

Now, $$m_1 r_1 = m_2 r_2$$

$$m_1 (r - r_2) = m_2 r_2$$

$$m_1 r - m_1 r_2 = m_2 r_2$$

$$m_1 r = m_1 r_2 + m_2 r_2$$

$$m_1 r = (m_1 + m_2) r_2$$

$$r_2 = \left(\frac{m_1}{m_1 + m_2}\right) r \qquad ...(4)$$

Now, $I = \sum_{i=1}^{2} m_i r_i^2$

For a diatomic system, $I = m_1 r_1^2 + m_2 r_2^2$

Using (3) and (4)

$$I = m_1 \left(\frac{m_2 r}{m_1 + m_2} \right)^2 + m_2 \left(\frac{m_1 r}{m_1 + m_2} \right)^2$$

$$I = \frac{m_1 m_2^2 r^2}{(m_1 + m_2)^2} + \frac{m_2 m_1^2 r^2}{(m_1 + m_2)^2}$$

$$I = \frac{m_1 m_2 \cancel{(m_1 + m_2)} r^2}{\cancel{(m_1 + m_2)}^2}$$

$$I = \left(\frac{m_1 m_2}{m_1 + m_2} \right) r^2$$

$$I = \mu r^2$$

where $\mu = \frac{m_1 m_2}{m_1 + m_2}$ = reduced mass of the system

(c) $L = I\omega$

Let v_1, v_2, v_3 ... v_n be the rotational velocities of particles

$$v_1 = r_1 \omega$$
$$v_2 = r_2 \omega$$
$$v_n = r_n \omega$$

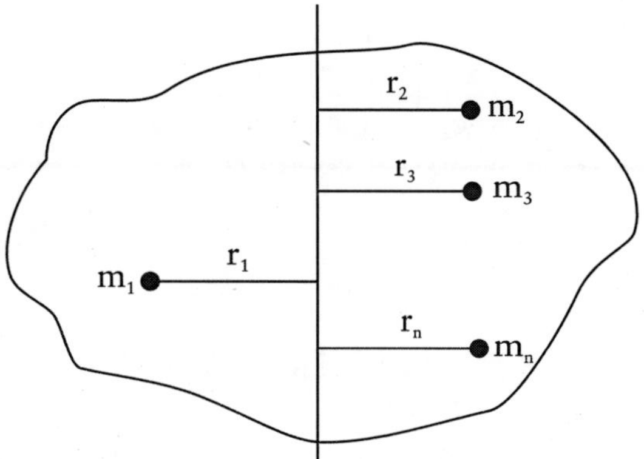

Angular momentum of particles

$$L_1 = m_1 v_1 r_1$$
$$L_1 = m_1 (r_1 \omega) r_1$$
$$L_1 = m_1 r_1^2 \omega$$

Similarly, $L_2 = m_2 r_2^2 \omega$

and $L_n = m_n r_n^2 \omega$

Total angular momentum would be:

$$L = L_1 + L_2 + L_3 + \text{.........} L_n$$

$$L = \left(m_1 r_1^2 + m_2 r_2^2 + m_3 r_3^2 + \ldots + m_n r_n^2 \right) \omega$$

$$L = I\omega$$

where, $$I = \sum_{i=1}^{n} m_i r_i^2$$

Q. 5. What quantum restrictions on the value of m and ℓ are introduced in a rigid rotator?

Ans. $\ell = 0, 1, 2, ...$

$m = 0, \pm 1, \pm 2, ... \pm \ell$

Q. 6. What is meant by Rigid rotator? Set up the Hamiltonian for a molecule AB undergoing rotational motion within this approximation. Explain briefly the terms in the expression of the Hamiltonian.

Ans. A rigid rotor is the simple model for a rotating diatomic molecule (AB) which consist of two atoms of mass 'm_A' and 'm_B' which are joined by a rigid bar of length 'r_{AB}'. The centre of mass of the system is fixed at point O and the system rotates around an axis perpendicular to its own axis and passing through point O. The bond length of the system does not change during rotation.

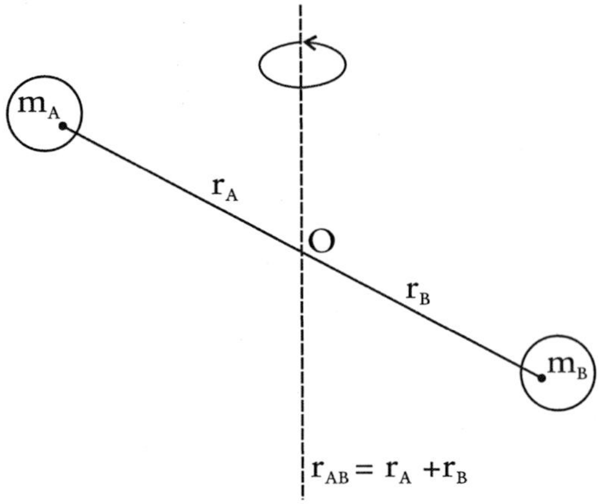

$$\text{Hamiltonian } (\hat{H}) = \hat{T} + \hat{V}$$

$$\hat{T} = \text{Kinetic energy} = \frac{\hat{L}^2}{2I}$$

$$\hat{V} = \text{Potential energy} = 0$$

$$\hat{H} = \hat{T} = \frac{\hat{L}^2}{2I} = \frac{\hat{L}^2}{2\mu r^2}$$

The angular momentum operator $(\hat{L}^2)$ which is in cartesian coordinates is converted into polar coordinates

$$\hat{L}^2 = -\hbar^2\left[\frac{1}{\sin\theta}\frac{\partial}{\partial\theta}\left(\sin\theta\frac{\partial}{\partial\theta}\right) + \frac{1}{\sin^2\theta}\frac{\partial^2}{\partial\phi^2}\right]$$

Therefore, Hamiltonian $$(\hat{H}) = \frac{-h^2}{8\pi^2\mu}\left[\frac{1}{r^2\sin\theta}\frac{\partial}{\partial\theta}\left(\sin\theta\frac{\partial}{\partial\theta}\right) + \frac{1}{r^2\sin^2\theta}\frac{\partial^2}{\partial\phi^2}\right]$$

where
$$x = r\sin\theta\cos\phi$$
$$y = r\sin\theta\sin\phi$$
$$z = r\cos\theta$$

3.5 Rotational Energy of the Rigid Rotator

We know that angular momentum is quantised whose values are given by

$$L = \sqrt{\ell(\ell+1)}\frac{h}{2\pi}$$

where, $\ell = 0, 1, 2, 3...$

$$E = \frac{L^2}{2I}$$

$$E = \ell(\ell+1)\frac{h^2}{4\pi^2} \times \frac{1}{2I}$$

$$E = \ell(\ell+1)\frac{h^2}{8\pi^2 I} \quad ...(3.14)$$

$$E = \beta\frac{h^2}{8\pi^2 I}$$

where, $\beta = \ell(\ell + 1)$

Thus, once again, we obtain a set of discrete energy levels and these are allowed energies by putting different values of ℓ, where, $\ell = 0, 1, 2, ...$ Thus, quantum mechanically it was proved that energy of rigid rotator is quantised. Rotational energy is quantised (like all other forms of molecular energy) means rotational energy is limited to certain definite values depending on the shape and size of the molecule.

- **Degeneracy:** From the above expression of energy (E) of the rigid rotator, i.e. equation (3.14) it is seen that energy of the rigid rotator does not depend on quantum number 'm' and depend only on quantum number 'ℓ'. Whereas wavefunction (Y) of rigid rotator depends on ℓ and m both. So we have degenerate states corresponding to different wavefunctions for a given level, i.e. different values of m, for a given value of ℓ.

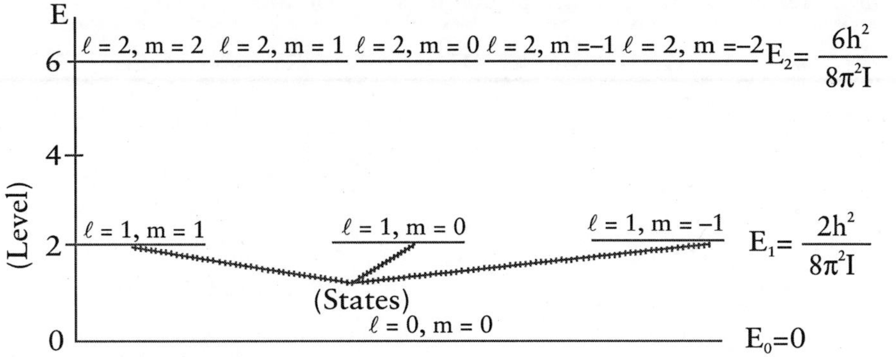

Fig. 3.3: Degeneracy of energy levels for a rigid rotator

- **Zero point energy:** Solution to Schrodinger wave equation of rigid rotator for kinetic energy is:

$$E = \ell(\ell+1)\frac{h^2}{8\pi^2 I}$$

The zero point energy of the rigid rotator is the lowest energy of the rigid rotator. The minimum value of ℓ is zero, therefore, the energy corresponding to $\ell = 0$ will be the zero point energy

$$E_0 = 0(0+1)\frac{h^2}{8\pi^2 I} = 0$$

Thus, for rigid rotator, zero point energy corresponds to $\ell = 0$, means no rotation. Also zero point energy is zero only for rotational motion but this is not possible for translational or vibrational motion.

i.e. $E_{\text{rot. zero point}} = 0$ (corresponds to no rotation)

$E_{\text{vib. zero point}} \neq 0$

$E_{\text{trans. zero point}} \neq 0$

According to Heisenberg uncertainty principle, kinetic energy of the system cannot be zero because if kinetic energy = 0, then, $\Delta x = 0$, which is not possible for microscopic objects. But in case of rotation, kinetic energy can be zero without violating Heisenberg uncertainty principle because rotation does not change position of the complete system.

Now, $E_{Total} = E_{Trans} + E_{vib} + E_{rot}$

If $E_{rot} = 0$ for $\ell = 0$, even then

$$\left.\begin{matrix} E_{vib} \\ E_{trans} \end{matrix}\right] \neq 0 \text{ and } E_{Total} \neq 0$$

Hence, Heisenberg uncertainty principle is not violated.

PROBLEMS BASED ON ROTATIONAL ENERGY AND DEGENERACY

Q. 1. Which has higher value: period of rotation or period of vibration for a diatomic molecule.

Ans. Frequency $(\nu) \propto \frac{1}{\text{time period}}$ ...(1)

and Energy (E) $\propto \nu$ $(\because\ E = h\nu)$ ...(2)

Comparing (1) and (2):

$\therefore$ Energy $(E) \propto \frac{1}{\text{time period}}$

Since, $E_{vib} > E_{rot}$

$\therefore$ Time period of vibration < Time period of rotation

Q. 2. (i) Write down the Hamiltonian for diatomic rotator and thus Schrodinger wave equation for rigid rotator.

(ii) Also write solution to this equation and thus calculate the zero point energy.

Ans. (i) For a diatomic rotator rotating freely, the potential energy (V = 0) is zero and hence its total energy is equal to its K.E.

$$E = T + V \quad \text{or} \quad \hat{H} = \hat{T} + \hat{V}$$

For V = 0,

$$E = T \quad \text{or} \quad \hat{H} = \hat{T}$$

Kinetic energy operator ($\hat{T}$) in terms of cartesian coordinates is

$$\hat{T} = \frac{-h^2}{8\pi^2 m}\nabla^2 = \frac{-h^2}{8\pi^2 m}\left(\frac{\partial^2}{\partial x^2} + \frac{\partial^2}{\partial y^2} + \frac{\partial^2}{\partial z^2}\right)$$

Putting the value of ∇^2 in terms of polar coordinates into expression of $\hat{T}$

$$\hat{T} = \frac{-h^2}{8\pi^2 m}\left[\frac{1}{r^2}\frac{\partial}{\partial r}\left(r^2\frac{\partial}{\partial r}\right) + \frac{1}{r^2 \sin\theta}\frac{\partial}{\partial\theta}\left(\sin\theta\frac{\partial}{\partial\theta}\right) + \frac{1}{r^2\sin^2\theta}\frac{\partial^2}{\partial\phi^2}\right]$$

Since the distance 'r' between the two masses of the rigid rotator remains constant, the derivative w.r.t. 'r' will vanish in the kinetic energy operator. Thus, the above equation for kinetic energy operator gets reduced to:

$$\hat{T} = \frac{-h^2}{8\pi^2 m}\left[\frac{1}{r^2 \sin\theta}\frac{\partial}{\partial\theta}\left(\sin\theta\frac{\partial}{\partial\theta}\right) + \frac{1}{r^2\sin^2\theta}\frac{\partial^2}{\partial\phi^2}\right]$$

For a microscopic diatomic rigid rotator, we consider reduced mass (μ) instead of individual masses m_1 and m_2. So, the above equation for kinetic energy operator and thus the Hamiltonian for diatomic rotator becomes

$$\hat{H} = \hat{T} = \frac{-h^2}{8\pi^2 \mu}\left[\frac{1}{r^2 \sin\theta}\frac{\partial}{\partial\theta}\left(\sin\theta\frac{\partial}{\partial\theta}\right) + \frac{1}{r^2\sin^2\theta}\frac{\partial^2}{\partial\phi^2}\right]$$

Therefore, the Schrodinger equation becomes

$$\hat{H}\psi = E\psi$$

$$\frac{-h^2}{8\pi^2 \mu}\left[\frac{1}{r^2 \sin\theta}\frac{\partial}{\partial\theta}\left(\sin\theta\frac{\partial}{\partial\theta}\right) + \frac{1}{r^2\sin^2\theta}\frac{\partial^2}{\partial\phi^2}\right]\psi = E\psi$$

Also the wavefunction of rigid rotator depends upon the angular orientation (θ and φ) and is independent of the radial part. Thus, the wavefunction ψ(r, θ, φ) has been replaced by a function Y(θ, φ) known as spherical harmonics, which depends only on θ and φ.

$$Y(\theta, \phi) = \Theta(\theta) \,.\, \Phi(\phi)$$

Thus, the Schrodinger wave equation for rigid rotator becomes

$$\frac{-h^2}{8\pi^2\mu}\left[\frac{1}{r^2\sin\theta}\frac{\partial}{\partial\theta}\left(\sin\theta\frac{\partial}{\partial\theta}\right)+\frac{1}{r^2\sin^2\theta}\frac{\partial^2}{\partial\phi^2}\right]Y(\theta,\phi)=EY(\theta,\phi)$$

(ii) The solution of Schrodinger wave equation of rigid rotator for kinetic energy is:

$$E=\ell(\ell+1)\frac{h^2}{8\pi^2 I}$$

The zero point energy of the rigid rotator is the lowest energy of the rigid rotator. The minimum value of 'ℓ' is zero, therefore, the energy corresponding to ℓ = 0 will be the zero point energy

$$E_0=0(0+1)\frac{h^2}{8\pi^2 I}=0 \qquad \text{(corresponding to no rotation)}$$

Q. 3. Formulate Schrodinger equation for rigid rotator. Find its eigen value and eigen function.

Ans. Schrodinger equation of rigid rotator is

$$\frac{-h^2}{8\pi^2\mu}\left[\frac{1}{r^2\sin\theta}\frac{\partial}{\partial\theta}\left(\sin\theta\frac{\partial}{\partial\theta}\right)+\frac{1}{r^2\sin^2\theta}\frac{\partial^2}{\partial\phi^2}\right]Y(\theta,\phi)=EY(\theta,\phi)$$

which on solving for kinetic energy gives

$$E=\ell(\ell+1)\frac{h^2}{8\pi^2 I}$$

Since the Schrodinger equation is an eigen value equation, i.e.

$$\hat{A}\psi = a\psi$$

Thus, on comparing Schrodinger equation for rigid rotator with eigen value equation:

Eigen function = $Y(\theta, \phi) = \Theta(\theta) . \Phi\phi$: Spherical harmonics

Eigen value = $E=\ell(\ell+1)\dfrac{h^2}{8\pi^2 I}$

Q. 4. Calculate the degeneracy of the rotating molecule with rotational quantum number J = 0, 1 and 2.

Ans. Refer Fig. 3.3, where J is just a replacement notation for 'ℓ'.

Q. 5. Show that the frequency of rotation per second (ν) for a rigid rotator is given by

$$\nu=\left(\frac{h}{4\pi^2 I}\right)\{J(J+1)\}^{1/2}$$

where, J = rotational quantum number

Ans. Energy expression for the rotational levels of a rigid rotator is given by

$$E=\frac{h^2}{8\pi^2 I}J(J+1) \qquad \text{(Quantum mechanics)}$$

Now, $E = \frac{1}{2}I\omega^2$ (Classical mechanics, equation 3.4)

where, ω = angular velocity in radians per second

$$\omega = 2\pi\nu$$

$$E = \frac{1}{2}I(2\pi\nu)^2 = 2I\pi^2\nu^2$$

$$\therefore \quad 2I\pi^2\nu^2 = \frac{h^2}{8\pi^2 I}J(J+1)$$

$$\nu = \frac{h}{4\pi^2 I}J\{J(J+1)\}^{1/2}$$

Q. 6. A rigid rotator consists of two particles of mass m_1 and m_2 joined by a rigid rod of length r. Based on classical considerations show that the total energy of a rigid rotator is $L^2/2I$, where L is the angular momentum and I is the moment of inertia. How does this result differ from quantum mechanical result?

Ans. The classical kinetic energy expression for linear systems can be written for rotating models as:

$$T = \frac{1}{2}I\omega^2$$

$$T = \frac{1}{2}I\omega^2 \times \frac{I}{I}$$

$$T = \frac{I^2\omega^2}{2I}$$

$$T = \frac{(I\omega)^2}{2I}$$

$$T = \frac{L^2}{2I}$$

where, L = Iω = angular momentum.

We know that angular momentum is quantised whose values are given by

$$L = \sqrt{\ell(\ell+1)}\frac{h}{2\pi}$$

where, $\ell = 0, 1, 2, 3...$

$$E = \frac{L^2}{2I}$$

$$E = \ell(\ell+1)\frac{h^2}{4\pi^2} \times \frac{1}{2I}$$

$$E = \ell(\ell+1)\frac{h^2}{8\pi^2 I}$$

$$E = \beta\frac{h^2}{8\pi^2 I}$$

where, $\beta = \ell\ (\ell + 1)$

Thus, according to classical mechanics, the kinetic energy of the rotator can have any value. However, quantum mechanically, we obtain a set of discrete energy levels and these are allowed energies by putting different value of ℓ, where ℓ = 0, 1, 2, ... Thus, quantum mechanically it was proved that the energy of rigid rotator is quantised.

3.6 Rotational Energy Levels

The rotational energy of a diatomic molecule during rotation in spectroscopy is given by:

$$E_J = \frac{h^2}{8\pi^2 I} J(J+1) = BJ(J+1) \qquad \text{[in Joules]} \qquad ...(3.15)$$

where, $B = \frac{h^2}{8\pi^2 I}$ = rotational constant [in Joules]

and J = 0, 1, 2, 3... = rotational quantum number

J is just a replacement notation for 'ℓ' i.e. azimuthal quantum number in the solution of rigid rotator. The permitted E_J values called **rotational energy** levels are calculated by putting J = 0, 1, 2, 3, ... in equation (3.15). The pattern of the rotational energy levels obtained will be as shown in Fig. 3.4.

The spacing between the energy levels increase as J increases. Only certain discrete rotational energy levels are allowed to the molecule, i.e. there is restriction on the 'J' values.

- In the rotational region, spectra are usually discussed in terms of wavenumber $(\overline{v})$, so we consider energies in cm^{-1}.

$$\epsilon_J = \frac{E_J}{hc} = \frac{h}{8\pi^2 Ic} J(J+1) \ \left[\text{in}\, cm^{-1}\right]$$

$$\epsilon_J = \overline{B} J(J+1) \qquad ...(3.16)$$

where, $\overline{B} = \frac{h}{8\pi^2 Ic}$ (in cm^{-1}) = rotational constant

The value of $\overline{B}$ is between 1 – 10 cm^{-1}.

Fig. 3.4: Rotational energy levels of a diatomic molecule

From Fig. 3.4, we can see that when J = 0, $\epsilon_J = 0$, means molecule is not rotating at all.

When J = 1, $\epsilon_J = 2\bar{B}$ and the rotating molecule then has its lowest angular momentum. Thus, we continue calculating ϵ_J with increasing J values. In principle, there is no limit to the rotational energy of that molecule. But in practice, there comes a point at which the centrifugal force of a rapidly rotating diatomic molecule is greater than the strength of the bond and the molecule is disrupted.

3.7 Rotational Selection Rules

For rigid diatomic molecule, we only consider transitions in which value of J changes by one unit. All other transitions being spectroscopically forbidden. Thus, during rotation, the transitions takes place between rotational energy levels for which

$$\Delta J = \pm 1$$

i.e. the change in the rotational quantum number is unity. Such a result is called **selection rule.**

The transition $\Delta J = +1$ corresponds to absorption and $\Delta J = -1$ corresponds to emission. All these transitions can take place with equal probability.

PROBLEM BASED ON SELECTION RULES

Q. 1. Give the selection rules for rotational spectra.

Ans. The transitions take place between rotational energy levels for which $\Delta J = \pm 1$

$$\Delta J = +1 \rightarrow \text{Absorption}$$
$$\Delta J = -1 \rightarrow \text{Emission}$$

3.8 Rotational Spectra of Rigid Diatomic Molecule

To discuss the rotational spectra, we consider the difference between the rotational energy levels. According to the rotational selection rules, only those rotational transitions are allowed for which $\Delta J = \pm 1$. Thus, a transition takes place from a lower rotational level with rotational quantum number 'J' to a higher rotational level with rotational quantum number $J' = J + 1$. The energy absorbed for the transition will be given by

$$\bar{v}_{J\to J+1} = \Delta \in_J = \bar{B}(J+1)(J+2) - \bar{B}J(J+1)$$

$$= \bar{B}(J+1)\left[\cancel{J} + 2 - \cancel{J}\right]$$

$$\bar{v}_{J\to J+1} = \Delta \in_J = 2\bar{B}(J+1)\ \text{cm}^{-1}$$

$\bar{v}$ represents the wavenumbers of the spectral lines which will be obtained as a result of the transitions between the rotational levels.

(a) For transition from level $J = 0 \to J = 1$,

$$\Delta\in_J = 2\bar{B}\,(0+1) = 2\bar{B}\ \text{cm}^{-1}$$

i.e. the absorption line will appear at $2\bar{B}$ cm^{-1}

(b) For transition from level $J = 1 \to J = 2$,

$$\Delta\in_J = 2\bar{B}\,(1+1) = 4\bar{B}\ \text{cm}^{-1}$$

i.e. the absorption line will appear at $4\bar{B}$ cm^{-1} and so on.

The most important feature of the pure rotational spectrum as shown in Fig. 3.5 is that every two successive lines have a constant difference of wavenumber equal to $2\bar{B}$ cm^{-1}. This is called **frequency separation** (or **wavenumber separation**). Thus, the various lines in the rotational spectra will be equally spaced.

Another important feature of the spectra is that the intensities of different transitions are not equal. The intensities first increases with increasing J, pass through a maximum and then decrease as J further increases.

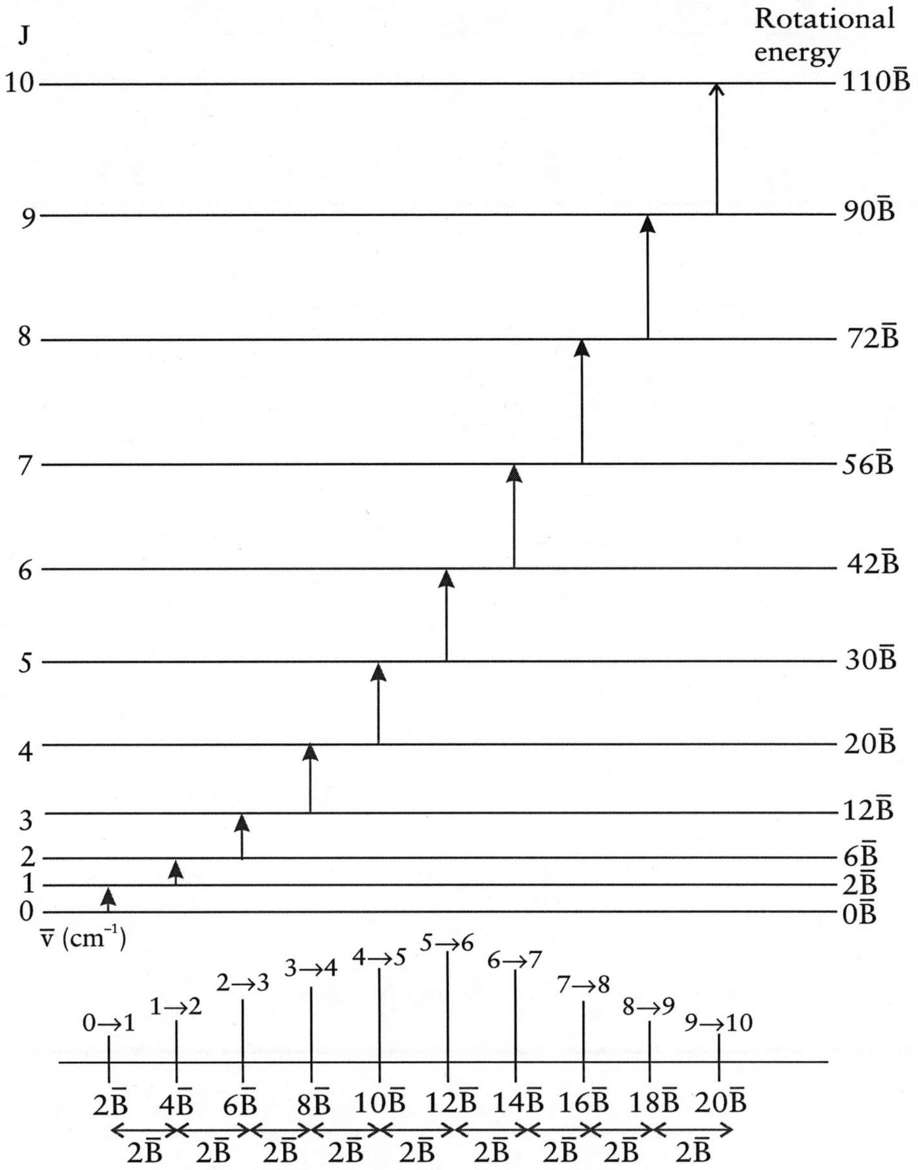

Fig. 3.5: Rotational spectrum

PROBLEMS BASED ON ROTATIONAL SPECTRA

Q. 1. (a) Draw the theoretical microwave spectrum for a heteronuclear diatomic molecule AB.

Ans.

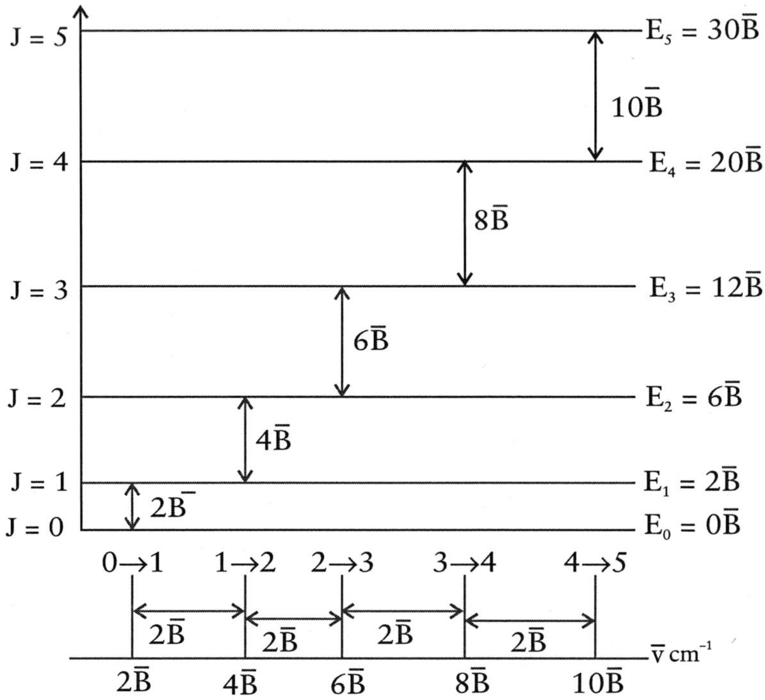

(b) What is the structural information that can be obtained from this spectrum?

Ans. (i) The spacing between different rotational level goes on increasing with increasing J value.

(ii) There is no zero point energy associated with rotation.

(iii) The spacing between different rotational lines give the value of $\bar{B}$.

(iv) For large molecule, the moment of inertia has large value and value of $\bar{B}$ will be small $\left(\bar{B} = \frac{h}{8\pi^2 Ic}\right)$. This signify that for large molecules the rotational levels are closely spaced in comparison to small molecules.

(v) Bond length of the molecule can be calculated.

Q. 2. Show that the difference between the adjacent lines in the rotation spectra of a diatomic molecule is constant.

Ans. To raise the molecule from state J to J + 1, the general expression is

$$\bar{v}_{J\to J+1} = \bar{B}(J+1)(J+2) - \bar{B}J(J+1)$$

$$\bar{\nu}_{J\to J+1} = \bar{B}\left(J^2 + 3J + 2 - J^2 - J\right)$$

$$\bar{\nu}_{J\to J+1} = 2\bar{B}(J+1)\ \text{cm}^{-1} \quad [\text{J = rotational constant for lower level}]$$

$$\bar{\nu}_{0\to 1} = 2\bar{B}(0+1) = 2\bar{B}$$

$$\bar{\nu}_{1\to 2} = 2\bar{B}(1+1) = 4\bar{B}$$

$$\bar{\nu}_{2\to 3} = 2\bar{B}(2+1) = 6\bar{B}$$

$$\bar{\nu}_{3\to 4} = 2\bar{B}(3+1) = 8\bar{B}$$

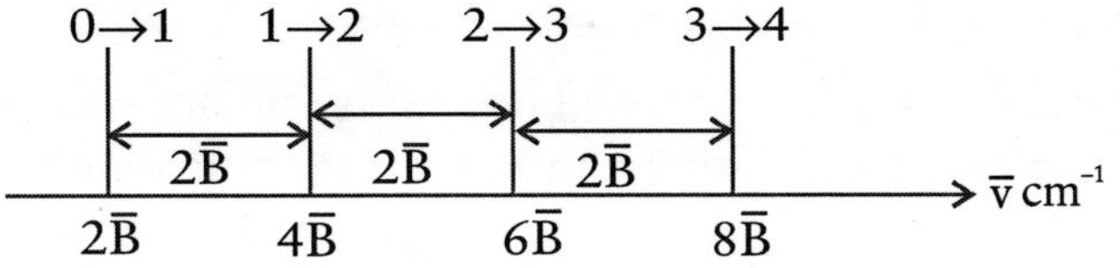

Therefore, the rotational transitions are separated by a constant value $2\bar{B}$.

Q. 3. Draw and discuss the allowed transitions between the rotational energy levels of a rigid diatomic molecule and the spectrum which arises from them.

Ans.

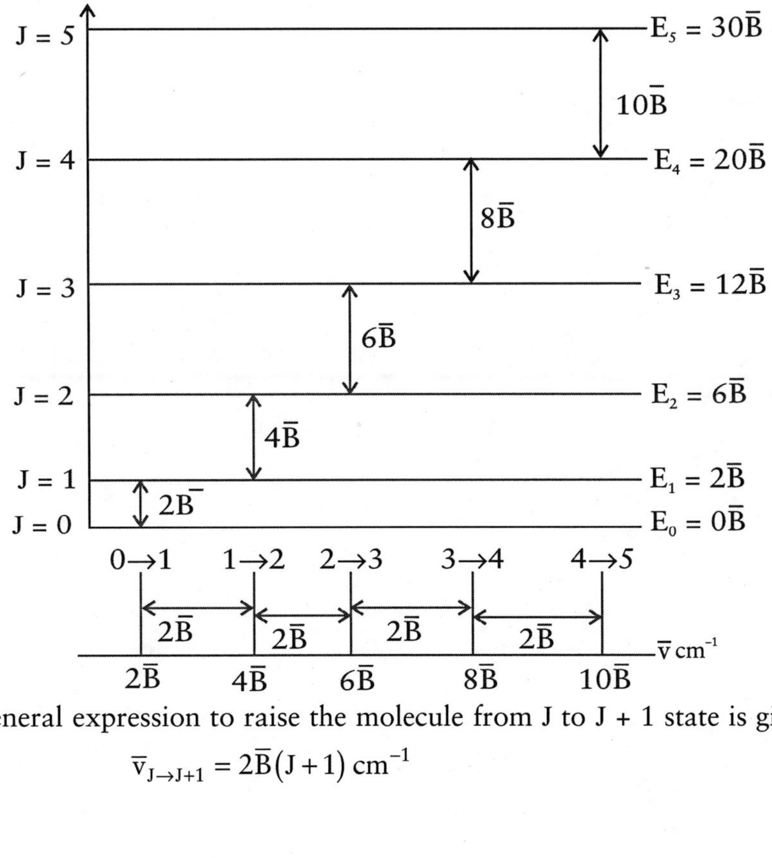

The general expression to raise the molecule from J to J + 1 state is given by

$$\bar{\nu}_{J\to J+1} = 2\bar{B}(J+1)\ \text{cm}^{-1}$$

Thus stepwise raising the rotational energy levels results in an absorption spectrum, having lines at $2\bar{B}$, $4\bar{B}$, $6\bar{B}$, $8\bar{B}$ .... cm^{-1}. Similarly, for the emission spectrum, similar spacing with lowering $\bar{B}$ value occurs. For deriving this spectrum, only those transitions are considered in which transition take place from a particular level to its next neighbour, either above or below, i.e. the selection rule followed is $\Delta J = \pm 1$, rest other transitions like $J = 0 \rightarrow J = 2 \rightarrow J = 4$ are considered forbidden.

Q. 4. A molecule XY is studied by microwave spectroscopy under the rigid rotator approximation. Draw the schematic stick line spectrum. What is the structural information that can be obtained from this spectrum.

3.9 Relative Intensities of Rotational Spectral Lines

Probability of all changes with $\Delta J = \pm 1$ is almost the same. But this does not mean that all spectral lines will be equally intense. Although the intrinsic probability that a single molecule in $J = 0$ will move to $J = 1$ is the same as that of a single molecule moving from $J = 1$ to $J = 2$ in an assemblage of molecules but there will be different number of molecules in each level to begin with and therefore different total number of molecules will carry out transition between the various levels. Thus, it is expected that the relative intensities of the spectral lines will depend upon the relative population of the energy levels. Greater is the population of an energy level, greater is the number of molecules that can be promoted to the next higher level and hence greater is the intensity of absorption.

The population of a rotational energy level with quantum number J (relative to that of the ground level for which $J = 0$) is governed by two factors:

(a) Boltzmann distribution factor

$$\frac{N_J}{N_0} = e^{-E_J/kT}$$

(b) Degeneracy of rotational energy level

$$g_J = (2J + 1)$$

Thus, the relative population of an energy level and hence intensity of the rotational lines will be proportional to the product of both the above factors, i.e.

$$\text{Population in } J_{th} \text{ level } \alpha(2J + 1)e^{-E_J/kT}$$

$$P_J \alpha(2J + 1)e^{-BJ(J + 1)/kT}$$

$$\bar{B} = \frac{B}{hc}$$

$$B = \bar{B}hc$$

$$P_J \alpha(2J + 1)e^{-\bar{B}hc\, J(J+1)/kT}$$

As J increases, the first factor, i.e. (2J + 1) increases, whereas the exponential factor decreases slowly in the beginning and then rapidly as the value of J becomes higher and higher. Hence, the plot of relative population versus J for a rigid diatomic molecule is obtained as shown in Fig. 3.6.

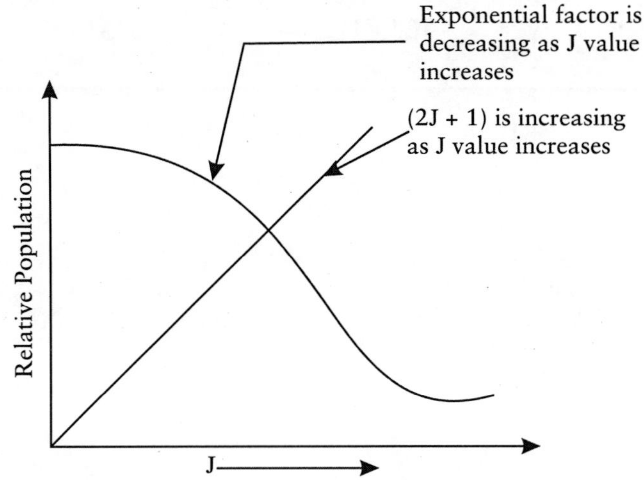

Fig. 3.6: Plot of relative population of rotational energy levels versus rotational quantum number (J)

The (2J + 1) factor is increasing and exponential factor is decreasing as J is increasing. Thus, the intensity of spectral lines for a particular value of J increases, then reaches a maximum value and then it decreases because the exponential factor predominates, shown in Fig. 3.7.

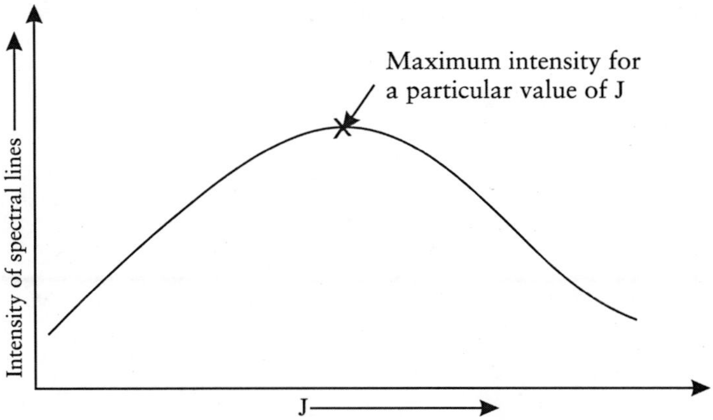

Fig. 3.7: Plot of intensity of spectral lines versus rotational quantum number (J)

Now, $\frac{dP_J}{dJ} = 0$ (maxima)

$$\frac{dP_J}{dJ} = \left[2e^{-\bar{B}hc\,J(J+1)/kT}\right] + \left\{(2J+1)e^{-\bar{B}hc\,J(J+1)/kT}\left[-\frac{\bar{B}hc(2J+1)}{kT}\right]\right\} = 0$$

$$= e^{-\bar{B}hc\,J(J+1)/kT}\left[2 - \frac{\bar{B}hc}{kT}(2J+1)^2\right] = 0$$

$$= 2 - \frac{\bar{B}hc}{kT}(2J+1)^2 = 0$$

$$(2J+1)^2 = \frac{2kT}{\bar{B}hc}$$

$$(2J+1) = \sqrt{\frac{2kT}{\bar{B}hc}}$$

$$J = \frac{1}{2}\left(\sqrt{\frac{2kT}{\bar{B}hc}} - 1\right)$$

$$J_{max} = \left(\sqrt{\frac{kT}{2\bar{B}hc}} - \frac{1}{2}\right)$$

J_{max} is the J value with maximum population.

$$\epsilon_{max} = \bar{B}hc\,J_{max}\left(J_{max} + 1\right)$$

$$\epsilon_{max} = \bar{B}hc\left(\sqrt{\frac{kT}{2\bar{B}hc}} - \frac{1}{2}\right)\left(\sqrt{\frac{kT}{2\bar{B}hc}} + \frac{1}{2}\right)$$

$$\epsilon_{max} = \bar{B}hc\left(\frac{kT}{2\bar{B}hc} - \frac{1}{4}\right)$$

$$\epsilon_{max} = \frac{kT}{2} - \frac{\bar{B}hc}{4}$$

Since $\bar{B}$ is small

$$\epsilon_{max} \approx \frac{kT}{2}$$

PROBLEMS BASED ON RELATIVE INTENSITIES OF ROTATIONAL SPECTRAL LINES

Q. 1. Show that the rotational level where quantum number is given by $J = \sqrt{\frac{kT}{2\bar{B}hc}} - \frac{1}{2}$ has maximum population.

Ans. $\epsilon_{max} = \bar{B}hc\,J_{max}(J_{max} + 1)$

$$\in_{max} = \bar{B}hc\left(\sqrt{\frac{kT}{2\bar{B}hc}} - \frac{1}{2}\right)\left(\sqrt{\frac{kT}{2\bar{B}hc}} - \frac{1}{2} + 1\right)$$

$$= \bar{B}hc\left(\sqrt{\frac{kT}{2\bar{B}hc}} - \frac{1}{2}\right)\left(\sqrt{\frac{kT}{2\bar{B}hc}} + \frac{1}{2}\right)$$

$$= \bar{B}hc\left(\frac{kT}{2\bar{B}hc} - \frac{1}{4}\right)$$

$$\in_{max} = \frac{kT}{2} - \frac{\bar{B}hc}{4}$$

Except at very low temperature, $\frac{kT}{2} \gg \frac{\bar{B}hc}{4}$

Therefore, the energy of the most populated level is, $\in_{max} \approx \frac{kT}{2}$

The relative population of an energy level will depend on:

$$\text{Population } \alpha(2J + 1) \exp(-E_J/kT) \quad ...(1)$$

When this is plotted against J, it will vary, at first it increases to maximum and then decreases. The differentiation of equation (1) show that population is maximum at the nearest integral J value

$$J_{max} = \sqrt{\frac{kT}{2\bar{B}hc}} - \frac{1}{2}$$

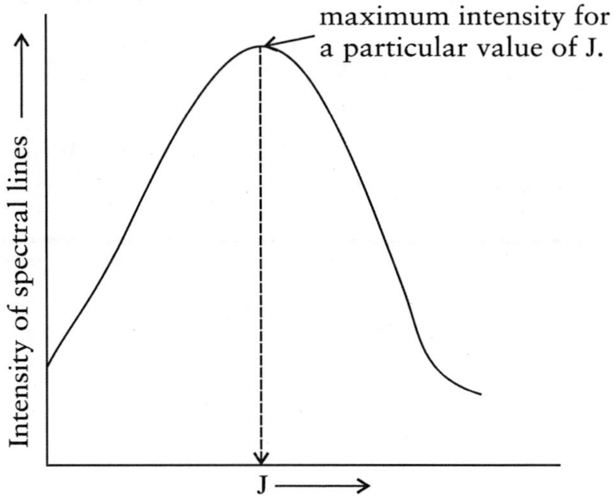

Q. 2. The transition corresponding to J = 0 to J = 1 is not necessarily the most intense line in microwave spectrum. Explain.

Ans. The spacing between nearest two signals in the rotational spectrum is constant and is equal to $2\bar{B}$. The intensity of all these transitions are not same and transition from

J = 0→J = 1 levels is not necessarily the most intense transition. The intensity of a transition depends on the number of molecules promoted from a particular level to the next higher level. So higher the population in a particular level, larger the number of molecules that can be promoted to next higher level and hence larger the intensity of absorption.

Q. 3. What are the various factors responsible for the different intensities of rotational spectral lines?

Ans. The intensities of rotational spectral lines depend on:

1. **Transition Probability**

   $\Delta J = \pm 1 \rightarrow$ Allowed Transitions

   Transitions corresponding to $\Delta J = \pm 1$ are allowed transitions and intense.

2. **Population of Rotational Level**

   Transitions with $\Delta J = \pm 1$ are equally probable and their intensity depends on the population of the molecule in the ground level. Population of rotational level is governed by two factors.

(a) **Boltzmann Distribution Law**

$$\frac{N_{upper}}{N_{lower}} = \exp\left(-\frac{\Delta E}{kT}\right) = \exp\left(\frac{-\bar{B}hc\ J(J+1)}{kT}\right)$$

(b) **Degeneracy of Energy States**

Degeneracy is the existence of two or more energy states which have exactly the same energy. In general, each spectral level is (2J + 1) fold degenerate.

Relative population at energy, $E \propto (2J+1)\exp\left(\frac{-\bar{B}hc\,J(J+1)}{kT}\right)$

Q. 4. Derive an expression for finding the rotational quantum number (J) of the state whose population is maximum.

(or)

Derive the expression for the rotational energy levels of rigid diatomic molecule corresponding to the most populated level. What will be the relative population of higher energy levels as J increases steadily?

(or)

For a diatomic molecule undergoing rotational motion

(i) Why does the population of the rotational level attain a maximum value at J_{max}.

(ii) Derive an expression for J_{max}.

(iii) Evaluate the rotational energy corresponding to J_{max} and show that under certain conditions this value is in accordance with classical theory.

(or)

Population of a rotational level is proportional to $(2J+1)\exp\left[-\bar{B}hc\,J(J+1)/kT\right]$ Derive an expression for J_{max}.

Ans. $$P_J \propto (2J+1)\exp\left[\frac{-\bar{B}hc\,J(J+1)}{kT}\right]$$

For maximum population, $\dfrac{dP_J}{dJ} = 0$

$$= \frac{\partial}{\partial J}(2J_{max}+1)\exp\left[\frac{-\bar{B}hc\,J_{max}(J_{max}+1)}{kT}\right] = 0$$

$$= (2J_{max}+1)\left(\frac{-\bar{B}hc}{kT}\right)(2J_{max}+1)\exp\left[\frac{-\bar{B}hc\,J_{max}(J_{max}+1)}{kT}\right] + 2\exp\left[\frac{-\bar{B}hc\,J_{max}(J_{max}+1)}{kT}\right] = 0$$

$$= \exp\left[\frac{-\bar{B}hc\,J_{max}(J_{max}+1)}{kT}\right]\left[\frac{-\bar{B}hc}{kT}(2J_{max}+1)^2 + 2\right] = 0$$

or $$\frac{-\bar{B}hc}{kT}(2J_{max}+1)^2 + 2 = 0$$

$$\not{-}(2J_{max}+1)^2 = \frac{\not{-}2kT}{\bar{B}hc}$$

$$2J_{max}+1 = \sqrt{\frac{2kT}{\bar{B}hc}}$$

$$J_{max} = \sqrt{\frac{kT}{\bar{B}hc}} - \frac{1}{2}$$

Population depends on two factors, the multiplicity factor (2J + 1) and exponential terms $\exp(-E_J/kT)$. As initially J increases, the multiplicity factor increases steadily, whereas the exponential factor decreases slowly in the beginning and more rapidly with higher value of J. As a consequence of these two factors, the number of molecules in the various levels increases in the beginning and after passing through the maxima it decreases

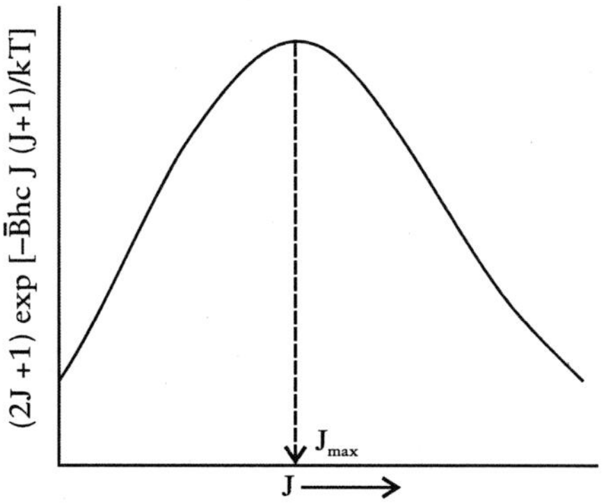

The rotational energy of the level which has highest population is

$$\in_{max} = J_{max}\left(J_{max}+1\right)\bar{B}hc$$

$$J_{max} = \sqrt{\frac{kT}{2\bar{B}hc}} - \frac{1}{2}$$

$$\in_{max} = \bar{B}hc\left[\left(\sqrt{\frac{kT}{2\bar{B}hc}} - \frac{1}{2}\right)\left(\sqrt{\frac{kT}{2\bar{B}hc}} - \frac{1}{2} + 1\right)\right]$$

$$\in_{max} = \bar{B}hc\left[\left(\sqrt{\frac{kT}{2\bar{B}hc}} - \frac{1}{2}\right)\left(\sqrt{\frac{kT}{2\bar{B}hc}} + \frac{1}{2}\right)\right]$$

$$\in_{max} = \bar{B}hc\left(\frac{kT}{2\bar{B}hc} - \frac{1}{4}\right)$$

$$\in_{max} = \frac{kT}{2} - \frac{\bar{B}hc}{4}$$

When temperature is moderately high, $\bar{B}$ being small is neglected $= \frac{kT}{2} \gg \frac{\bar{B}hc}{4}$

$$\in_{max} \approx \frac{kT}{2}$$

This kT/2 value is the same as desired from classical theory using equipartition principle. Each degree of freedom is ascertained a value of kT/2.

Q. 5. Determine the value of rotational quantum number 'J' in the most populated rotational level of ICl at 300 K with $\bar{B}$ = 0.114 cm^{-1}.

Ans. $J_{max} = \sqrt{\frac{kT}{2\bar{B}hc}} - \frac{1}{2}$

T = 300 K, k = 1.38 × 10^{-23} JK^{-1}

$\overline{B}$ = 0.114 cm^{-1} = 11.4 m^{-1}

c = 3 ×10^8 ms^{-1}

$$J_{max} = \sqrt{\frac{(1.38\times10^{-23})\times(300)}{2\times(11.4)\times(6.626\times10^{-34})\times(3\times10^{8})}} - \frac{1}{2}$$

$$= \sqrt{\frac{414\times10^{-23}}{453.21\times10^{-26}}} - \frac{1}{2}$$

$$= \sqrt{\frac{414000}{453.21}} - \frac{1}{2}$$

$$= \sqrt{913.48} - \frac{1}{2}$$

$$= 30.22 - 0.5$$

$$J_{max} = 29.7 \approx 30$$

The nearest whole number is 30.

Q. 6. The intensity of rotational absorption spectral lines, I, is related to the rotational quantum number J by the relation.

$$I = 2\overline{B}(J+1)(2J+1)e^{-\overline{B}hc\,J(J+1)/kT}$$

The variation of I vs J is shown in figure.

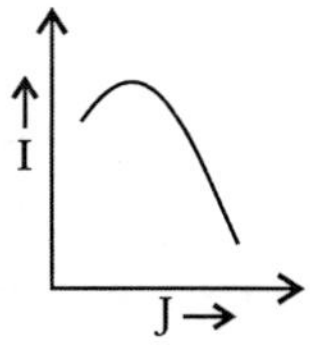

Assuming 4J + 3 ~ 4J + 4, calculate the value of J for which it will show maximum intensity.

Ans. $$\frac{dI}{dJ} = 2\overline{B}\left[(2J+1)\times1+2(J+1)\right]e^{-\overline{B}hc\,J(J+1)/kT}$$

$$+2\overline{B}(J+1)(2J+1)e^{-\overline{B}hc\,J(J+1)/kT}\times\left[\frac{-\overline{B}hc(2J+1)}{kT}\right]=0$$

$$\frac{dI}{dJ} = 2\overline{B}e^{-\overline{B}hc\,J(J+1)/kT}\times\left[(4J+3)-\frac{\overline{B}hc(2J+1)^2(J+1)}{kT}\right]=0$$

$$(4J+3)-\frac{\overline{B}hc(2J+1)^2(J+1)}{kT}=0$$

$$\frac{(2J+1)^2(J+1)}{kT}=\frac{4J+3}{\overline{B}hc}$$

$$\frac{(2J+1)^2(J+1)}{kT} = \frac{4J+4}{\bar{B}hc} \quad \{\text{Assuming} : (4J+3) \sim (4J+4)\}$$

$$\frac{(2J+1)^2 \cancel{(J+1)}}{kT} = \frac{4\cancel{(J+1)}}{\bar{B}hc}$$

$$(2J+1) = \sqrt{\frac{4kT}{\bar{B}hc}}$$

$$2J = 2\sqrt{\frac{kT}{\bar{B}hc}} - 1$$

$$\cancel{2}J = \cancel{2}\sqrt{\frac{kT}{\bar{B}hc}} - \frac{1}{2}$$

$$J_{max} = \sqrt{\frac{kT}{\bar{B}hc}} - \frac{1}{2}$$

Therefore, at $= \sqrt{\frac{kT}{\bar{B}hc}} - \frac{1}{2}$, will show maximum intensity.

3.10 Microwave Active Molecules or Types of Molecules Showing Rotational Spectra

To show pure rotational spectra, the molecule must have a permanent dipole moment so that it can interact with the electromagnetic radiation.

Thus, only polar or asymmetric (heteronuclear) molecules like H_2O, N_2O, HCl, CO, etc. show pure rotational spectra or are microwave active because they have permanent dipole moment and show dipole component change during rotation on interaction with radiation. Whereas homonuclear diatomic molecules (like N_2, O_2, etc.) and symmetrical linear molecules (like CO_2, C_6H_6, etc.) are not microwave active because there will be no dipole component change during the rotation and hence no interaction with radiation.

3.10.1 How do Microwave Radiation Interact with the Rotating Molecule Having a Permanent Dipole?

The rotating molecule generates fluctuating electrical field (similar to the field generated by electromagnetic radiation) and this field is responsible for interaction between microwave radiation and rotating molecule. The electrical component of the radiation oscillates in the plane perpendicular to the direction of propagation of the wave. The frequency of the oscillating electric component of radiation when becomes equal to the fluctuating electric field frequency from molecule, then resonance takes place and two interact with each other. Hence, the energy transfer from radiation to molecule or vice-versa takes place. For producing fluctuating electrical field by rotation, the molecule must possess a permanent dipole. **That is why, molecule with permanent dipole moment only shows rotational spectra.**

On rotation, the positive and negative ends of dipolar molecule change place periodically and thus the component of dipole moment in a given direction such as the vertical component in the plane of the paper fluctuates.

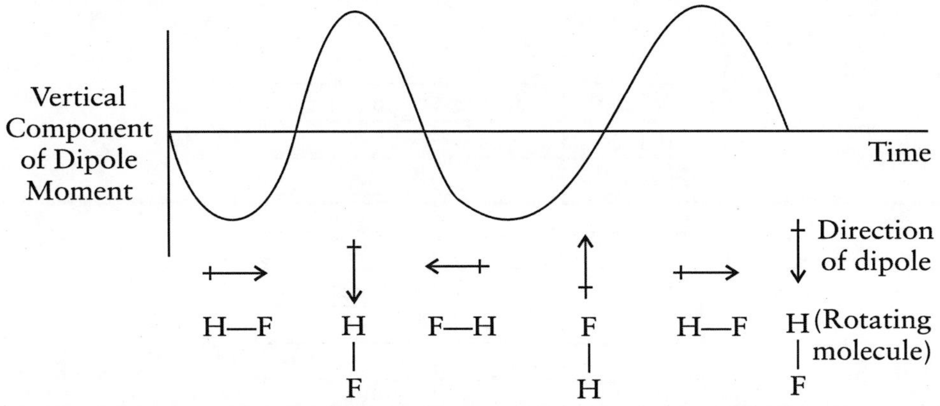

Fig. 3.8: Dipole changes with rotation in HF molecule

This sinusoidal component of the dipole moment enables interaction with the incident electromagnetic radiation.

A molecule with no permanent dipole moment does not generate fluctuating electric field on rotation, hence, such molecules are microwave inactive, i.e. they do not produce pure rotational spectra.

PROBLEMS BASED ON MICROWAVE ACTIVE MOLECULES

Q. 1. What are the fundamental requirements for a molecule to show rotational spectra or for a molecule to be microwave active?

Ans. The fundamental requirement for a molecule to show rotational spectra is that it should have a permanent dipole moment.

Q. 2. Why it is not possible to obtain the pure rotational spectrum of a homonuclear diatomic molecule?

Ans. For obtaining the rotational spectrum the fluctuating electric field of the radiation should interact with the similar field generated by the molecule. The electric component of radiation oscillates in a plane perpendicular to the direction of propagation of wave. The frequency of this oscillating electric component is equal to frequency of radiation. The oscillating electric field is generated by molecules which have permanent dipole moment. Since homonuclear diatomic molecule do have permanent dipole moment, thus their rotational spectrum cannot be recorded.

Q. 3. Which of the molecules are microwave active/inactive and why?

Ans.

Molecule	Active	Inactive
HCl	yes	—
$H_2/Cl_2/I_2$	—	yes
CH_4	—	yes
CH_3Cl	yes	—
O_2	—	yes
SF_6	—	yes
C_2H_6	—	yes
CO_2	—	yes
HI	yes	—
ICl	yes	—

Q. 4. (a) HF gives a rotational spectrum whereas H_2 does not. Comment.

Ans. HF is a heterodiatomic molecule and has permanent dipole moment, hence microwave active. Whereas, H_2 molecules has no permanent dipole moment and hence microwave inactive.

(b) Diagrammatically explain the interaction of HF molecule with incident electromagnetic radiation.

Ans. Refer Fig. 3.8.

3.11 Classification of Molecules on the Basis of Moment of Inertia

A body has three principal moments of inertia, one about each axis, designated as I_a, I_b and I_c. Molecules may be classified into groups according to the relative values of their three principal moments of inertia.

Classification

- Spherical top molecule or Spherical symmetrical molecule
- Linear molecule
- Symmetric top molecule
  - Prolate top molecule
  - Oblate top molecule
- Asymmetric top molecule

A. Spherical Top Molecule or Spherical Symmetrical Molecule

Octahedral (SF_6) and Tetrahedral (CH_4) molecules are spherically symmetrical molecules. For these molecules, $I_a = I_b = I_c$. These molecules have zero dipole moment, thus their spectra is not important and none of them is microwave active

$$E_{rot} = \frac{L_a^2}{2I_a} + \frac{L_b^2}{2I_b} + \frac{L_c^2}{2I_c}$$

$$L^2 = L_a^2 + L_b^2 + L_c^2$$

Since $I_a = I_b = I_c$

$$E_{rot} = \frac{L_a^2 + L_b^2 + L_c^2}{2I_b}$$

$$E_{rot} = \frac{L^2}{2I_b}$$

B. Linear Molecule

Molecules like HCl, CO, OCS, CO_2, CS_2, etc. are linear molecules.

For these molecules,

$I_a = 0$ (no rotation about bond axis)

and $I_b = I_c$

Linear molecules (like CS_2, CO_2, etc.) have zero dipole moment, thus they are microwave inactive. Whereas linear molecules (like HCl, CO, OCS, etc.) have dipole moment value thus they are microwave active.

$$E_{rot} = \frac{L_b^2}{2I_b} + \frac{L_c^2}{2I_c}$$

$$E_{rot} = \frac{L^2}{2I_b}$$

C. Symmetric Top Molecule

Molecule having C_3 axis or atleast C_4 axis are called symmetric top molecules. For these molecules $I_a \neq 0$ and $I_b = I_c$. These are of two types:

1. **Prolate Top Molecules**

   For these, $I_a < (I_b = I_c)$

   Like: CH_3F

   This molecule has C_3 axis of symmetry. The moment of inertia about the C–F bond axis is not negligible because it involves the rotation of three comparatively massive hydrogen atoms off the axis, therefore $I_a = 0$.

   But the value of I_a is very small as none of heavier atom is making contribution to I_a of C_3 axis

   $\therefore \quad I_a < (I_b = I_c)$

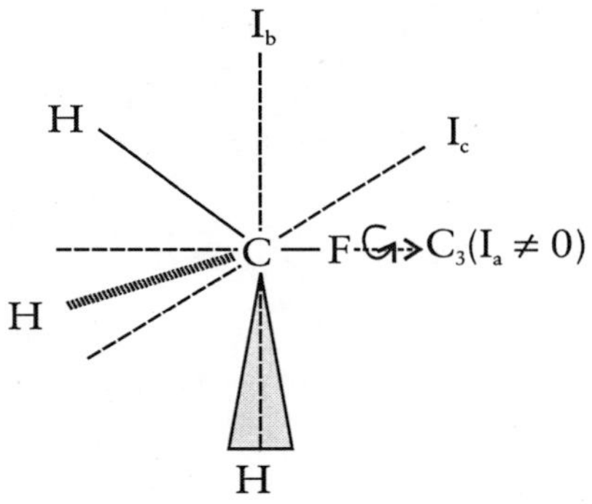

$$E_{rot} = \frac{L_a^2}{2I_a} + \frac{L_b^2 + L_c^2}{2I_b} = \frac{L_a^2}{2I_a} + \frac{L^2 - L_a^2}{2I_b}$$

2. **Oblate Top Molecules**

For these, $(I_a = I_b) < I_c$

Like: BCl_3

$$E_{rot} = \frac{L_a^2}{2I_a} + \frac{L_b^2}{2I_b} + \frac{L_c^2}{2I_c}$$

$$= \frac{L_a^2 + L_b^2}{2I_b} + \frac{L_c^2}{2I_c}$$

$$= \frac{L^2 - L_c^2}{2I_b} + \frac{L_c^2}{2I_c}$$

D. **Asymmetric Top Molecule**

For these molecules, $I_a = I_b = I_c$

Like: ethane, vinyl chloride.

3.12 Rotation about Bond Axis Rejected in Linear Rigid Rotator

The rotation about bond axis is rejected in linear rigid rotator because

(a) For linear molecules ($I_a = 0$ and $I_b = I_c$), the moment of inertia is very small about the bond axis, thus energy levels are extremely widely spaced $\left(E \propto \frac{1}{I}\right)$. Thus, molecule requires a great deal of energy to be raised from J = 0 to J = 1 and such transitions do not occur under normal spectroscopic conditions. Thus, all linear diatomic molecules are in the J = 0 state for rotation about the bond axis and may be said to be not rotating.

(b) Even if such a transition occur, there will be no dipole moment change and no spectra obtained.

3.13 Parameters Calculated from Rotational Spectra

1. From the rotational spectra, we can determine the value of $\overline{B}$.
2. Using the value of $\overline{B}$, it is possible to calculate the moment of inertia (I) of the molecule using the equation

$$\overline{B} = \frac{h}{8\pi^2 Ic}$$

3. Then knowing the masses of the atoms present in the diatomic molecule, we can calculate reduced mass (μ) of the molecule

$$\mu = \frac{m_1 m_2}{m_1 + m_2}$$

4. Using the value of moment of inertia (I) and reduced mass (μ), it is possible to calculate distance between the atoms, r_0 (internuclear distance or bond length) using the equation

$$I = \mu r_0^2$$

PROBLEMS BASED ON CALCULATION OF $\bar{B}$, I AND r_0

Q. 1. The first line in the microwave spectrum of carbon monoxide is positioned at 3.84235 cm^{-1}. Calculate the moment of inertia of the molecule and the internuclear distance.

Ans.

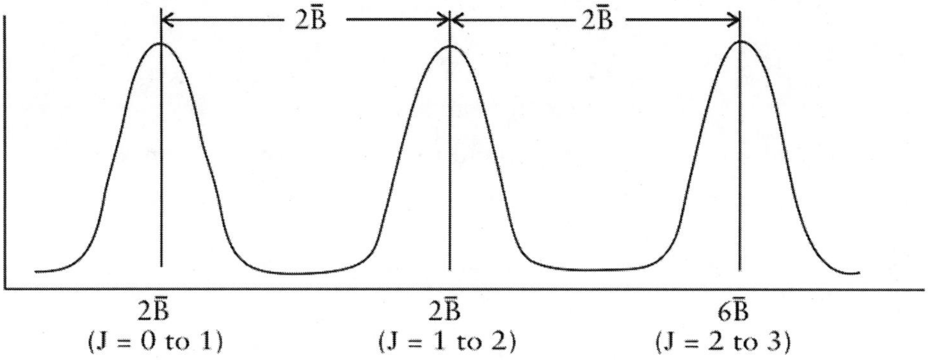

(a) The first rotational line corresponds to the rotational transition, J = 0 to 1.

$$\bar{v}_{J\to J+1} = 2\bar{B}(J+1)\ cm^{-1}$$

$$\bar{v} = 2\bar{B}(0+1)\ cm^{-1}$$

$$\bar{v} = 2\bar{B}\ cm^{-1}$$

$$2\bar{B} = 3.84235$$

$$\bar{B} = 1.92118\ cm^{-1}$$

(b)
$$\bar{B} = \frac{h}{8\pi^2 Ic}$$

$$I = \frac{h}{8\pi^2 \bar{B}c}$$

$$I = \frac{\left(6.6\times10^{-27}\,g\ cm^2\,s^{-1}\right)}{8\times(3.14)^2\times\left(1.92118\ cm^{-1}\right)\times\left(3\times10^{10}\,cm\,s^{-1}\right)}$$

$$I = 0.02\times10^{-37}\,gcm^2$$

(c)
m_1 = 12.0000 g (carbon) and
m_2 = 15.9949 g (oxygen)

$$\text{Reduced mass} = \mu = \frac{m_1 m_2}{m_1 + m_2}\times\frac{1}{N_A}$$

$$\mu = \frac{(12 \times 15.9949)}{(12 + 15.9949)} \times \frac{1}{6.023 \times 10^{23}}$$

$$\mu = 1.14 \times 10^{-23} \text{ g}$$

(d)

$$I = \mu r_0^2$$

$$r_0^2 = \frac{I}{\mu}$$

$$r_0 = \sqrt{\frac{I}{\mu}} \text{ cm}$$

$$r_0 = 1.31 \times 10^{-8} \text{ cm}$$

Q. 2. The rotational spectrum of HI is found to contain a series of lines with a separation of 12.8 cm^{-1}. Calculate I and r_0.

Ans. Separation between spectral lines = $2\bar{B}$

(a)

$$2\bar{B} = 12.8 \text{ cm}^{-1}$$

$$\bar{B} = 6.4 \text{ cm}^{-1}$$

(b)

$$\bar{B} = \frac{h}{8\pi^2 Ic}$$

$$I = \frac{h}{8\pi^2 \bar{B}c}$$

$$I = \frac{6.6 \times 10^{-27}}{8 \times (3.14)^2 \times (6.4) \times (3 \times 10^{10})}$$

$$I = 4.372 \times 10^{-40} \text{ g cm}^2$$

(c)

m_1 = 1.0000 g (hydrogen)

m_2 = 127.00 g (iodine)

$$\mu = \frac{(1 \times 127)}{(1 + 127)} \times \frac{1}{(6.023 \times 10^{23})}$$

$$\mu = \frac{127}{128} \times \frac{1}{6.023 \times 10^{23}}$$

$$\mu = 1.65 \times 10^{-25} \text{ g}$$

(d) $$I = \mu r_0^2$$

$$r_0^2 = \frac{I}{\mu}$$

$$r_0^2 = \frac{4.372 \times 10^{-40}}{1.65 \times 10^{-25}} cm^2$$

$$r_0^2 = 2.6254 \times 10^{-16} cm^2$$

$$r_0 = 1.62 \times 10^{-8} cm^2$$

Q. 3. Calculate the most populated level. (J_{max}) for CO molecule at 25°C, if the internuclear distance is 0.133 nm.

Ans. For CO molecule at 25°C (298 K), the most populated state can be found, if $\bar{B}$ is known. The $\bar{B}$ for CO can be calculated as:

For CO, r_{CO} = 1.33 nm = 0.133×10^{-9} m

$$\mu_{CO} = \frac{m_c m_0}{m_c + m_0}$$

$$\mu_{CO} = \frac{(0.012 kg / 6.023 \times 10^{23})(0.016 kg / 6.023 \times 10^{23})}{(0.012 kg / 6.023 \times 10^{23}) + (0.016 / 6.023 \times 10^{23})}$$

$$\mu_{CO} = 1.14 \times 10^{-26} kg$$

As, $$I = \mu r$$

So, $$I = (1.14 \times 10^{-26} kg)(0.113 \times 10^{-9} m)^2$$

$$I = 14.5 \times 10^{-49} kg\ m^2$$

We know that, $$\bar{B} = \frac{h}{8\pi^2 Ic}$$

$$\bar{B} = \frac{6.6 \times 10^{-34} Js}{8 \times (3.14)^2 \times (14.5 \times 10^{-47} kg m^2)(3 \times 10^8 ms^{-1})}$$

$$\bar{B} = 1.932 \times 10^2 J kg^{-1} m^{-3} s^2$$

Now, $$1J = 1\ kg\ m^2 s^{-2}$$

So, $$\bar{B} = 1.932 \times 10^2 m^{-1}$$

Substitute this value in J_{max},

$$J_{max} = \sqrt{\frac{kT}{2\bar{B}hc}} - \frac{1}{2}$$

$$J_{max} = \left\{\frac{1.38\times10^{-23}\,JK^{-1}\times298K}{2\left(1.932\times10^{2}\,m^{-1}\right)\times\left(6.6\times10^{-34}\,Js\right)\times\left(3\times10^{8}\,ms^{-1}\right)}\right\}^{1/2} - \frac{1}{2}$$

$$J_{max} = 6.83 \approx 7$$

Q. 4. Pure rotational spectrum of gaseous CN molecule exhibit that spacing between two adjacent lines is 3.7989 cm^{-1}. Calculate internuclear distance between C and N.

Ans. Spacing = $2\bar{B} = 3.7878\,cm^{-1}$

$$\bar{B} = 1.8989\,cm^{-1} = 1.8989\times10^{-2}\,m^{-1}$$

As, $\bar{B} = \dfrac{h}{8\pi^2 Ic}$

$$I = \frac{h}{8\pi^2\bar{B}c}$$

$$I = \frac{6.6\times10^{-34}\,Js}{8\times(3.14)^2\times\left(1.8989\times10^{-2}\,m^{-1}\right)\times\left(3\times10^{8}\,ms^{-1}\right)}$$

$$= 1.4742\times10^{-46}\,kg\,m^2$$

$$\mu = \frac{m_1 m_2}{(m_1+m_2)N_A} = \frac{(12\times14)}{(12+14)}\times\frac{1}{6.023\times10^{23}}$$

$$\mu = 1.07\times10^{-23}\,g$$

$$\mu = 1.07\times10^{-26}\,kg$$

We know that $r_0 = \left(\dfrac{I}{\mu}\right)^{1/2} = \left(\dfrac{1.4742\times10^{-46}\,kg\,m^2}{1.07\times10^{-26}\,kg}\right)^{1/2}$

$$= 1.17\times10^{-10}\,m$$

Q. 5. Calculate the average period of rotation of HCl when it is in the J = 2 state. The bond length of the molecule is 1.274 Å. (Avogadro number = 6.023×10^{23})

Ans. In classical mechanics, the rotational energy of a rotator is $I\omega^2/2$, where I is the moment of inertia about the axis of rotation and ω is the angular frequency.

Rotational energy is quantum mechanics, when the molecule is in the J state, is

$$E_J = \frac{h^2}{8\pi^2 I}J(J+1)$$

Equating these two expressions for energy of the rotator, we have

$$\frac{1}{2}I\omega^2 = \frac{h^2}{8\pi^2 I}J(J+1)$$

$$\omega = \frac{h\sqrt{J(J+1)}}{2\pi I}$$

Mass of hydrogen atom m_H and mass of chlorine atom m_{Cl} are

$$m_H = \frac{1}{6.023 \times 10^{23}} \text{ and } m_{Cl} = \frac{35.5}{6.023 \times 10^{23}}$$

Thus, the reduces mass μ is

$$\mu = \frac{m_H m_{Cl}}{m_H + m_{Cl}} \text{ and } m_{Cl} = \frac{1 \times 35.5}{36.5 \times (6.023 \times 10^{23})}$$

$$= 1.6151 \times 10^{-24} \text{g}$$

The moment of inertia I about the axis of rotation is

$$r_0 = 1.274 \text{Å} = 1.274 \times 10^{-10} \text{ m} = 1.274 \times 10^{-8} \text{ cm (given)}$$

$$I = \mu r_0^2 = (1.6151 \times 10^{-24}) \times (1.274 \times 10^{-8})^2 = 2.621 \times 10^{-40} \text{g cm}^2$$

Thus, the angular frequency ω in J = 2 state is:

$$\omega = \frac{h\sqrt{J(J+1)}}{2\pi I} = \frac{6.6 \times 10^{-27}\sqrt{6}}{2 \times 3.14 \times 2.621 \times 10^{-40}}$$

$$= 9.854 \times 10^{12} \text{ s}^{-1}$$

The time period of rotation T is

$$T = \frac{2\pi}{\omega} = \frac{2 \times 3.14}{9.854 \times 10^{12}} = 6.38 \times 10^{-13} \text{s}$$

Q. 6. For $^1H^{35}Cl$, the rotational constant $\bar{B}$ is 10.6 cm^{-1}. Find the period of rotation in a rotational level characterized by $\ell = 4$.

Ans. $E = BJ(J + 1)$ Joules, (B in Joules)

$$\in = \frac{E}{hc} = \frac{BJ(J+1)}{hc}$$

$$\in = \bar{B}J(J+1)\text{cm}^{-1} \quad (\bar{B} \text{ in cm}^{-1})$$

$$E = \in \times hc$$

$$E = \bar{B}J(J+1) \times hc$$

J is same as ℓ for rigid rotator.

Here, $\bar{B}$ = 10.6 cm^{-1}, J = 4, h = 6.6 × 10^{-27} gcm^{2}s^{-1} and c = 3 × 10^{10} cms^{-1}

$$E = h\nu_{rot} \text{ and } \nu_{rot} = \frac{1}{\text{period of rotation}}$$

$$\therefore \quad E = \frac{h}{\text{period of rotation}}$$

Thus, period of rotation $= \frac{h}{E}$

Period of rotation $= \frac{h}{\bar{B} \cdot J(J+1) \times hc}$

$$= \frac{1}{(10.6) \times 4(4+1) \times (3 \times 10^{10})}$$

$$= \frac{1}{10.6 \times 20 \times (3 \times 10^{10})}$$

Period of rotation $= 0.0015 \times 10^{-20}$ s.

Q. 7. Rotational lines from HCl gas were found at the following positions ($\mu_{HCl} = 1.6266 \times 10^{-27}$ g):

83.32, 104.13, 124.73, 145.35, 165.89, 186.23, 200.60, 226.86 cm^{-1}

Find the moment of inertia and bond length of the molecule.

Ans. Separation between spectral lines $= 2\bar{B}$

$$(104.13 - 83.32)\ cm^{-1} = 2\bar{B}$$

$$20.81\ cm^{-1} = 2\bar{B}$$

$$\bar{B} = 10.41\ cm^{-1}$$

(or)

$\Delta \in (J) = 83.32\ cm^{-1} = 2\bar{B}\ (J + 1)$ ...(1)

$\Delta \in (J + 1) = 104.13\ cm^{-1} = 2\bar{B}\ (J + 2)$ ...(2)

Divide equation (1) by (2)

$$\frac{2\bar{B}(J+1)}{2\bar{B}(J+2)} = \frac{83.32}{104.13}$$

$$20.81J = 62.51$$

$$J = 3.004.$$

Substitute the value of J in equation (1)

$$2\bar{B}(3.004 + 1) = 83.32$$

$$\bar{B} = 10.41\ cm^{-1}$$

(i) $\bar{B} = \frac{h}{8\pi^2 Ic}$ or $I = \frac{h}{8\pi^2 \bar{B}c}$

$$I = \frac{(6.626\times10^{-27})}{8\times(3.14)^2(10.41)\times(3\times10^{10})}$$

$$I = 2.69\times10^{-40}\ gcm^2$$

(ii) $I = \mu r_0^2$

$\mu = 1.6266 \times 10^{-27}$ g (given)

$$2.69\times10^{-40} = (1.6266\times10^{-27})\times r_0^2$$

$$r_0^2 = 1.66\times10^{-13} = 0.166\times10^{-12}$$

$$r_0 = 0.407\times10^{-6}\ cm$$

Q. 8. (a) The wave number of pure rotational lines of NH molecule are represented by equation.

$$\bar{v} = 16.7(J+1)cm^{-1}$$

Calculate the moment of inertia and bond length of NH molecule.

Ans. $\Delta \in_J = \bar{v} = 2\bar{B}(J+1)cm^{-1}$

We have, $\bar{v} = 16.7(J+1)cm^{-1}$

Thus, $2\bar{B} = 16.7\ cm^{-1}$

$\bar{B} = 8.35\ cm^{-1}$

$$\bar{B} = \frac{h}{8\pi^2 Ic} \quad \text{or} \quad I = \frac{h}{8\pi^2\bar{B}c}$$

$$I = \frac{6.626\times10^{-27}}{8\times(3.14)^2\times(8.35)\times(3\times10^{10})}$$

$$= 3.3\times10^{-40}\ gcm^2$$

$$\mu_{NH} = \left(\frac{1\times14}{1+14}\right)\times\frac{1}{6.023\times10^{23}} = \frac{14}{90.345\times10^{23}}$$

$$\mu_{NH} = 1.55\times10^{-24}\ g$$

$I = \mu r_0^2$

$3.3 \times 10^{-40} = (1.55 \times 10^{-24}) \times r^2$

On solving, we get

$r_0^2 = 2.12 \times10^{-16}$ cm

$r_0 = 1.47 \times10^{-8}$ cm

(b) Calculate and draw the rotational energy diagram and thus show the first four spectral transitions on this energy diagram and thus generate the position of rotational lines for NH molecule.

Ans. For transition from level J = 0 → J = 1

$$\bar{v} = 16.7\,(J+1)\ cm^{-1}$$

$$\bar{v}_1 = 16.7\,(0+1)\ cm^{-1}$$

$$\bar{v}_1 = 16.7\ cm^{-1} = 2\bar{B}$$

For transition from level J = 1 → J = 2

$$\bar{v}_2 = 16.7\,(1+1)$$

$$\bar{v}_2 = 33.4\ cm^{-1} = 4\bar{B}$$

For transition from level J = 2 → J = 3

$$\bar{v}_3 = 16.7\,(2+1)$$

$$\bar{v}_3 = 50.1\ cm^{-1} = 6\bar{B}$$

For transition from level J = 3 → J = 4

$$\bar{v}_4 = 16.7\,(3+1)$$

$$\bar{v}_4 = 66.8\ cm^{-1} = 8\bar{B}$$

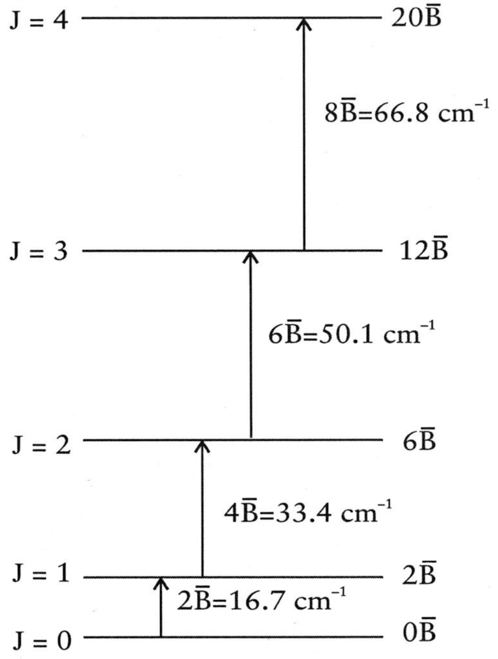

Rotational Energy Diagram for NH Molecule

(c) For which transition will the absorption spectra on NH molecule show the maximum intensity at 27°C?

Ans. $$J_{max} = \sqrt{\frac{kT}{2\bar{B}hc}} - \frac{1}{2}$$ Now, k = 1.38 × 10^{16} erg K^{-1}

$$J_{max} = \sqrt{\frac{(1.38\times10^{-16})\times(300)}{2\times(8.35)\times(6.626\times10^{-27})\times(3\times10^{10})}} - \frac{1}{2}$$

$$J_{max} = 3.53 - \frac{1}{2} = 3.03 \approx 3$$

$$J_{max} = 3$$

Thus for the fourth transition from J = 3 to J = 4, maximum intensity will be seen at 27°C.

Q. 9. The HCl molecule shows pure rotational lines at the following frequencies (cm^{-1})

20.7, 41.5, 62.0, 83.0 and 103.8

(a) Assign theses lines to the rotational transitions, J → J + 1

Ans. $\Delta\in (J) = 20.7 = 2\bar{B}\ (J + 1)$ ...(1)

$\Delta\in (J + 1) = 41.5 = 2\bar{B}\ (J + 2)$ ...(2)

Divide equation (1) by (2)

$$\frac{2\bar{B}(J+1)}{2\bar{B}(J+2)} = \frac{20.7}{41.5}$$

$$20.8J = -0.1$$

$$J = -0.004 \approx 0$$

Substitute the value of J in equation (1)

$$2\bar{B}\ (J + 1) = 20.7$$

$$2\bar{B} = 20.7\ cm^{-1}$$

∴ $2\bar{B} = 20.7\ cm^{-1}$ represents J = 0 → J = 1 transition

$4\bar{B} = 41.5\ cm^{-1}$ represents J = 1 → J = 2 transition

$6\bar{B} = 62.0\ cm^{-1}$ represents J = 2 → J = 3 transition

$8\bar{B} = 83.0\ cm^{-1}$ represents J = 3 → J = 4 transition

$10\bar{B} = 103.8\ cm^{-1}$ represents J = 4 → J = 5 transition

(b) Also calculate the bond distance of HCl (m_H = 1.0 and m_{Cl} = 35.5)

Ans. $$2\bar{B} = 20.775\ cm^{-1}$$

$$\bar{B} = 10.3875\ cm^{-1}$$

$$\bar{B} = \frac{h}{8\pi^2 Ic} \quad \text{or} \quad I = \frac{h}{8\pi^2 \bar{B} c}$$

$$I = \frac{(6.6 \times 10^{-27})}{8 \times (3.14)^2 \times (10.3875) \times (3 \times 10^{10})}$$

$$= 2.6956 \times 10^{-40}\, gcm^2$$

Now,

$$\mu = \left(\frac{1 \times 35.5}{1 + 35.5}\right) \times \frac{1}{6.023 \times 10^{23}} = \frac{35.5}{36.5} \times \frac{1}{6.023 \times 10^{23}}$$

$$\mu = \frac{35.5}{219.8395} \times 10^{-23} = 0.161 \times 10^{-23}\, g$$

$$I = \mu r_0^2$$

$$r_0^2 = \sqrt{\frac{I}{\mu}} = \sqrt{\frac{2.6956 \times 10^{-40}}{0.161 \times 10^{-23}}} = \sqrt{1.674 \times 10^{-16}}$$

$$r_0 = 1.3010 \times 10^{-8}\ cm$$

Q. 10. The rotational constant for $^1H^{35}Cl$ is observed to be 10.5909 cm^{-1}. Calculate the value of $\bar{B}$ for $^1H^{37}Cl$. Find the $^1H^{37}Cl$ bond distance. [Take the relative atomic weights of ^1H, ^{35}Cl and ^{37}Cl to be 1, 35 and 37 respectively].

Absolute mass of H–atom = 1.67×10^{-27} kg

Ans. (i) For $^1H^{35}Cl$: $\bar{B}$ = 10.5909 cm^{-1}

$$\bar{B} = \frac{h}{8\pi^2 Ic} \quad \text{or} \quad I = \frac{h}{8\pi^2 \bar{B} c}$$

$$I = \frac{(6.626 \times 10^{-27})}{8 \times (3.14)^2 \times (10.5909) \times (3 \times 10^{10})}$$

$$I = 2.64 \times 10^{-40}\, g\,cm^2$$

Now,

$$\frac{\bar{B}}{\bar{B}'} = \frac{I'(^1H^{37}Cl)}{I(^1H^{35}Cl)} = \frac{\mu'_{red}(^1H^{37}Cl)}{\mu_{red}(^1H^{35}Cl)}$$

$$\mu_{red}(^1H^{35}Cl) = \frac{(1 \times 35)}{(1 + 35)} \times \frac{1}{6.023 \times 10^{23}} = 1.62 \times 10^{-24}\, g$$

$$\mu'_{red}(^1H^{37}Cl) = \frac{(1 \times 37)}{(1 + 37)} \times \frac{1}{6.023 \times 10^{23}} = 1.63 \times 10^{-24}\, g$$

Now, $$\frac{\bar{B}}{\bar{B}'} = \frac{\mu'_{red}}{\mu_{red}}$$

$$\frac{10.5909}{\bar{B}'} = \frac{1.63\times10^{-24}}{1.62\times10^{-24}} \Rightarrow \bar{B}' = 10.5277\text{cm}^{-1}$$

(ii) $$\frac{\bar{B}}{\bar{B}'} = \frac{I'}{I}$$

$$\frac{10.5909}{10.5277} = \frac{I'}{2.64\times10^{-40}}$$

$$I' = 2.66\times10^{-40}\,\text{g cm}^2$$

$$I' = \mu' r_0^2$$

$$r_0 = \sqrt{\frac{I'}{\mu'}} = \sqrt{\frac{2.66\times10^{-40}}{1.63\times10^{-24}}}\ \text{cm}$$

$$r_0 = \sqrt{1.63\times10^{-16}}\,\text{cm}$$

$$r_0 = 1.276\times10^{-8}\,\text{cm}$$

Q. 11. The microwave spectrum of $^{12}C^{14}N$ radical shows a series of lines. The spacing between the lines is 3.978 cm^{-1}. Calculate the bond length of the radical.

Ans. $2\bar{B} = 3.978\ cm^{-1}$

$\bar{B} = 1.989\ cm^{-1}$

Now, $$\bar{B} = \frac{h}{8\pi^2 Ic} \quad \text{or} \quad I = \frac{h}{8\pi^2 \bar{B}c}$$

$$I = \frac{6.626\times10^{-27}}{8\times(3.14)^2\times(1.989)\times(3\times10^{10})}$$

$$I = 1.4\times10^{-39}\,\text{gcm}^2$$

$$\mu_{CN} = \left(\frac{12\times14}{12+14}\right)\times\frac{1}{6.023\times10^{-23}} = \frac{168}{156.598\times10^{23}}$$

$$\mu_{CN} = 1.07\times10^{-23}\,\text{g}$$

Now, $$I = \mu r_0^2$$

$$r_0 = \sqrt{\frac{I}{\mu}} = \sqrt{\frac{1.4\times10^{-39}}{1.07\times10^{-23}}} = \sqrt{1.308\times10^{-16}}$$

$$r_0 = 1.145\times10^{-8}\ \text{cm}$$

Q. 12. For $^{1}H^{79}Br$, the rotational constant, $\bar{B}$ is found to be 8.45 cm^{-1}

(a) Draw the rotational energy level diagram for J = 1, 2, 3, 4.

(b) Predict the position of the first four lines in the pure microwave spectrum.

(c) Evaluate the equilibrium internuclear distance for this molecule.

Ans. $\bar{B}$ = 8.45 cm^{-1}

$$\bar{B} = \frac{h}{8\pi^2 Ic} \quad \text{or} \quad I = \frac{h}{8\pi^2 \bar{B}c}$$

$$I = \frac{(6.626\times10^{-27})}{8\times(3.14)^2\times(8.45)\times(3\times10^{10})}$$

$$I = 3.316\times10^{-40}\,\text{gcm}^2$$

$$\mu = \left(\frac{1\times79}{1+79}\right)\times\frac{1}{6.023\times10^{-23}} = \frac{79}{80}\times\frac{1}{6.023\times10^{23}} = \frac{79\times10^{-23}}{481.84}$$

$$\mu = 0.163\times10^{-23}\,\text{g}$$

$$I = \mu r_0^2$$

$$r_0 = \sqrt{\frac{I}{\mu}} = \sqrt{\frac{3.316\times10^{-40}}{0.163\times10^{-23}}} = \sqrt{2.034\times10^{-16}}$$

$$r_0 = 1.42\times10^{-8}\,\text{cm}$$

Now, $2\bar{B}$ = 16.9 cm^{-1} represents J = 0 → J = 1 transition

$4\bar{B}$ = 33.8 cm^{-1} represents J = 1 → J = 2 transition

$6\bar{B}$ = 50.7 cm^{-1} represents J = 2 → J = 3 transition

$8\bar{B}$ = 67.6 cm^{-1} represents J = 3 → J = 4 transition

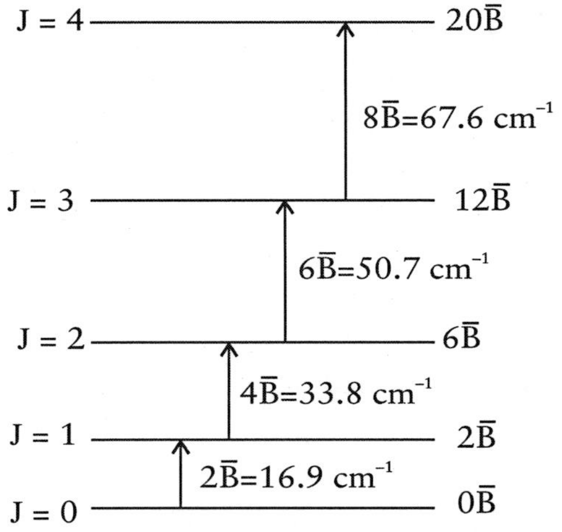

Rotational Energy Diagram for $^{1}H^{79}Br$ Molecule

Q. 13. The rotational constant for $^{1}H^{35}Cl$ is found to be 10.5909 cm^{-1}. Calculate moment of inertia and bond length. (Atomic mass of ^{1}H = 1.673× 10^{-27} kg and ^{35}Cl = 0.58120 × 10^{-27} kg.)

Ans. (i) $\bar{B} = 10.5909\ cm^{-1}$

$$\bar{B} = 10.5909 \times 100\ m^{-1} = 1050.09\ m^{-1}$$

$$\bar{B} = \frac{h}{8\pi^2 Ic} \quad \text{or} \quad I = \frac{h}{8\pi^2 \bar{B} c}$$

$$I = \frac{(6.626\times10^{-34})}{8\times(3.14)^2\times(1050.09)\times(3\times10^8)}$$

$$= 2.6439\times10^{-47}\ kgm^2$$

(ii) $$\mu = \frac{m_H \cdot m_{Cl}}{m_H + m_{Cl}} = \frac{(1.673\times10^{-27})\times(0.58120\times10^{-27})}{(1.673\times10^{-27})+(0.58120\times10^{-27})}$$

$$\mu = 4.313\times10^{-26}\ kg$$

$$I = \mu r_0^2$$

$$r_0 = \sqrt{\frac{I}{\mu}} = \sqrt{\frac{2.6439\times10^{-47}}{4.313\times10^{-26}}} = \sqrt{6.1301\times10^{-22}}$$

$$r_0 = 2.4759\times10^{-11}\ m$$

Q. 14. Calculate the bond length in a covalent molecule PQ, if the frequency separation in its microwave spectrum is 20.80 cm^{-1}. Its reduced mass is 1.6×10^{-27} kg.

Ans. $2\bar{B} = 20.80\ cm^{-1}$

$$\bar{B} = 10.40\ cm^{-1}$$

$$\bar{B} = 10.40 \times 10^2\ m^{-1} = 1040\ m^{-1}$$

$$\bar{B} = \frac{h}{8\pi^2 Ic} \quad \text{or} \quad I = \frac{h}{8\pi^2 \bar{B} c}$$

$$I = \frac{6.626\times10^{-34}}{8\times(3.14)^2\times(1040)\times(3\times10^8)}$$

$$= 2.6940\times10^{-47}\ kgm^2$$

Now, $I = \mu r_0^2$

$$r_0 = \sqrt{\frac{I}{\mu}} = \sqrt{\frac{2.6940\times10^{-47}}{1.6\times10^{-27}}}$$

$$r_0 = \sqrt{1.6837\times10^{-20}}$$

$$r_0 = 1.2975\times10^{-10}\ m$$

Q. 15. In a very bad rotational spectra of HCl^+ following lines were observed in cm^{-1}

30, 45, 90, 105

(i) Calculate the rotational energy levels for the HCl^+ molecule

Ans. Separation between the spectral lines.

$$2\bar{B} = (45 - 30)\ cm^{-1}$$

$$2\bar{B} = 15\ cm^{-1}$$

$$\bar{B} = 7.5\ cm^{-1}$$

Thus, the various rotational energy levels are: $\in_J = \bar{B}J(J+1)\ cm^{-1}$

$\in_0 = 0$ (J = 0)

$\in_1 = 7.5 \times 1(1+1) = 15\ cm^{-1}$ (J = 1)

$\in_2 = 7.5 \times 2(2+1) = 45\ cm^{-1}$ (J = 2)

$\in_3 = 7.5 \times 3(3+1) = 90\ cm^{-1}$ (J = 3)

(ii) Assign the observed lines to various transitions.

$$\Delta \in_J = \bar{v} = 2\bar{B}(J+1)\ cm^{-1}$$

$\bar{v}_1 = 2\bar{B}(0+1) = 15\,cm^{-1}$ $(J = 0 \rightarrow J = 1)$

$\bar{v}_2 = 2\bar{B}(1+1) = 30\,cm^{-1}$ $(J = 1 \rightarrow J = 2)$

$\bar{v}_3 = 2\bar{B}(2+1) = 45\,cm^{-1}$ $(J = 2 \rightarrow J = 3)$

$\bar{v}_4 = 2\bar{B}(3+1) = 60\,cm^{-1}$ $(J = 3 \rightarrow J = 4)$

$\bar{v}_5 = 2\bar{B}(4+1) = 75\,cm^{-1}$ $(J = 4 \rightarrow J = 5)$

$\bar{v}_6 = 2\bar{B}(5+1) = 90\,cm^{-1}$ $(J = 5 \rightarrow J = 6)$

(iii) Calculate the line position of missing lines.

Ans. $\bar{v}_1 = 2\bar{B}(0+1) = 15\,cm^{-1}$ $(J = 0 \rightarrow J = 1)$

$\bar{v}_4 = 2\bar{B}(3+1) = 60\,cm^{-1}$ $(J = 3 \rightarrow J = 4)$

$\bar{v}_5 = 2\bar{B}(4+1) = 75\,cm^{-1}$ $(J = 4 \rightarrow J = 5)$

(iv) Calculate the internuclear distance in HCl^+ molecule.

Ans. $\bar{B} = 7.5\ cm^{-1}$

$$I = \frac{h}{8\pi^2 \bar{B} c}$$

$$I = \frac{(6.6 \times 10^{-27})}{8 \times (3.14)^2 \times (7.5) \times (3 \times 10^{10})}$$

$$= 3.6811 \times 10^{-39}\,gcm^2$$

Now, $\mu = \frac{(1\times 35.5)}{(1+35.5)} \times \frac{1}{6.023\times 10^{23}} = \frac{35.5}{36.5} \times \frac{1}{6.023\times 10^{23}} = 1.6150\times 10^{-24}\,g$

$$I = \mu r_0^2$$

$$r_0 = \sqrt{\frac{I}{\mu}} = \sqrt{\frac{3.811\times 10^{-39}}{1.6150\times 10^{-24}}}$$

$$r_0 = \sqrt{2.2793\times 10^{-15}}$$

$$r_0 = 4.7742\times 10^{-7}\,cm$$

Q. 16. The bond length of $^{12}C^{16}O$ is 113 pm. Calculate the frequency separation in m^{-1} in its microwave spectrum.

Ans. r_0 = 113 pm = 113 $\times 10^{-12}$ m = 113 $\times 10^{-10}$ cm

$$\mu = \frac{(12\times 16)}{(12+16)} \times \frac{1}{6.023\times 10^{23}}$$

$$\mu = 1.1386\times 10^{-23}\,g$$

Now, $I = \mu r_0^2$

$$I = \left(1.1386\times 10^{-23}\right)\times\left(113\times 10^{-10}\right)^2$$

$$I = 1.4538\times 10^{-39}\,g\,cm^2$$

$$\bar{B} = \frac{h}{8\pi^2 Ic}$$

$$\bar{B} = \frac{\left(6.626\times 10^{-27}\right)}{8\times(3.14)^2\times\left(1.4538\times 10^{-39}\right)\times(3\times 10)^{10}}$$

$$\bar{B} = 192.6088\ cm^{-1}$$

Frequency separation = $2\bar{B}$ = 2 × (192.6088) cm^{-1}
= 385.21 cm^{-1}
= 38521 m^{-1}

Q. 17. Show that the rotational transition of a diatomic molecule occurs in the microwave or IR region. Use an atomic mass of 40 × 10^{-27} kg for atoms and a bond length of 200 pm.

Ans. μ = 40 × 10^{-27} kg
r_0 = 200 pm = 200 × 10^{-12} m = 2 × 10^{-10} m
$I = \mu r_0^2$ = (40 × 10^{-27} kg) (2 × 10^{-10} m)^2
= 16 × 10^{-46} kgm^2

$$E = h\nu = \frac{\cancel{h}c}{\lambda} = \frac{h^{\cancel{2}}}{8\pi^2 I}$$

$$\lambda = \frac{8\pi^2 Ic}{h}$$

$$\lambda = \frac{8\times(3.14)^2\times\left(16\times10^{-46}\,\text{kgm}^2\right)\left(3\times10^8\,\text{ms}^{-1}\right)}{\left(6.626\times10^{-34}\,\text{Js}\right)}$$

$$\lambda = 5.70\times10^{-2}\ \text{m}$$

(which lies in microwave region)

Q. 18. The frequency separation in the rotation spectrum of BrF is found to be 71.43 m^{-1}. Calculate the bond length in metres, if the reduced mass is 2.54×10^{-26} kg.

Ans. Frequency separation = $2\bar{B}$ = 71.43 m^{-1}

$$\bar{B} = 35.715\ \text{m}^{-1}$$

$$I = \frac{h}{8\pi^2\bar{B}c}$$

$$I = \frac{6.626\times10^{-34}}{8\times(3.14)^2\times(35.715)\times\left(3\times10^8\right)}$$

$$I = 7.8402\times10^{-46}\ \text{kg}\,\text{m}^2$$

$$I = \mu r_0^2$$

$$r_0 = \sqrt{\frac{I}{\mu}} = \sqrt{\frac{7.8402\times10^{-46}}{2.54\times10^{-26}}}$$

$$r_0 = \sqrt{3.0866\times10^{-20}}$$

$$r_0 = 1.7568\times10^{-10}\,\text{m}$$

Q. 19. If the wavenumber of J = 2 → 3 transition (rotational) of $^1H^{35}Cl$ considered as a rigid rotator is 63.56 cm^{-1}, what is the

(i) moment of inertia

(ii) bond length

Ans. $\bar{\nu}_{J\to J+1} = 2\bar{B}(J+1) = 63.56\ \text{cm}^{-1}$ (J = rotational quantum number of lower state)

so, $2\bar{B}(2+1) = 63.56$

or, $\bar{B} = \frac{63.56}{6} = 10.59\ \text{cm}^{-1}$

(i) $\text{Moment to Inertia (I)} = \frac{h}{8\pi^2\bar{B}c}$

$$I_{HCl} = \frac{\left(6.626\times10^{-27}\right)}{8\times(3.14)^2\times(10.59)\times\left(3\times10^{10}\right)} = 2.644\times10^{-40}\ \text{gcm}^2$$

(ii) $I_{HCl} = \mu_{HCl}\left(r_{HCl}\right)^2$

$$r_{HCl} = \sqrt{\frac{I_{HCl}}{\mu_{HCl}}}$$

$$\mu = \frac{(1 \times 35)}{(1+35)} \times \frac{1}{6.023 \times 10^{23}}$$

$$\mu = 1.6144 \times 10^{-24}\,g$$

$$r_0 = \sqrt{\frac{I}{\mu}} = \sqrt{\frac{2.644 \times 10^{-40}}{1.6144 \times 10^{-24}}}$$

$$r_0 = 1.2797 \times 10^{-8}\,cm$$

Q. 20. Gilliam et al. (1950) measured the first line in the rotational spectrum of CO as 3.84235 cm^{-1}. Calculate the rotational constant, moment of inertia and find out which rotational state of CO would be most populated at 300 K?

Ans. (i) $2\bar{B} = 3.84235\ cm^{-1}$

$\bar{B} = 1.921175\ cm^{-1}$

Rotational constant, $\bar{B} = 1.921175 \times 10^2\ m^{-1}$

(ii) $$\bar{B} = \frac{h}{8\pi^2 Ic}$$

$$I = \frac{h}{8\pi^2 \bar{B}c}$$

$$I = \frac{(6.626 \times 10^{-34})}{8 \times (3.14)^2 \times (1.921175 \times 10^2) \times (3 \times 10^8)}$$

$$I = 14.5752 \times 10^{-47}\,kgm^2$$

(iii) $$J_{max} = \sqrt{\frac{kT}{2\bar{B}hc}} - \frac{1}{2}$$

$$J_{max} = \sqrt{\frac{(1.38 \times 10^{-23}) \times 300}{2 \times (1.921175 \times 10^2) \times (6.626 \times 10^{-34}) \times (3 \times 10^8)}} - \frac{1}{2}$$

$$J_{max} = \sqrt{54.2} - 0.5$$

$$J_{max} = 6.86 \approx 7$$

3.14 Isotopic Effect

When a particular atom in a molecule is replaced by its isotope, the resultant molecule is found to be chemically identical to the original molecule. In particular, there is no appreciable change in internuclear distance and electronic distribution of the molecule on isotopic substitution. However, there is a change in total mass, moment of inertia and $\bar{B}$ value of the molecule.

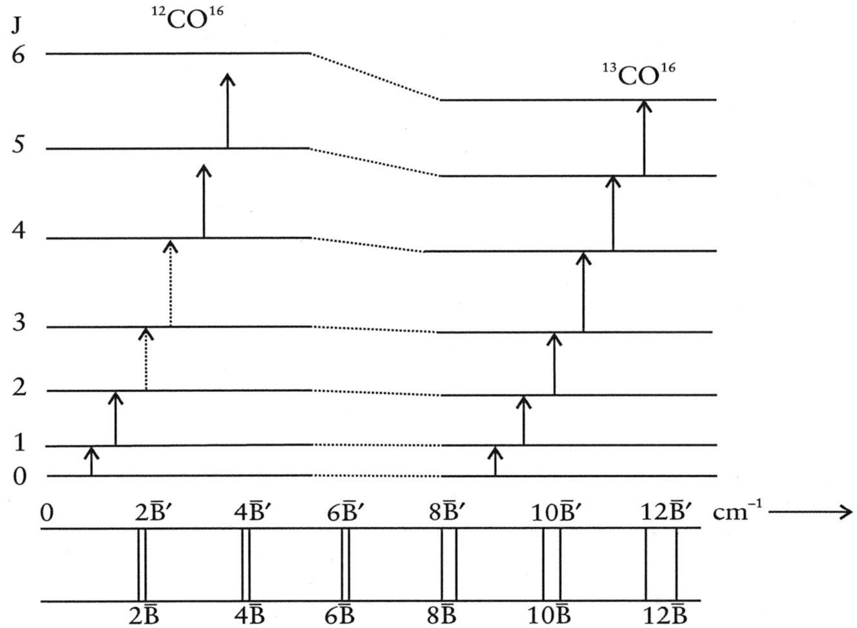

Fig. 3.9: The isotopic effect on the energy levels and rotational spectrum of CO molecule

Like: For CO molecule, on isotopic substitution ($^{12}C^{16}O \rightarrow {}^{13}C^{16}O$), there is increase in mass of carbon atoms, hence there is increase in the value of moment of inertia (I) and thus leads to decrease in the value of $\bar{B}\left(\bar{B} \propto \frac{1}{I}\right)$. There is relative lowering of the ^{13}C levels with respect to those of ^{12}C levels. ($\bar{B} > \bar{B}'$), i.e. the energy levels become closer to one another on going from $^{12}C \rightarrow {}^{13}C$. Thus, the spectra of the heavier species will show a smaller separation between the spectral lines ($2\bar{B}'$) than that of lighter one ($2\bar{B}$).

For $^{12}C^{16}O$, $2\bar{B} = 3.84235\ cm^{-1}$

$\bar{B} = 1.92118\ cm^{-1}$

$^{13}C^{16}O$, $2\bar{B}' = 3.67337\ cm^{-1}$

$\bar{B}' = 1.83669\ cm^{-1}$

The effect of isotopic substitution on the energy levels and spectrum obtained is shown in Fig. 3.9.

3.14.1 Application of Isotopic Substitution

Microwave studies can give directly an estimate of the abundance of isotopes by comparison of absorption intensities.

Like: Consider CO molecule,

for $^{12}C\ ^{16}O;\ \bar{B} = 1.9118\ cm^{-1}$

$^{13}C\ ^{16}O;\ \bar{B}' = 1.83669\ cm^{-1}$

$$\frac{\bar{B}}{\bar{B}'} = \frac{h}{8\pi^2 Ic} \times \frac{8\pi^2 I'c}{h}$$

$$\frac{\bar{B}}{\bar{B}'} = \frac{I'}{I} = \frac{\mu'}{\mu} = 1.046$$

Now, mass of ^{16}O = 15.9949 g

mass of ^{12}C = 12.0000 g

mass of ^{13}C = m′

$$\frac{\mu'}{\mu} = \frac{(15.9949 \times m')}{(15.9949 + m')} \times \frac{(15.9949 + 12)}{(15.9949 \times 12)} = 1.046$$

$$m' = 13.0007$$

This is 0.02% of the best value in other ways.

PROBLEMS BASED ON ISOTOPIC SUBSTITUTION

Q. 1. What is the effect of isotopic substitution on the rotational spectra of the molecule?

Q. 2. How will the rotational spectra change when ^{12}C in $^{12}C^{16}O$ is replaced by ^{13}C?

Q. 3. The first absorption in rotational spectrum of ^{12}CO molecule is found to occur at 3.8424 cm^{-1}. The first absorption in rotational spectrum of isotopic substituted (^{13}CO) is found to occur at 3.6734 cm^{-1}. Calculate atomic mass of ^{13}C. (Given: ^{12}C = 12.600 g).

Ans. $2\bar{B}(^{12}CO) = 3.8424\ cm^{-1}$

$2\bar{B}(^{13}CO) = 3.6734\ cm^{-1}$

$\bar{B}(^{12}CO) = 1.9212\ cm^{-1}$

$\bar{B}(^{13}CO) = 1.8367\ cm^{-1}$

$$\frac{\bar{B}(^{12}CO)}{\bar{B}(^{13}CO)} = \frac{h/8\pi^2 I_{(^{12}CO)}c}{h/8\pi^2 I_{(^{13}CO)}c}$$

$$= \frac{I_{(^{13}CO)}}{I_{(^{12}CO)}} = \frac{\mu_{(^{13}CO)}}{\mu_{(^{12}CO)}}$$

$$= \frac{m_{13C} m_{16O}}{(m_{13C} + m_{16O})/N_A} \times \frac{(m_{12C} m_{16O})/N_A}{(m_{12C} + m_{16O})}$$

$$= \frac{m_{13C}(m_{12C} + m_{16O})}{m_{12C}(m_{13C} + m_{16O})}$$

$$\text{So,} \ \frac{1.9212\,\text{cm}^{-1}}{1.8367\,\text{cm}^{-1}} = \frac{m_{13C}(12.000+15.9994)}{12.000(m_{13C}+15.9994)}$$

$$m_{13C} = 13.0007\,\text{g}.$$

3.15 Non-Rigid Rotator

Spectral lines for CO, other than the first would not have shown the constant $2\bar{B}$ separation. The separation between successive lines (and hence the apparent $\bar{B}$ value) decreases steadily with increasing value of J. This is because so far we have considered that the bond in a molecule to be a rigid bond. But this is only an approximation. However, all bonds are elastic to some extent. The bond length (r_0) increases with increasing value of J. This reflects that more quickly a diatomic molecule rotates, the greater is the centrifugal force tending to move the atoms apart.

The consequences of elasticity are:

1. When the bond is elastic, a molecule may have vibrational energy, i.e. the bond will stretch or compress periodically with a certain frequency depending upon the masses of the atoms and the force constant of the bond.
2. Due to elasticity, the quantities 'r_0' and '$\bar{B}$' vary during a vibration. When these quantities are measured by microwave technique, many hundreds of vibrations occur during a rotation and hence the measured value is average.

3.15.1 Rotational Energy for Non-Rigid Rotator

For a non-rigid rotator, Schrodinger wave equation yields the rotational energy as:

$$E_J = \frac{h^2}{8\pi^2 I}J(J+1) - \frac{h^4}{32\pi^4 I^2 r^2 k}J^2(J+1)^2 \text{ Joules}$$

$$\epsilon_J = \frac{E_J}{hc} = \frac{h}{8\pi^2 Ic}J(J+1) - \frac{h^3}{32\pi^4 I^2 r^2 kc}J^2(J+1)^2 \text{ cm}^{-1}$$

$$\epsilon_J = \bar{B}J(J+1) - \bar{D}J^2(J+1)^2 \text{ cm}^{-1}$$

where, $\bar{B}$ is the rotational constant, whose value is,

$$\bar{B} = \frac{h}{8\pi^2 Ic}\text{cm}^{-1}$$

and $\bar{D}$ is the centrifugal distortion constant whose value is

$$\bar{D} = \frac{h^3}{32\pi^4 I^2 r^2 kc}\text{cm}^{-1}$$

In the expression of $\bar{D}$, the k is force constant whose value is

$$k = 4\pi^2\bar{\omega}^2 c^2 \mu \text{ Nm}^{-1}$$

where, $\bar{\omega}$ is the vibrational frequency, expressed in cm^{-1}.

Now, using expressions of k and $\bar{B}$ in the expression of $\bar{D}$, we have

$$\bar{D} = \frac{4\bar{B}^3}{\bar{\omega}^2}$$

Usually, $\bar{\omega} \sim 10^3$ cm^{-1}

$\bar{B} \sim 10$ cm^{-1}

Thus, $$\bar{D} \sim \frac{(10)^3}{(10^3)^2} = \frac{10^3}{10^6} \sim 10^{-3}\,cm^{-1}$$

Since $\bar{D}$ is very small compared to $\bar{B}$, thus, for small J values, the correction term $\bar{D}J^2(J + 1)^2$ is almost negligible. While for J value of 10 or more, it may become appreciable.

Knowledge of $\bar{D}$ enables us to determine the vibrational frequency ($\bar{\omega}$) of a diatomic molecule

$$\bar{D} = \frac{4\bar{B}^3}{\bar{\omega}^2}$$

$$\bar{\omega}^2 = \frac{4\bar{B}^3}{\bar{D}}\,cm^{-1}$$

3.15.2 Selection Rule for Non-Rigid Rotator

For non-rigid rotator, we only consider transitions in which the value of J changes by one unit. All other transitions being spectroscopically forbidden. Thus, during rotation, the transitions takes place between rotational energy levels for which

$$\Delta J = \pm 1$$

i.e. the change in rotational quantum number is unity.

3.15.3 Rotational Spectra of Non-Rigid Rotator

To discuss the rotational spectra, we consider the difference between the rotational energy levels. According to the rotational selection rules, only those rotational transitions are allowed for which $\Delta J = \pm 1$. Thus, a transition takes place from lower rotational level with rotational quantum number J to a higher rotational level with rotational quantum number $J' = J + 1$. The energy absorbed for the transition is given by

$$\bar{v}_{J\to J+1} = \Delta \epsilon_J = \bar{B}\left[(J+1)(J+2) - J(J+1)\right]$$
$$-\bar{D}\left[(J+1)^2(J+2)^2 - J^2(J+1)^3\right]\,cm^{-1}$$

$$\Delta \epsilon_J = \left[2\bar{B}(J+1) - 4\bar{D}(J+1)^3\right]\,cm^{-1}$$

From this equation, it follows that the spectrum of elastic rotator (or non-rigid rotator) is almost similar to that of a rigid rotator except that each spectral line is displaced slightly to the low frequency value as shown in Fig. 3.10.

Table 3.1: Wavenumbers in cm^{-1} for CO

J	Rigid rotator	Non-rigid rotator	Shift
0	3.8620	3.8620	0.0000
1	7.7240	7.7238	0.0002
2	11.5860	11.5853	0.0007
3	15.4480	15.4464	0.0016
4	19.3100	19.3069	0.0030
5	23.1720	23.1667	0.0053
6	27.0340	27.0256	0.0084
7	30.8960	30.8835	0.0125
8	34.7580	34.7402	0.0178

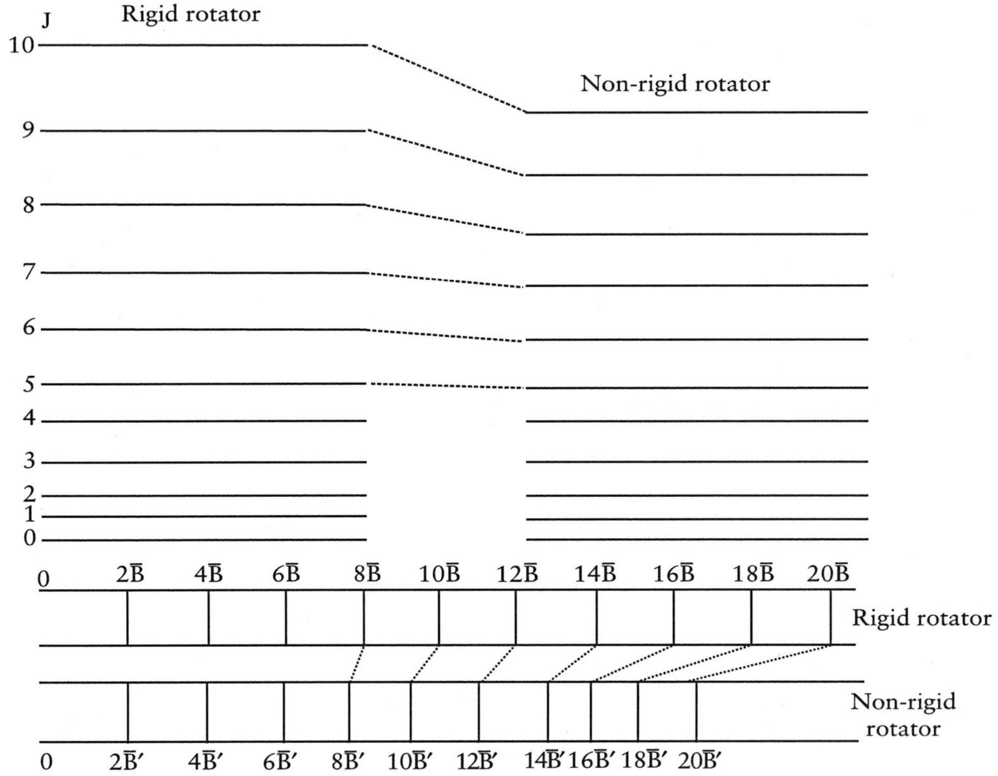

Fig. 3.10: Spectrum of Non-rigid rotator

PROBLEM BASED ON NON-RIGID ROTATOR

Q. 1. In the rotational spectrum of a diatomic molecule two consecutive lines are observed at 84.544 and 101.355 cm^{-1}. Calculate the values of the rotational constant $\overline{B}$, distortional constant $\overline{D}$ and the rotational quantum J for these transitions.

Ans. For the transition $J \rightarrow J + 1$, the wavenumber is

$$\bar{\nu}_J = 2\overline{B}(J+1) - 4\overline{D}(J+1)^3$$

Thus, we have
$$84.544 = 2\overline{B}(J + 1) - 4\overline{D}(J + 1)^3 \qquad ...(1)$$

and
$$101.355 = 2\overline{B}(J + 2) - 4\overline{D}(J + 2)^3 \qquad ...(2)$$

From these equations, we have

$$16.811 = 2\overline{B} - 4\overline{D}[(J + 2)^3 - (J + 1)^3]$$

Since $\overline{D} << \overline{B}$, thus $\overline{B}$ can be obtained by neglecting $\overline{D}$ in comparison to $\overline{B}$, Thus, we have $2\overline{B} = 16.811$. Using the value of $2\overline{B}$ in equation (1) and again neglecting $\overline{D}$, we have

$$84.544 = 16.811(J+1)$$

$$J = \frac{84.544}{16.811} - 1 = 4.03$$

But J must be an integer, therefore, correct value of J is 4. Hence, the line 84.544 cm^{-1} corresponds to the transition $J = 4 \rightarrow 5$ and the line 101.355 cm^{-1} correspond to the transition $J = 5 \rightarrow 6$. Using J = 4 in the equations, we have

$$84.544 = 2\overline{B} \times 5 - 4\overline{D} \times 125$$

and
$$101.355 = 2\overline{B} \times 6 - 4\overline{D} \times 126$$

Solution of these two relations gives $\overline{B} = 8.473\ cm^{-1}$ and $\overline{D} = 3.7 \times 10^{-4}\ cm^{-1}$.

3.16 Application of Microwave Radiation: Microwave Oven

In recent years, the microwave oven has become a very familiar and useful cooking device in the kitchen. Its mode of operation depends upon the absorption by the food of the microwave radiation in which it is placed. The water molecules in the food absorb the microwave radiation and are thereby raised to the higher rotational states. The biological molecules in the food, on the other hand, are far too big to be able to rotate. The heating in the microwave oven is entirely internal. In contrast to the conventional oven where the food is cooking by external heating. In internal heating, the water molecules in the food are excited and the excess rotational energy is re-emitted as heat with the help of which the food gets cooked. Unfortunately, the microwave radiation affects the human body.

4

Vibrational Motion and Infrared (IR) Spectroscopy

4.1 Introduction

Linear harmonic oscillator is the simplest system to understand the vibrational motion of microscopic particles, therefore, it is also called as simple harmonic oscillator. For the sake of simplicity, we treat here only the simple harmonic motion of a single particle which oscillates to and fro about its mean position. Thus, oscillations take place periodically in one dimension only, like ball of mass m, tied to a spring through a support shown in Fig. 4.1.

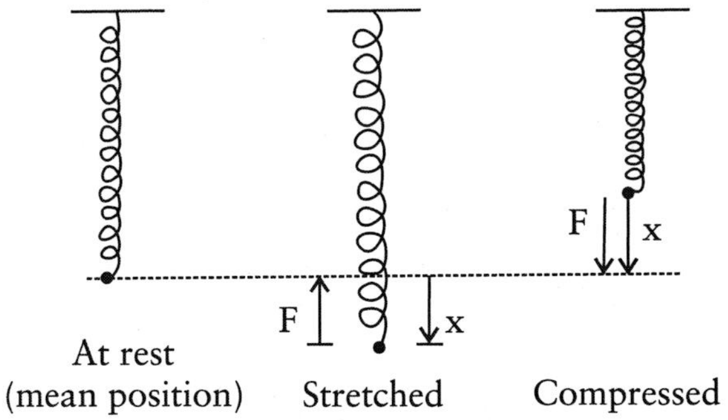

Fig. 4.1: Schematic diagram of linear harmonic oscillator

This kind of a vibration motion obeys Hooke's law, according to which the restoring force (F) acting on the particle is proportional to the displacement (x) from the mean position.

$$F \propto x$$

$$F = -k_f x \qquad ...(4.1)$$

where, k_f is called the force constant.

It is characteristic of the spring and is a measure of the stiffness of the spring. A strong inflexible spring will have a larger value of k_f and a weaker spring will have a lesser value of k_f. In chemical system, value of k_f depends on the strength of the bond. Also the negative sign in the above expression implies that the restoring force (F) always acts in the opposite direction of the displacement (x).

4.2 Classical Treatment of Linear Harmonic Oscillator

In classical mechanics, the above vibrational motion can be described in terms of Newton's second law of motion, according to which, we have

$$F = ma \qquad \text{(a = acceleration)}$$

$$\text{Velocity (v) = rate of change of displacement (x)} = \frac{dx}{dt}$$

$$\text{Acceleration (a) = rate of change of velocity (v)} = \frac{dv}{dt}$$

$$a = \frac{d}{dt}(v) = \frac{d}{dt}\left(\frac{dx}{dt}\right) = \frac{d^2x}{dt^2}$$

$$F = ma = m\frac{d^2x}{dt^2} \qquad ...(4.2)$$

Equating equations (4.1) and (4.2)

$$ma = -k_f x \qquad ...(4.3)$$

$$m\frac{d^2x}{dt^2} = -k_f x$$

$$\frac{d^2x}{dt^2} = \frac{-k_f x}{m}$$

$$\frac{d^2x}{dt^2} + \frac{k_f x}{m} = 0$$

It is a differential equation with general solution:

$$x = A \sin(2\pi\nu t) \qquad ...(4.4)$$

where, x = maximum displacement of the vibration or amplitude of the vibration.

ν = frequency of the vibration.

4.2.1 Frequency of Linear Harmonic Oscillator

The expression of frequency (ν) for a linear harmonic oscillator moving about its mean position can be evaluated as follows:

$$x = A \sin(2\pi\nu t) \qquad ...(4.4)$$

Evaluating the second differential of equation (4.4)

$$\frac{dx}{dt} = A\cdot(2\pi\nu)\cdot\cos(2\pi\nu t) \qquad ...[4.4(a)]$$

$$\frac{d^2x}{dt^2} = -A\cdot(2\pi\nu)^2\cdot\sin(2\pi\nu t) \qquad ...[4.4(b)]$$

Substituting equation (4.4) and equation [4.4(b)] in equation (4.3)

$$m\frac{d^2x}{dt^2} = -k_f x$$

$$m\left[\cancel{-}A\cdot(2\pi\nu)^2\cdot\cancel{\sin(2\pi\nu t)}\right] = \cancel{-}k_f\left[A\cancel{\sin(2\pi\nu t)}\right]$$

$$m(2\pi\nu)^2 = k_f$$

$$\nu = \frac{1}{2\pi}\sqrt{\frac{k_f}{m}}\ \text{sec}^{-1}$$

$$\bar{\nu} = \frac{1}{2\pi c}\sqrt{\frac{k_f}{m}}\ \text{cm}^{-1} \qquad ...(4.5)$$

where, $k_f = 4\pi^2\nu^2 m = 4\pi^2 c^2 \bar{\nu}^2 m$

Equation (4.5) is the required characteristic frequency at which atoms of a diatomic molecule undergo harmonic vibration.

Thus, frequency of oscillation is a characteristic of the system which depends only on k_f and m, and not on the amplitude of vibration. Thus, whatever may be the amplitude, the particle will oscillate with the same frequency.

4.2.2 Potential Energy of Linear Harmonic Oscillator

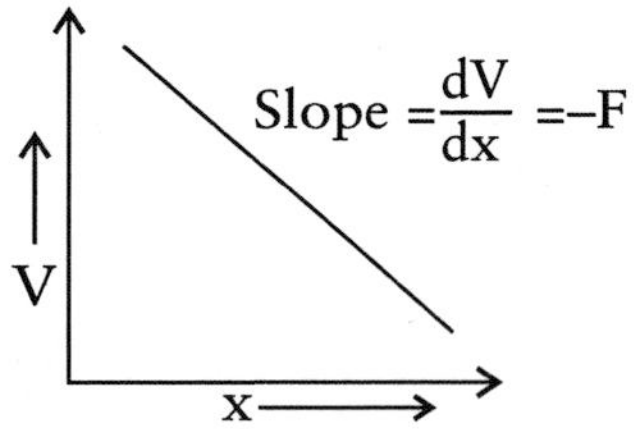

Linear harmonic oscillator is the simplest system to understand the vibrational motion of microscopic particle whose potential energy is not zero (or assumed zero like in case of particle in a box or rigid rotator). In linear harmonic oscillator, the potential energy (V) is the potential gradient of the force needed to extend or compress the spring

$$F = -k_f x \text{ (From Hooke's law)}$$

Also, $$F = -\frac{dV}{dx}$$

Hence, $$\cancel{-}\left(\frac{dV}{dx}\right) = \cancel{-}k_f x$$

$$dV = k_f \, . \, x\, dx$$

Integrating both the sides:

$$\int_0^V dV = \int_0^x k_f x\, dx$$

$$[V]_0^V = k_f\left[\frac{x^2}{2}\right]_0^x$$

$$V = \frac{1}{2}k_f x^2 \qquad ...(4.6)$$

Thus, the potential energy of the particle, besides depending on k_f depends on the square of the displacement. At equilibrium position (x = 0), the potential energy is zero, so the lowest vibrational energy is zero in case of classical vibrator.

The variation in potential energy with displacement 'x' is parabolic in nature.

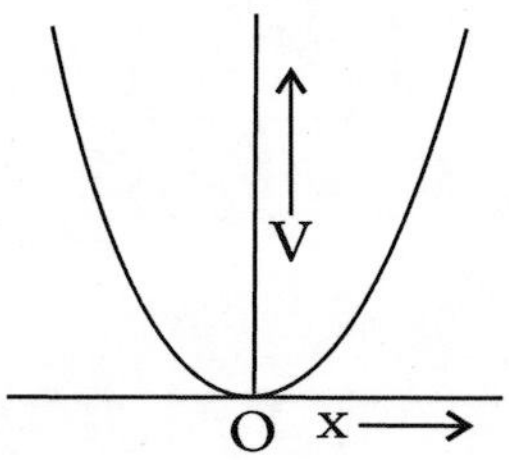

Fig. 4.2: Parabolic potential energy curve for the linear harmonic oscillator

Classically speaking, the molecule can vibrate with any amount of potential energy, i.e. the potential energy of the molecule is a continuous function of the displacement from the equilibrium position.

4.2.3 Total Energy of the Linear Harmonic Oscillator

The momentum of the particle at any instant is

$$p = mv = m\frac{dx}{dt} \qquad \text{(x = displacement)}$$

Using equation [4.4(a)]

$$p = m\left[A\cdot(2\pi\nu)\cdot\cos(2\pi\nu t)\right] \qquad \text{...(4.7)}$$

Total energy of the particle at any instant is

$$E = T + V$$

$$E = \frac{1}{2}mv^2 + \frac{1}{2}k_f x^2 \qquad \text{[Using equation (4.6)]}$$

$$E = \frac{1}{2m}m^2v^2 + \frac{1}{2}k_f x^2$$

$$E = \frac{p^2}{2m} + \frac{1}{2}k_f x^2$$

Using equation (4.4) and equation (4.7):

$$E = \frac{\left[mA\cdot(2\pi\nu)\cos(2\pi\nu t)\right]^2}{2m} + \frac{1}{2}k_f\left[A\sin(2\pi\nu t)\right]^2$$

$$E = \frac{m^2A^2\cdot(2\pi\nu)^2\cos^2(2\pi\nu t)}{2m} + \frac{1}{2}k_f\cdot A^2\cdot\sin^2(2\pi\nu t)$$

Using equation (4.5),

$$\nu = \frac{1}{2\pi}\sqrt{\frac{k_f}{m}}$$

$$2\pi\nu = \sqrt{\frac{k_f}{m}}$$

$$(2\pi\nu)^2 = \frac{k_f}{m}$$

Putting this expression for $(2\pi\nu)^2$ in the expression of energy (E), we have

$$E = \frac{m^2A^2\left(\frac{k_f}{m}\right)\cos^2(2\pi\nu t)}{2m} + \frac{1}{2}k_fA^2\sin^2(2\pi\nu t)$$

$$E = \frac{1}{2}k_fA^2\cos^2(2\pi\nu t) + \frac{1}{2}k_fA^2\sin^2(2\pi\nu t)$$

$$E = \frac{1}{2}k_fA^2\left[\cos^2(2\pi\nu t) + \sin^2(2\pi\nu t)\right]$$

$$E = \frac{1}{2}k_fA^2 \qquad \text{...(4.8)}$$

$[\therefore \cos^2 x + \sin^2 x = 1]$

Thus, total energy of the particle besides depending of k_f, depends on the square of the amplitude (A) of the vibration. Since the oscillating particle can possess any amplitude, it is obvious that the total energy of the particle will vary in a continuous manner. Thus, according to classical mechanics, the energy of the linear harmonic oscillator can have any value.

However, quantum mechanically, it was proved that energy of linear harmonic oscillator is quantised and thus cannot possess any value.

- **Total energy of linear harmonic oscillator is conserved:**

  Since the total energy of an oscillator depends only on the amplitude of the vibration (equation 4.8), it is obvious that the sum of kinetic energy and potential energy remains the same at any instant of vibration. The potential energy(equation 4.6) increases as the displacement 'x' is increased and it attains a maximum value equal to the total energy at the two extremes of the vibration. Consequently, the kinetic energy of the particle decreases as the displacement 'x' is increased and it has zero value at the two extremes of the vibration.

  At equilibrium position ($x = 0$), the potential energy is zero and the particle carries only kinetic energy equal to the total energy.

  Thus, we have:

  At $x = 0$; kinetic energy is maximum or velocity of the oscillator is maximum.

  At $x = A$; kinetic energy is zero or velocity of the oscillator is zero.

  Since the velocity of the oscillator at $x = 0$ is maximum and is zero at the extremes of vibration, it is expected that the particle will spend maximum time at the extremes and the minimum at the equilibrium position during the course of its vibration.

4.2.4 Effect of Isotopic Substitution on Force Constant and Vibrational Frequency

- Force constant is independent of isotopic substitution.
- Vibrational frequency is inversely proportional to the reduced mass (equation 4.5). On isotopic substitution, the reduced mass of the molecule increases and the value of vibrational frequency decreases.

PROBLEMS BASED ON HARMONIC OSCILLATOR

Q. 1. Write short note on Harmonic Oscillator.

Q. 2. Describe the effect of vibrational frequency when one or both H-atoms are replaced by deuterium D in H_2 molecule.

Ans. $\nu = \dfrac{1}{2\pi}\sqrt{\dfrac{k_f}{\mu}}$

$$\mu_{D-D} > \mu_{H-D} > \mu_{H-H}$$

Therefore, $\nu_{H-H} > \nu_{H-D} > \nu_{D-D}$

Q. 3. Calculate the force constant for HCl^{35} when its fundamental vibrational frequency is 8.667×10^{13} s^{-1} and compare with the force constant of HCl^{37}.

Ans.

$$\nu = \frac{1}{2\pi}\sqrt{\frac{k_f}{\mu}}$$

ν = fundamental vibrational frequency (s^{-1})

k_f = force constant

μ = reduced mass

$k_f = 4\pi^2\nu^2\mu$

For HCl^{35}

$$\mu = \frac{m_1 m_2}{m_1 + m_2} \times \frac{1}{N_A}$$

$m_1 = 1.008$; $m_2 = 35.5$

$$\mu = \left(\frac{1.008 \times 35.5}{1.008 + 35.5}\right) \times \frac{1}{6.023 \times 10^{23}}$$

$$= \frac{35.784}{36.508} \times \frac{1}{6.023 \times 10^{23}}$$

$$= 0.1626 \times 10^{-23}\text{ g}$$

$$= 1.626 \times 10^{-27}\text{ kg}$$

$$k_f = 4\pi^2\nu^2\mu = 4 \times \left(\frac{22}{7}\right)^2 \times \left(8.668 \times 10^{13}\right)^2 \times (1.626 \times 10^{-27})$$

$$= 481.7\text{ kg/s}^2$$

$$= 481.7\text{ N/m}$$

Force constant is independent of isotopic substitution, so force constant will be same for both HCl^{35} and HCl^{37}.

Q. 4. Calculate the force constant for the bond in HCl from the fact that the fundamental frequency is 8.667×10^{13} s^{-1}.

Ans. m_1 (for H-atom) = 1.088 g

m_2 (for Cl-atom) = 35.5 g

$\therefore$ Reduced mass $\mu = \dfrac{m_1 \times m_2}{m_1 + m_2} \times \dfrac{1}{N_A} = 1.6277 \times 10^{-27}\,kg$

Also it is given that

$$\nu = 8.667 \times 10^{13}\ s^{-1}$$

$$\therefore \quad k_f = 4\pi^2\nu^2\mu$$

$$= 4 \times \left(\frac{22}{7}\right) \times \left(8.667 \times 10^{13}\right)^2 \times (1.6277 \times 10^{-27})$$

$$= 483.2\ Nm^{-1}$$

Q. 5. The reduced mass of a diatomic molecule is 2.5×10^{-26} kg and its vibrational frequency is 2900 cm^{-1}. Calculate force constant.

Ans. $\bar{\nu} = \dfrac{1}{2\pi c}\sqrt{\dfrac{k_f}{\mu}}$

$$k_f = 4\pi^2 c^2 \bar{\nu}^2 \mu$$

$$= 4 \times (3.14)^2 \times \left(3 \times 10^8\,ms^{-1}\right)^2 \times \left(2900 \times 10^2\,m^{-1}\right)^2 \times (2.5 \times 10^{-26}\,kg)$$

$$= 7476.7\,Nm^{-1}\left(\text{Because } 1\ N = 1\,kg\,ms^{-2}\right)$$

Q. 6. In each of the parts below, arrange the moieties in increasing orders of vibrational frequency giving reasons:

(i) $-\overset{|}{\underset{|}{C}}-Br \quad -\overset{|}{\underset{|}{C}}-F \quad -\overset{|}{\underset{|}{C}}-CI \quad -\overset{|}{\underset{|}{C}}-I \quad -\overset{|}{\underset{|}{C}}-H$

(ii) $-C\equiv C- \quad >C=C< \quad -\overset{|}{\underset{|}{C}}-\overset{|}{\underset{|}{C}}-$

Ans. For one type of the group, the vibrational frequency increases with the increase in value of force constant and decreases with increase in value of reduced mass

(i) Reduced mass order:

$$\mu_{C\text{–}I} > \mu_{C\text{–}Br} > \mu_{C\text{–}Cl} > \mu_{C\text{–}F} > \mu_{C\text{–}H}$$

Therefore, frequency order

$$\nu_{C\text{–}I} < \nu_{C\text{–}Br} < \nu_{C\text{–}Cl} < \nu_{C\text{–}F} < \nu_{C\text{–}H}$$

i.e. vibrational frequency increases in order of,

$$-\overset{|}{\underset{|}{C}}-I < -\overset{|}{\underset{|}{C}}-Br < -\overset{|}{\underset{\underset{C}{|}}{C}}-Cl < -\overset{|}{\underset{|}{C}}-F < -\overset{|}{\underset{|}{C}}-H$$

(ii) In this case the reduced mass is same, thus the force constant follows the relation

$$k_{c\equiv c} > k_{c=c} > k_{c-c}$$

Therefore, vibrational frequency increases in order

$$\nu_{c\equiv c} > \nu_{c=c} > \nu_{c-c}$$

or, $H—C\equiv C—H \quad > \quad H_2C=CH_2 \quad > \quad H_3C—CH_3$

4.3 Quantum Mechanical Treatment: Schrödinger Wave Equation for Linear Harmonic Oscillator

Since the classical laws do not hold good for particles of very small masses, equation (4.6) will not be applicable to diatomic molecules. The potential energy expression for such a case has to be derived quantum mechanically by solving the appropriate Schrödinger equation.

The Linear Harmonic Oscillator can be solved for energy using Schrödinger wave equation

$$\hat{H}\psi = E\psi$$

In this system; total energy (E) = K.E. + P.E.

Thus, the Hamiltonian will comprise the sum of the operators for kinetic energy and potential energy, i.e.

$$\hat{H} = \hat{T} + \hat{V}$$

$$\hat{H} = -\frac{h^2}{8\pi^2 m}\nabla^2 + \frac{1}{2}k_f x^2 \qquad \text{[using equation (4.6)]}$$

Since we are considering that the linear harmonic oscillator is oscillating periodically in one dimensional only, thus,

$$\hat{H} = -\frac{h^2}{8\pi^2 m}\frac{\partial^2}{\partial x^2} + \frac{1}{2}k_f x^2$$

Hence, the Schrödinger wave equation for linear harmonic oscillator consisting of a particle of mass 'm' moving with frequency 'ν' can be given as:

$$\hat{H}\psi_v = E_v\psi_v$$

$$\left[-\frac{h^2}{8\pi^2 m}\frac{\partial^2}{\partial x^2} + \frac{1}{2}k_f x^2\right]\psi_v = E_v\psi_v$$

$$-\frac{h^2}{8\pi^2 m}\frac{\partial^2\psi_v}{\partial x^2} + \frac{1}{2}k_f x^2\psi_v = E_v\psi_v \qquad ...(4.9)$$

Since linear harmonic oscillator is a vibrational system, thus the wavefunction (ψ_v) gives the vibrational energy (E_v).

PROBLEMS BASED ON HAMILTONIAN OPERATOR

Q. 1. Write Hamiltonian operator for a particle executing simple harmonic motion at any instant.

Q. 2. Write the expression for the Hamiltonian operator for a single particle executing simple harmonic motion.

4.4 Solution of Schrödinger Wave Equation: Vibrational Energy and Wavefunction of the Linear Harmonic Oscillator

In the above equation (4.9), the potential energy term is an additional term compared to the Schrödinger wave equation of particle in one-dimensional box. This additional term makes the Schrödinger wave equation of linear harmonic oscillator difficult to solve. Though we can solve it by either of the two methods:

1. Factorization method
2. Power series method

Both the methods give same result for the system. However, factorization method is a recently developed method for solving Schrödinger wave equation for linear harmonic oscillator.

4.4.1 Factorization Method

In factorization method, the Schrödinger equation is converted into an appropriate form and then factorization method is applied to it.

$$-\frac{h^2}{8\pi^2 m}\frac{\partial^2\psi_v(x)}{\partial x^2}+\frac{1}{2}k_f x^2\cdot\psi_v(x)=E_v\psi_v(x) \quad ...(4.9)$$

↓ Transformation of Schrödinger wave equation to an appropriate form (not discussed here)

$$\left(\frac{\partial^2}{\partial y^2}-y^2\right)\psi_v(y)=-\lambda\psi_v(y)$$

↓ Application of Factorization method on appropriate form to solve for energy and wavefunction (not discussed here)

$$E_v = h\nu\left(v+\frac{1}{2}\right) \quad ...(4.10)$$

$v = 0, 1, 2, 3, ...$

$$\psi_v(x) = N_v H_v(y) e^{-y^2/2} \quad ...(4.11)$$

where,

$$N_v = \sqrt{\frac{1}{\alpha 2^v v! \sqrt{\pi}}}$$: **Normalisation constant**

and

$$H_v = (-1)^v e^{y^2}\left(\frac{\partial^v}{\partial y^v}e^{-y^2}\right)$$: **Hermite polynomial**

4.4.2 Power Series Method

In the power series method, the Schrödinger equation is converted to the same appropriate form and then power series method is applied using asymptotic solution.

$$-\frac{h^2}{8\pi^2 m}\frac{\partial^2\psi_v(x)}{\partial x^2}+\frac{1}{2}k_f x^2\cdot\psi_v(x)=E_v\psi_v(x) \quad ...(4.9)$$

↓ Transformation of Schrödinger wave equation to an appropriate form

$$\left(\frac{\partial^2}{\partial y^2}-y^2\right)\psi_v(y)=-\lambda\psi_v(y)$$

↓ Application of Power series method on appropriate form to solve for energy and wavefunction

$$E_v = h\nu\left(v+\frac{1}{2}\right) \quad ...(4.10)$$

$v = 0, 1, 2, 3, ...$

$$\psi_v(x)=N_v H_v(y)e^{-y^2/2} \quad ...(4.11)$$

where,

$$N_v = \sqrt{\frac{1}{\alpha 2^v v!\sqrt{\pi}}}$$: **Normalisation constant**

and

$$H_v = (-1)^v e^{y^2}\left(\frac{\partial^v}{\partial y^v}e^{-y^2}\right)$$: **Hermite polynomial**

PROBLEMS BASED ON VIBRATIONAL ENERGY

Q. 1. A diatomic molecule AB undergoes vibrational motion according to the harmonic oscillator model. Write the mathematical expression for the Hamiltonian, Schrödinger equation and the vibrational energy, E_{vib}, associated with this system.

Q. 2. What is meant by simple harmonic oscillator? A diatomic molecule e.g., HCl can be considered so, as a first approximation. Give the relation which gives the energy of such a molecule when the Schrödinger equation is employed. Are these energies quantized?

4.5 Wavefunction Plots for Linear Harmonic Oscillator

Wavefunction for linear harmonic oscillator can be written in the generalized form as equation (4.11)

$$\psi_v(x) = N_v \cdot H_v(y) \cdot e^{-y^2/2}$$

where,

$$\begin{aligned} v = 0;\ \psi_0(x) &= N_0(1)e^{-y^2/2} \\ v = 1;\ \psi_1(x) &= N_1(2y)e^{-y^2/2} \\ v = 2;\ \psi_2(x) &= N_2(4y^2 - 2)e^{-y^2/2} \end{aligned} \qquad \left\{ \begin{array}{l} \text{Because} \\ v = 0,\ H_v(y) = 1 \\ v = 1,\ H_v(y) = 2y \\ v = 2,\ H_v(y) = 4y^2 - 2 \end{array} \right\}$$

The function $e^{-y^2/2}$ when plotted gives bell shaped curve with no nodes (shown in Fig. 4.3).

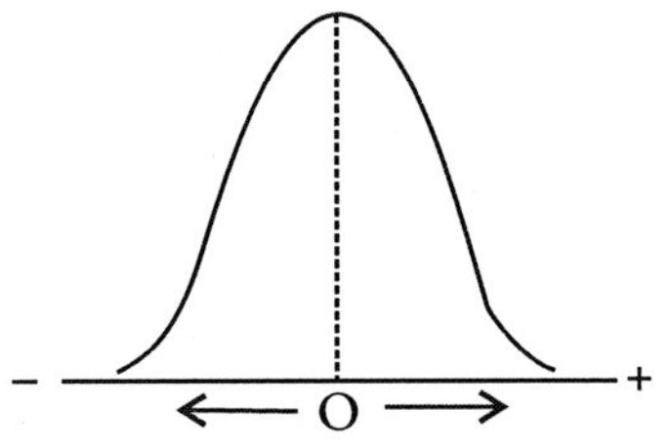

Fig. 4.3: Plot for function $e^{-y^2/2}$

Wavefunction plots for linear harmonic oscillator:

(a) $v = 0;\ \psi_0(x) = N_0(1)e^{-y^2/2}$

N_0 = Normalisation constant

In Fig. 4.4(a), number of nodes = 0

where, $\psi_0(x)$ depends only on $e^{-y^2/2}$ (as variable), therefore it has similar plot as $e^{-y^2/2}$.

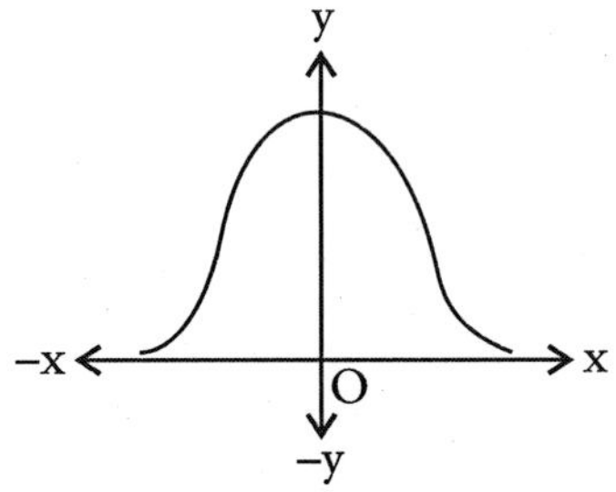

Fig. 4.4(a): Wavefunction plot for v = 0

(b) $v = 1;\ \psi_1(x) = N_1(2y)e^{-y^2/2}$

N_1 = Normalisation constant

In Fig. 4.4(b), number of nodes = 1 where, $\psi_1(x)$ depends on y. $e^{-y^2/2}$ (as variable). Hence, for negative value of y the $\psi_1(x)$ also becomes negative.

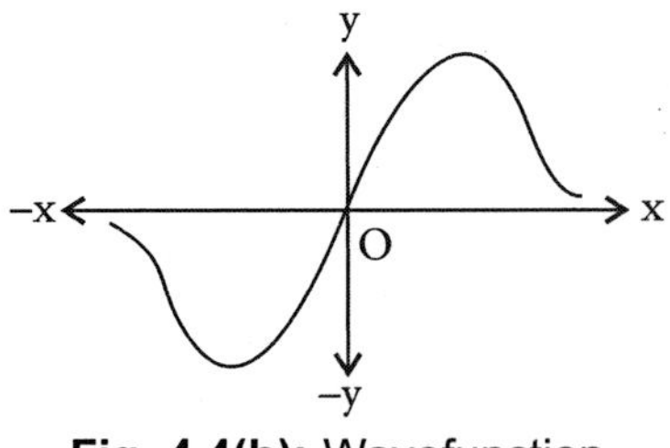

Fig. 4.4(b): Wavefunction plot for v = 1

(c) $v = 2; \psi_2(x)$

$= N_2(4y^2 - 2)e^{-y^2/2}$

N_2 = Normalisation constant

In Fig. 4.4(c), number of nodes = 2

where, $\psi_2(x)$ depends on $(4y^2 - 2)\ e^{-y^2/2}$ (as variable).

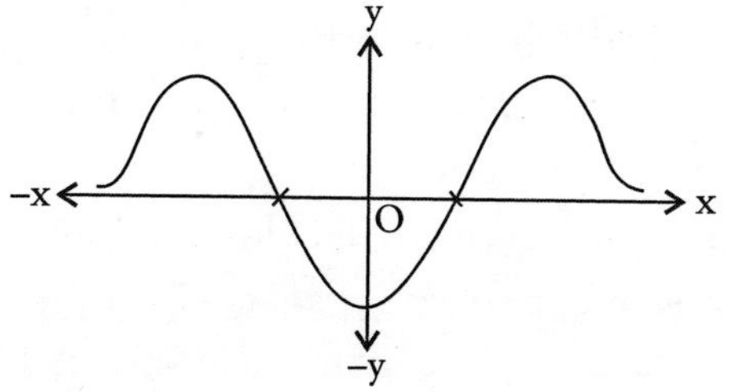

Fig. 4.4(c): Wavefunction plot for v = 2

The nodes ($\psi = 0$) in the wavefunction plot occur due to the Hermite polynomial of the order 'y'. As the order of Hermite polynomial in the wavefunction equation increases, the number of nodes increases. Also, as the order increases, the vibrational quantum number 'v' also increases. Thus, as the value of vibrational quantum number increases, the number of nodes increases in the wavefunction plot.

The number of nodes in the vibrational wavefunction plot = v.

4.6 Probability Plots for Linear Harmonic Oscillator

The probability distribution is achieved by squaring the wavefunction plots. Squaring the wavefunction plots shift the curve from the negative quadrant to the positive quadrant of the graph.

(a) $\psi\ (v = 0)$

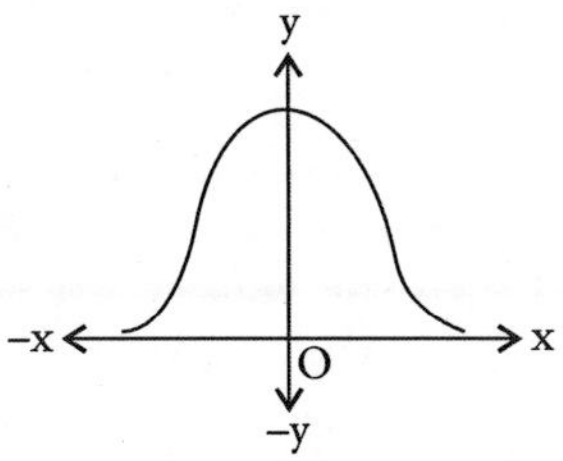

(b) $\psi_1^2(v = 1)$

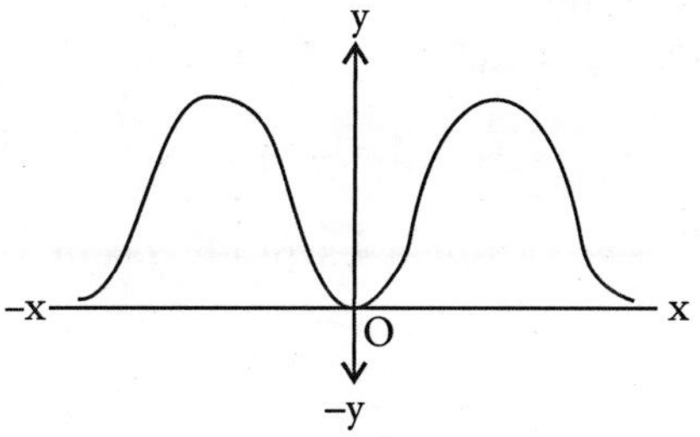

(c) $\psi_2^2(v = 2)$

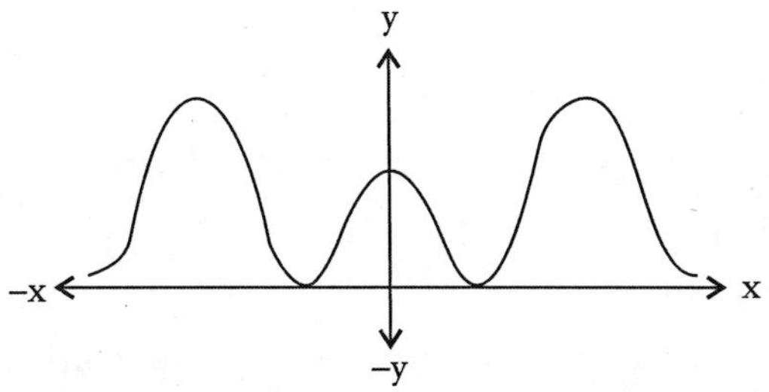

Fig. 4.5: Probability plots for linear harmonic oscillator

4.7 Symmetry of the Vibrational Wavefunction

- The symmetry is assigned to a function, f(x), according to:

  Symmetric function: f(x) = –f(x) → odd functions are symmetric

  Antisymmetric function: f(x) = –f(–x) → even functions are antisymmetric

- When two functions are multiplied, their symmetry follows certain rules:

  (i) (odd function) × (odd function) = Even function

  (ii) (even function) × (even function) = Even function

  (iii) (even function) × (odd function) = Odd function

  (iv) (odd function) × (even function) = Odd function

  From these rules, we conclude that, whenever two functions of same symmetry are multiplied (i.e. both odd or both even), the resultant function is an even function. However, functions of different symmetry (i.e. one odd and other even) give odd function as resultant.

- The symmetry of the vibrational wavefunction (whether odd or even) depends upon the Hermite polynomial. The Hermite polynomials are odd functions for odd values of 'v' and they are even functions for even value of 'v'.
- The symmetry of the vibrational wavefunction is important for calculation of integrals while solving the properties of the system.

  Like: $\int_{-\infty}^{+\infty} f(x)\,dx = 0$ [if f(x) is an odd function]

4.8 Vibrational Energy Levels and Zero Point Energy for Linear Harmonic Oscillator

The absorption spectrum in the infra-red region corresponds to the promotion of a molecule from its lowest vibrational level to the next higher one. The correlation of experimental absorption with the theoretical difference in energy levels yields information on the flexibility of the bond undergoing the vibrational motion. By flexibility, we mean the relative easiness with which the bond length and bond angle of the molecule can be distorted from the equilibrium configuration.

For a linear harmonic oscillator, the expression of the energy is given as equation (4.10).

$$E_v = h\nu\left(v + \frac{1}{2}\right) \text{ (in Joules)} \qquad ...(4.10)$$

$$\epsilon_v = \frac{E_v}{hc} = \frac{\nu}{c}\left(v + \frac{1}{2}\right) = \bar{\nu}\left(v + \frac{1}{2}\right) \text{ in cm}^{-1} \qquad ...(4.12)$$

where, ν = vibrational frequency or classical frequency of vibration (in s^{-1})

$\bar{\nu}$ = vibrational frequency (in cm^{-1})

v = vibrational quantum number with allowed values of 0, 1, 2, 3 ... etc.

Thus, once again, we obtain a set of discrete energy levels (shown in Fig. 4.6) and these are allowed energies by putting different values of v, where v = 0, 1, 2, 3, ... Thus, quantum mechanically it was proved that energy of linear harmonic oscillator is quantized, i.e. the molecule cannot vibrate with any vibrational arbitrary amount of energy but with only certain discrete vibrational energies.

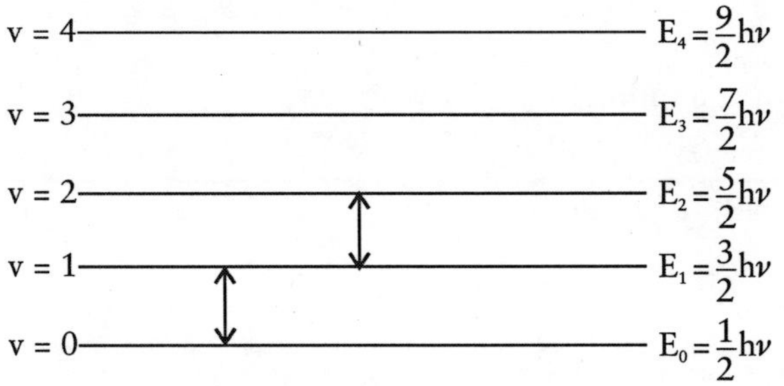

Fig. 4.6: Vibrational energy levels of linear harmonic oscillator

- **Zero Point Energy:** The lowest vibrational energy (v = 0) is not zero as in the case of classical vibrator but has a value of $1/2h\nu$.

For the lowest vibrational level, v = 0, the energy for the level will be

$$E_0 = h\nu\left(0+\frac{1}{2}\right) = \frac{1}{2}h\nu \text{ Joules}$$

This is called **zero point energy**. It indicates that the diatomic molecule can never have zero vibrational energy or the atoms can never be completely at rest relative to each other. Even at absolute zero when all translational and rotational motion ceases in crystal, the residual energy of vibration (E_0) still remains i.e., the vibrational motion still exists.

This fact is in agreement with the Heisenberg uncertainty principle. If the atoms were standstill, the uncertainty in the position would be zero and hence, according to the uncertainty principle, the uncertainty in momentum would have been infinite.

In order to avoid this, the atoms must have uncertainty in position, i.e. the atoms will have to have some vibrational energy.

In classical mechanics, the minimum vibrational energy which a diatomic molecule can possess is zero, i.e. the atoms can be at rest relative to each other.

PROBLEM BASED ON VIBRATIONAL ENERGY AND ZERO POINT ENERGY

Q. 1. What is zero point energy of a simple harmonic oscillator?

4.9 Vibrational Selection Rules for Linear Harmonic Oscillator

When a molecule absorbs electromagnetic radiation, under vibrational spectroscopy (i.e., IR region), it gets excited to higher vibrational levels. However, excitation to higher vibrational levels follows certain selection rules. Those transitions which follow selection rules are called as **allowed transitions** and which do not follow selection rules are called **forbidden transitions.** According to vibrational selection rules, only those vibrational transitions are allowed for which $\Delta v = \pm 1$.

Quantum mechanically, transition probability for all allowed transitions is non zero and for forbidden transitions it is zero. Transition probability can be determined by finding the average transition dipole moment integral ($\mu_{v'v}$).

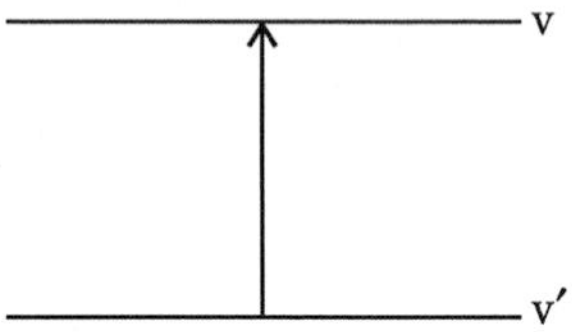

$$\mu_{v'v} = \int_{-\infty}^{+\infty} \psi_{v'} \cdot \hat{\mu}\psi_v d\tau$$

where,

$\hat{\mu}$ = dipole moment operator = $e \cdot \hat{x}$

e = charge

$\hat{x}$ = operator (distance of separation)

$$\mu_{v'v} = \int_{-\infty}^{+\infty} \psi_{v'} \cdot (e \cdot \hat{x}) \cdot \psi_v d\tau$$

$$= e\int_{-\infty}^{+\infty} \psi_{v'} \cdot \hat{x} \cdot \psi_v d\tau$$

If, $\mu_{v'v} = 0$ (forbidden transition)

and $\mu_{v'v} \neq 0$ (allowed transition)

For a molecule to be vibrational active, it should experience change in its dipole moment during vibration. If $\mu_{v'v} \neq 0$, then there is change in dipole moment. Therefore, transitions are allowed and hence molecule is vibrational active. If $\mu_{v'v} = 0$, then there is no change in dipole moment. Therefore, transitions are forbidden and hence molecule is vibrational inactive.

1. Transition between v′ = 0 and v = 1:

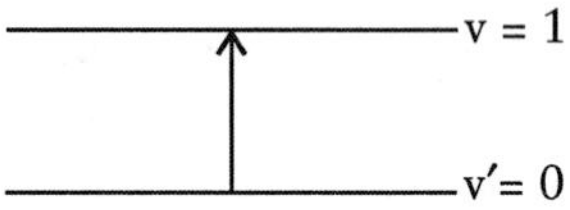

$$\mu_{0,1} = e\int_{-\infty}^{+\infty} \psi_0(x) \cdot x \cdot \psi_1(x)dx$$

$$= e\int_{-\infty}^{+\infty} N_0(1) \cdot e^{-y^2/2} \cdot x \cdot N_1(2y)e^{-y^2/2}dx$$

$$(\because\ x = \alpha y \text{ and } dx = \alpha dy)$$

$$= e\int_{-\infty}^{+\infty} N_0 e^{-y^2/2} \cdot (\alpha y) N_1(2y) \cdot e^{-y^2/2} \cdot \alpha dy$$

$$= 2eN_0 N_1 \alpha^2 \int_{-\infty}^{+\infty} y^2 \cdot e^{-y^2} dy$$

Using standard integral:
$$\int_{-\infty}^{+\infty} y^{2n} \cdot e^{-ay^2} dy = 2\left[\frac{1.3.5....(2n-1)}{2^{n+1} \cdot a^n}\sqrt{\frac{\pi}{a}}\right]$$

$$\mu_{0,1} = 2e \cdot N_0 N_1 \cdot \alpha^2 \frac{\sqrt{\pi}}{2} \neq 0$$

Therefore, vibrational transition from v′ = 0 to v = 1 is allowed.

2. Transition between any two even levels: (Let v′ = 0 and v = 2):

$$\mu_{0,2} = e\int_{-\infty}^{+\infty} \psi_0(x)\cdot x\cdot \psi_2(x)dx$$

$$= e\int_{-\infty}^{+\infty} N_0(1)e^{-y^2/2}\cdot x\cdot N_2(4y^2-2)\cdot e^{-y^2/2}dx$$

$$(\because x = \alpha y \text{ and } dx = \alpha\, dy)$$

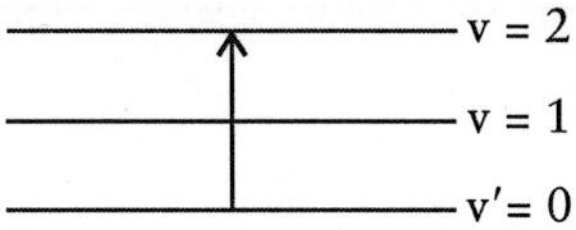

$$\mu_{0,2} = e\int_{-\infty}^{+\infty} N_0\cdot(\alpha y)\cdot N_2(4y^2-2)\cdot e^{-y^2}\cdot dy$$

$$= e\cdot N_0N_2\alpha^2\int_{-\infty}^{+\infty} y\left(4y^2-2\right)e^{-y^2}dy$$

$$= e\cdot N_0N_2\alpha^2\left[\int_{-\infty}^{+\infty} 4y^3\cdot e^{-y^2}dy - \int_{-\infty}^{+\infty} 2ye^{-y^2}dy\right]$$

> Using standard integral:
>
> $$\int_{-\infty}^{+\infty} y^{2n}\, e^{-ay^2}dy = 2\left[\frac{1.3.5....(2n-1)}{2^{n+1}\cdot a^n}\sqrt{\frac{\pi}{a}}\right]$$

$$\mu_{0,2} = eN_0N_2\alpha^2\left[2\sqrt{\pi}-2\sqrt{\pi}\right] = 0$$

Therefore, vibrational transition from v′ = 0 to v = 2 is forbidden.

3. Transition between v' = 0 and v = 3:

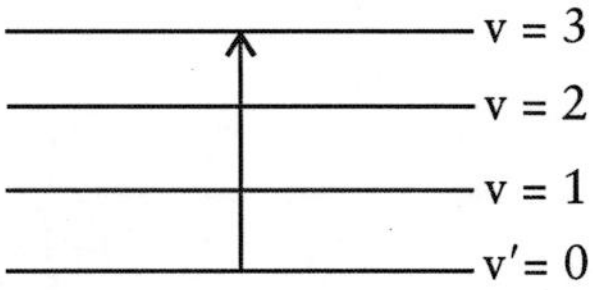

$$\mu_{0,3} = e\int_{-\infty}^{+\infty} \psi_0(x)\cdot x\cdot \psi_3(x)dx$$

$$\mu_{0,3} = e\int_{-\infty}^{+\infty} N_0(1)\cdot e^{-y^2/2}\cdot(\alpha y)\cdot N_3\left(8y^3-12y\right)\cdot e^{-y^2/2}\cdot \alpha\, dy$$

$$(\because x = \alpha y \text{ and } dx = \alpha dy)$$

$$= eN_0N_3\cdot\alpha^2\int_{-\infty}^{+\infty} y\left(8y^3-12y\right)\cdot e^{-y^2}dy$$

$$= eN_0N_3\cdot\alpha^2\left[\int_{-\infty}^{+\infty} 8y^4\cdot e^{-y^2}dy - \int_{-\infty}^{+\infty} 12y^2\cdot e^{-y^2}dy\right]$$

> Using standard integral:
>
> $$\int_{-\infty}^{+\infty} y^{2n}\, e^{-ay^2}dy = 2\left[\frac{1.3.5....(2n-1)}{2^{n+1}\cdot a^n}\sqrt{\frac{\pi}{a}}\right]$$

$$\mu_{0,3} = eN_0N_3\alpha^2\left[6\sqrt{\pi}-6\sqrt{\pi}\right] = 0$$

Therefore, vibrational transition from v′ = 0 to v = 3 is forbidden.

4. Transition between v′ = 1 and v = 2:

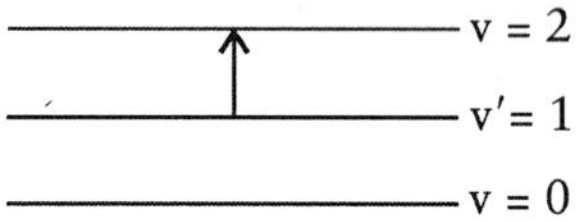

$$\mu_{1,2} = e\int_{-\infty}^{+\infty} \psi_1(x)\cdot x \cdot \psi_2\, dx$$

$$= e\int_{-\infty}^{+\infty} N_1 \cdot (2y)\cdot e^{-y^2/2} \cdot x \cdot N_2(4y^2 - 2)e^{-y^2/2} dx$$

$$= e\int_{-\infty}^{+\infty} N_1\,(2y)e^{-y^2/2} \cdot (\alpha y)\cdot N_2(4y^2 - 2)\cdot e^{-y^2/2} \cdot \alpha\, dy$$

$$(\because\ x = \alpha y \text{ and } dx = \alpha \cdot dy)$$

$$= eN_1N_2\alpha^2 \int_{-\infty}^{+\infty} 2y^2\left(4y^2 - 2\right)e^{-y^2} dy$$

$$= eN_1N_2\alpha^2 \left[\int_{-\infty}^{+\infty} 8y^4 \cdot e^{-y^2} dy - \int_{-\infty}^{+\infty} 4y^2 \cdot e^{-y^2} \cdot dy\right]$$

> Using standard integral:
>
> $$\int_{-\infty}^{+\infty} y^{2n} \cdot e^{-ay^2} dy = 2\left[\frac{1.3.5....(2n-1)}{2^{n+1}\cdot a^n}\sqrt{\frac{\pi}{a}}\right]$$

$$\mu_{1,2} = eN_1N_2\alpha^2\left[6\sqrt{\pi} - 2\sqrt{\pi}\right]$$

$$= e\cdot N_1N_2\alpha^2 4\sqrt{\pi} \neq 0$$

Therefore, vibrational transition from v′ = 1 to v = 2 is allowed.

5. Transition between any two odd levels (let v′ = 1 to v = 3):

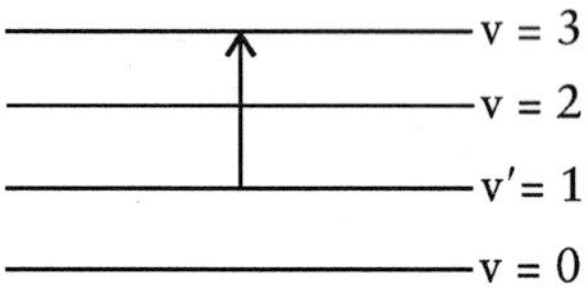

$$\mu_{1,3} = e\int_{-\infty}^{+\infty} \psi_1(x)\cdot x \cdot \psi_3\, dx$$

$$= e\int_{-\infty}^{+\infty} N_1 \cdot (2y)\cdot e^{-y^2/2} \cdot x \cdot N_3(8y^3 - 12y)e^{-y^2/2} dx$$

$$= e\int_{-\infty}^{+\infty} N_1\,(2y)\cdot(\alpha y)\cdot N_3(8y^3 - 12y)e^{-y^2} \cdot \alpha\, dy$$

$$(\because\ x = \alpha y \text{ and } dx = \alpha\, dy)$$

$$= eN_1N_3\alpha^2 \int_{-\infty}^{+\infty} 2y^2\left(8y^3 - 12y\right)e^{-y^2} dy$$

$$= eN_1N_3\alpha^2 \left[\int_{-\infty}^{+\infty} 16y^5 \cdot e^{-y^2} dy - \int_{-\infty}^{+\infty} 24y^3 \cdot e^{-y^2} dy\right]$$

> Using standard integral:
>
> $$\int_{-\infty}^{+\infty} y^{2n} \cdot e^{-ay^2} dy = 2\left[\frac{1.3.5....(2n-1)}{2^{n+1}\cdot a^n}\sqrt{\frac{\pi}{a}}\right]$$

$$\mu_{1,3} = 0$$

Therefore, vibrational transition from v′ = 1 to v = 3 is forbidden.

6. Transition between v' = 1 and v = 4:

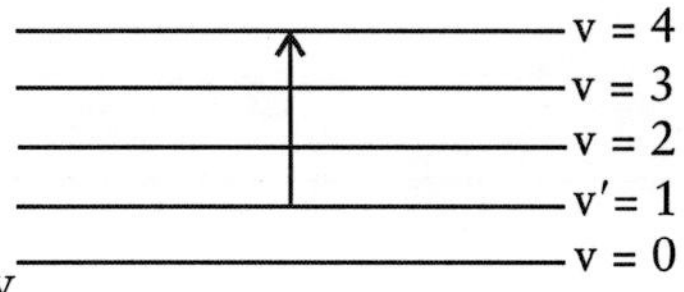

$$\mu_{1,4} = e\int_{-\infty}^{+\infty} \psi_1(x)\cdot x\cdot \psi_4\, dx$$

$$= e\int_{-\infty}^{+\infty} N_1(2y)e^{-y^2/2}\cdot x\cdot N_4(16y^4 - 48y^2 + 12)e^{-y^2/2}dx$$

$$(\because\ x = \alpha y \text{ and } dx = \alpha\, dy)$$

$$\mu_{1,4} = e\int_{-\infty}^{+\infty} N_1(2y)e^{-y^2/2}\cdot(\alpha y)\cdot N_4(16y^4 - 48y^2 + 12)e^{-y^2/2}\cdot \alpha\, dy$$

$$= eN_1N_4\cdot\alpha^2\int_{-\infty}^{+\infty} 2y^2\left(16y^4 - 48y^2 + 12\right)e^{-y^2}dy$$

$$= eN_1N_4\cdot\alpha^2\left[\int_{-\infty}^{+\infty} 32y^6\cdot e^{-y^2}dy - \int_{-\infty}^{+\infty} 96y^4\cdot e^{-y^2}\cdot dy + \int_{-\infty}^{+\infty} 24y^2\cdot e^{-y^2}dy\right]$$

Using standard integral:

$$\int_{-\infty}^{+\infty} y^{2n}\cdot e^{-ay^2}dy = 2\left[\frac{1.3.5....(2n-1)}{2^{n+1}\cdot a^n}\sqrt{\frac{\pi}{a}}\right]$$

$$\mu_{1,4} = e\cdot N_1\cdot N_4\alpha^2\cdot\left[\frac{15}{4} - \frac{9}{2} + \frac{3}{4}\right] = 0$$

$$\mu_{1,4} = 0$$

Therefore, vibrational transition from v′ = 1 to v = 4 is forbidden.

Conclusion

Therefore, we can conclude that according to selection rule only the transitions following $\Delta v = \pm 1$ are allowed. But from $v' = 0 \rightarrow v = 2$ or $v' = 1 \rightarrow v = 3$, i.e. $\Delta v = \pm 2$ are forbidden. Also $\Delta v = \pm 3$ ($v' = 0 \rightarrow v = 3$ and $v' = 1 \rightarrow v = 4$) are also forbidden transitions.

Hence, quantum mechanically it was proved that selection rule for allowed vibrational transitions follow as $\Delta v = \pm 1$. Transitions for which $\Delta v = +1$ corresponds to absorption and those with $\Delta v = -1$ (corresponds to emission).

PROBLEMS BASED ON SELECTION RULES

Q. 1. Give the selection rules for vibrational spectra.

Q. 2. What is the selection rule for simple harmonic oscillator?

4.10 Vibrational Spectra of Linear Harmonic Oscillator

To discuss the vibrational spectra, we consider the difference between the vibrational energy levels. According to the vibrational selection rules, only those vibrational transitions are allowed for which $\Delta v = \pm 1$. Thus, a transition takes place from a lower vibrational level with vibrational quantum number 'v' to a higher vibrational level with vibrational quantum number $v' = v + 1$. The energy absorbed for the transition will be given by

$$\Delta E_v = E_{v+1} - E_v$$

$$\Delta E_v = h\nu\left(v + 1 + \frac{1}{2}\right) - h\nu\left(v + \frac{1}{2}\right)$$

$$\Delta E_v = h\nu\left(v + \frac{3}{2} - v - \frac{1}{2}\right)$$

$$\Delta E_v = h\nu \text{ Joules}$$

$$\Delta E_v = \bar{\nu}_{v \to v+1} = \bar{v} \text{ cm}^{-1}$$

According to this, it has been proved that vibrational energy levels are quantized with energy spacing equal to hν, as shown in Fig. 4.7.

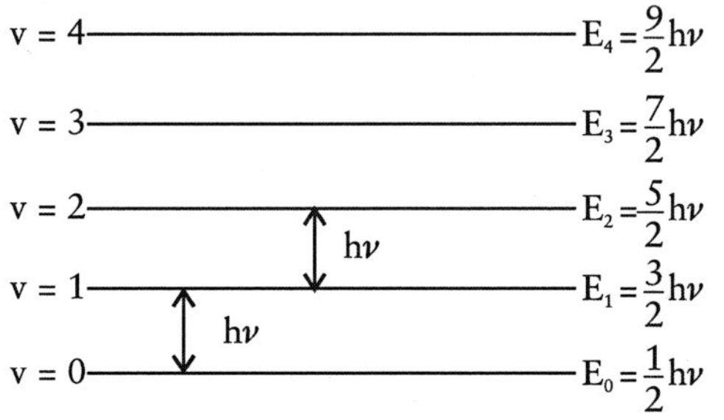

Fig. 4.7: Vibrational energy spacing of linear harmonic oscillator

Thus, the transition between any two neighbouring states will give rise to same energy change of hν. So the vibrational energy levels are equally spaced. Thus, only one absorption line will be obtained in the vibrational spectrum whose wavenumber ($\bar{v}$) is equal to the vibrational frequency ($\bar{\nu}$) of the diatomic molecule (shown in Fig. 4.8).

Also, as most of the molecules are in the lowest (ground state) vibrational level with v = 0, therefore the transitions taking place are from v = 0 → v = 1. The frequency thus observed is called the **fundamental frequency**.

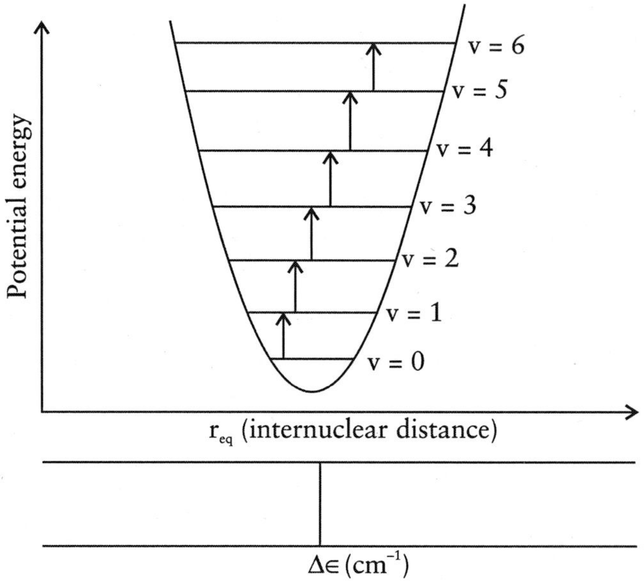

Fig. 4.8: Vibrational spectra of linear harmonic oscillator

PROBLEMS BASED ON VIBRATIONAL SPECTRA OF HARMONIC OSCILLATOR

Q. 1. HCl absorbs at 2890 cm^{-1}, calculate force constant.

Ans. $\Delta \in_v = \bar{\nu}\ cm^{-1}$

$$\bar{\nu} = 2890\ cm^{-1} = 2890 \times 10^2\ m^{-1}$$

$$\mu_{HCl} = \left(\frac{1.008 \times 35.5}{1.008 + 35.5}\right) \times \frac{1}{6.023 \times 10^{23}} = \frac{35.784}{36.508} \times \frac{1}{6.023 \times 10^{23}} = 0.1626 \times 10^{-23} g$$

$$= 1.626 \times 10^{-27} kg$$

so, $$k_f = 4\pi^2 c^2 \mu (\bar{\nu})^2$$

$$= 4 \times (3.14)^2 \times (3 \times 10^8)^2 \times (1.626 \times 10^{-27}) \times (2890 \times 10^2)^2$$

$$= 4.781 \times 10^2 kg s^{-1} \text{ or } Nm^{-1}$$

The force constant, k_f, is a good measure of strength of bond. Also, higher the value of k_f, higher is absorption frequency or wave number and vice-versa. It is also seen that as bond order increases, the value of force constant also increases and so is their absorption frequency/wavenumber. This is found to be true for bond stretching absorptions of C$\equiv$C, C$=$C, C$-$C follows order in the same manner.

$$\bar{\nu}_{C\equiv C} > \bar{\nu}_{C=C} > \bar{\nu}_{C-C}$$

These variations are helpful in determining the structures of various molecules.

Q. 2. The force constant of CO is 1840 Nm^{-1}. Calculate $\bar{\nu}$, vibrational frequency (cm^{-1}) for fundamental vibration and spacing between the vibrational energy levels.

(μ for CO is 11.4×10^{-27} kg)

Ans.
$$\bar{\nu} = \frac{1}{2\pi c}\sqrt{\frac{k_f}{\mu}}$$

$$= \frac{1}{2 \times 3.14 \times (3 \times 10^8)} \left(\frac{1840 kg s^{-2}}{11.4 \times 10^{-27} kg}\right)^{1/2}$$

$$= 2140\ cm^{-1}$$

The spacing between energy levels

$$\Delta E = E_{v+1} - E_v = h\nu \text{ Joules}$$

$$\Delta \in = \frac{h\nu}{hc}$$

$$\Delta \in = \frac{\nu}{c} = \bar{\nu}\ cm^{-1}$$

$$\Delta \in = 2140\ cm^{-1}$$

Q. 3. The fundamental stretching vibration for $^{1}HCl^{35}$ is observed at 2886 cm^{-1}.

(a) Draw an energy diagram indicating the vibration energy levels.

(b) Calculate the force constant and zero point energy.

Ans. Hint:

$$\Delta E_v = \bar{\nu}\ cm^{-1} = 2886\ cm^{-1}$$

$$k_f = 4\pi^2 c^2 \mu (\bar{\nu})^2\ Nm^{-1}$$

Zero point energy

$$E_0 = \frac{1}{2} h\nu$$

$$\epsilon_0 = \frac{1}{2}\frac{h\nu}{hc} = \frac{1}{2}\bar{\nu}\ cm^{-1}$$

Q. 4. A harmonic oscillator diatomic molecule vibrates with a frequency of ν. Write down

(i) In what way it is related to vibrational energy level E?

(ii) Draw the vibration energy levels for v = 0, 1, 2, 3.

(iii) Using the selection rule, $\Delta v = \pm 1$, calculate the transition energies between these levels.

(iv) Show these transitions on the energy level diagram in part (ii) and thus draw the vibrational spectrum.

Ans. (Refer sec. 4.10)

Q. 5. The $^{1}H^{35}Cl$ molecule shows a vibrational absorption line corresponding to a wavelength of 3.465×10^{-6} m. Calculate the force constant of the bond.

Ans.

$$\mu_{HCl} = 1.626 \times 10^{-27}\ kg$$

$$\lambda = 3.465 \times 10^{-6}\ m$$

$$k_f = 4\pi^2 c^2 \bar{\nu}^2 \mu$$

$$k_f = 4\pi^2 c^2 \frac{1}{\sqrt{\lambda^2}} \mu$$

$$k_f = \frac{4 \times (3.14)^2 \times (3 \times 10^8\ ms^{-1})^2 \times (1.626 \times 10^{-27}\ kg)}{(3.465 \times 10^{-6}\ m)^2}$$

$$k_f = 480.35\ Nm^{-1}$$

Q. 6. For CO molecule, $\Delta E = 4.24 \times 10^{-21}$ J. Calculate the force constant of the bond.

Ans. $\Delta E = h\nu$ Joules

$$\nu = \frac{\Delta E}{h}$$

Now, $$k_f = 4\pi^2\mu\nu^2 = 4\pi^2\mu\left(\frac{\Delta E}{h}\right)^2$$

$$\mu_{co} = \frac{12\times16}{(12+16)}\times\frac{1}{6.023\times10^{23}} = 1.138\times10^{-25}\,g = 1.38\times10^{-28}\,kg$$

$$\Delta E = 4.24\times10^{-20}\,J$$

$$k_f = 4\pi^2\mu\left(\frac{\Delta E}{h}\right)^2$$

$$k_f = 4\times(3.14)^2\times\left(1.138\times10^{-28}\right)\times\left(\frac{4.24\times10^{-20}}{6.626\times10^{-34}}\right)^2$$

$$k_f = 1.838\times10^3\,kg\,s^{-2} = 1.838\times10^3\,Nm^{-1}$$

4.11 Relative Population of Energy Levels

The relative populations of energy levels are given by Boltzmann equation is

$$N_i = N_0\exp(-\Delta E_i/kT)$$

where ΔE_i is the energy difference between the ith level and the zeroeth level. The term N_i/N_0, i.e. relative population depends on the ratio $\Delta E_i/kT$. From the experimental data, it is found that in most diatomic molecules $\Delta E_{0\to1}$ is much higher than the thermal energy kT.

For example, CO and Cl_2 molecules we have

Carbon monoxide	$\Delta E_{0\to1} = 42\times10^{-21}$ J
Chlorine	$\Delta E_{0\to1} = 11\times10^{-21}$ J

These values may be compared with the thermal energy 4.1×10^{-21} J at 298 K. On computing the ratios N_1/N_0, N_2/N_0 for these molecules, we find that

For CO

$$\frac{N_1}{N_0} = \exp\left\{\left(-42\times10^{-21}\,J\right)/\left(4.1\times10^{-21}\,J\right)\right\} = \exp(-10.244) = 3.6\times10^{-5}$$

$$\frac{N_2}{N_0} = \exp\left\{\left(-84\times10^{-21}\,J\right)/\left(4.1\times10^{-21}\,J\right)\right\} = \exp(-20.488) = 3.1\times10^{-9}$$

For Cl_2

$$\frac{N_1}{N_0} = \exp\left\{\left(-11\times10^{-21}\,J\right)/\left(4.1\times10^{-21}\,J\right)\right\} = \exp(-2.683) = 6.8\times10^{-2}$$

$$\frac{N_2}{N_0} = \exp\left\{\left(-22\times10^{-21}\,J\right)/\left(4.1\times10^{-21}\,J\right)\right\} = \exp(-5.366) = 4.7\times10^{-3}$$

From the above calculated values, we observe that the population in the higher vibrational levels are negligible in comparison to the population in the ground level. Thus, we may conclude that the excitation of molecules from v = 0 level will be much more intense than from any of the higher levels.

PROBLEMS BASED ON RELATIVE POPULATION

Q. 1. Calculate the relative number of bromine molecules in the second and first excited vibrational state at 800 K if it is given that

$\bar{v}$ (v = 1 → 2) = 321 cm^{-1}

Ans. $\Delta\in = \bar{v}$ cm^{-1}

$$\Delta\in = \bar{v} \times hc \text{ Joules}$$

$$\frac{N_{v=2}}{N_{v=1}} = \exp\left(-\Delta E/_{kT}\right)$$

$$= \exp\left(\frac{-\bar{v} \times hc}{kT}\right)$$

$$= \exp\left(\frac{-321\,\text{cm}^{-1} \times 6.626 \times 10^{-34}\,\text{Js} \times 3 \times 10^{10}\,\text{cm s}^{-1}}{1.381 \times 10^{-23}\,\text{JK}^{-1} \times 800\,\text{K}}\right)$$

$$\frac{N_{v=2}}{N_{v=1}} = 0.561$$

Q. 2. Determine the ratio of population of the v = 1 level to that of the v = 0 level of the CO molecule at 300 K. The fundamental vibration frequency is 2170 cm^{-1}. Also calculate the same for 1000 K.

Ans. $\Delta\in = \bar{\nu}$ cm^{-1} = $\bar{\nu}$ × hc Joules

$\bar{\nu}$ = 2170 cm^{-1}

$$\frac{N_{v=1}}{N_{v=0}} = \exp\left(-hc\bar{\nu}/_{kT}\right)$$

$$= \exp\left(\frac{-6.626 \times 10^{-34}\,\text{Js} \times 3 \times 10^{10}\,\text{cm s}^{-1} \times 2170\,\text{cm}^{-1}}{1.381 \times 10^{-23}\,\text{JK}^{-1} \times 300\,\text{K}}\right)$$

$$= \exp(-10.40)$$

$$= 0.00003$$

At $T = 1000$ K

$$\frac{N_{v=1}}{N_{v=0}} = \exp\left(\frac{-6.626 \times 10^{-34}\,\text{Js} \times 3 \times 10^{10}\,\text{cm}^{-1} \times 2170\,\text{cm}^{-1}}{1.381 \times 10^{-23}\,\text{JK}^{-1} \times 1000\,\text{K}}\right)$$

$$= \exp(-3.12) = 0.044$$

From the results, it is seen that the increase in temperature raises the population of the excited state vibrational level.

4.12 Infrared Active Molecules or Types of Molecules Showing Vibrational Spectra

The interaction between the radiation and the vibrating molecule is through the fluctuating electric field of the radiation and the similar field generated by the vibration of the molecule. The electrical component of the radiation oscillates in a plane perpendicular to the direction of propagation of the wave. The frequency of this oscillating electrical component is equal to the frequency of the radiation. On the other hand, an oscillating electrical field from a molecule is generated only when there occurs a change in dipole moment during the vibration of the molecule.

In other words, for a molecule to be IR active, the dipole moment of the molecule must change when the atoms are displaced due to vibration. Actually, while vibrating there is some change in dipole moment and this oscillating dipole moment interacts with the electromagnetic radiation and then molecule starts vibrating more.

Thus, the condition required is only the change in dipole moment and the molecule need not have a permanent dipole moment. For example, in case of homonuclear diatomic molecules like H_2, O_2, N_2, etc., we have only one type of vibration, i.e. symmetrical stretching vibration, where the molecule is alternatively stretched and compressed. It is expected that the change in dipole moment during the vibration of a diatomic molecule will occur only if the molecule has a permanent dipole moment. This is shown in Fig. 4.9.

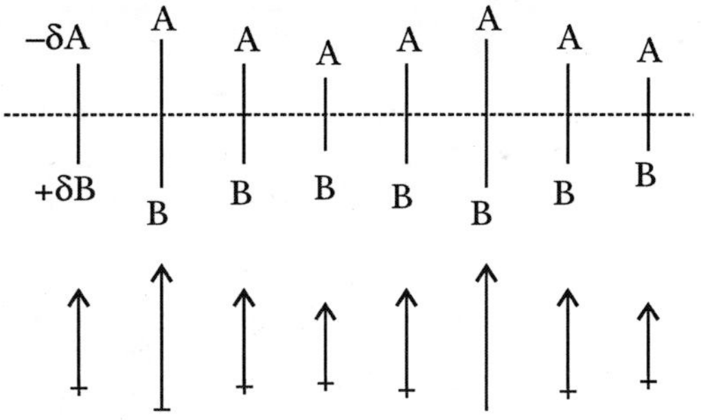

Fig. 4.9: Dipole moment change on vibration

Homonuclear diatomic molecules have zero dipole moment and also there occurs no change in dipole moment during the symmetrical stretching vibration, and hence such molecules are infrared inactive, i.e. they do not show any vibrational spectra, whereas heteronuclear diatomic molecules like HCl, CO, NO, etc. and polyatomic molecules like CO_2, H_2O, CH_4, C_2H_4, etc. do have a permanent dipole moment and they also show change in dipole moment in some mode of vibration and hence give vibrational spectra and said to be infrared-active.

PROBLEMS BASED ON MOLECULES SHOWING VIBRATIONAL SPECTRA

Q. 1. What are the fundamental requirements for a molecule to show IR spectra?

Q. 2. State the conditions for a molecule to be rotationally and vibrationally active.

Q. 3. HCl exhibits IR spectrum, whereas H_2 does not do so. Explain.

Q. 4. Show the nature of interaction between H-Cl and electromagnetic radiation.

Q. 5. What are the essential conditions, for a molecule to show IR spectra? Which of the following will be IR active: O_2, CO_2, CO and SO_2?

Ans 5. The condition is that molecular vibration should bring about a change in dipole moment.

CO_2, CO and SO_2 are IR active.

4.13 Amplitude of Vibration or Maximum Displacement of the Vibration or Vibrational Amplitude

From the classical mechanics, we find that at the extremes of vibration, whole of the energy is present in the form of potential energy. We can find the maximum displacement, x_{max}, from the equilibrium bond length by equating the classical expression of potential energy $\left(\frac{1}{2}k_f x_{max}^2\right)$ with the vibrational energy.

$$\frac{1}{2}k_f\, x_{max}^2 = h\nu\left(v+\frac{1}{2}\right)$$

$$\frac{1}{\not 2}k_f\, x_{max}^2 = \frac{h}{\not 2\pi}\sqrt{\frac{k_f}{m}}\left(v+\frac{1}{2}\right); \quad \left[\nu = \frac{1}{2\pi}\sqrt{\frac{k_f}{m}}\right]$$

$$x_{max}^2 = \frac{h}{\pi k_f}\sqrt{\frac{k_f}{m}}\left(v+\frac{1}{2}\right)$$

$$x_{max} = \left\{\frac{h}{\pi}\sqrt{\frac{1}{k_f m}}\left(v+\frac{1}{2}\right)\right\}^{1/2}$$

PROBLEM BASED ON AMPLITUDE OF VIBRATION

Q. 1. Calculate the vibrational amplitude of HCl molecule in the lowest vibrational level. The k_f of the molecule is 4.781×10^2 kg s^{-2}.

Ans. For HCl molecule, lowest vibrational level is v = 0.

$\mu = 1.615 \times 10^{-27}$ kg

$k_f = 4.781 \times 10^2$ kg s^{-2}

$$x_{max} = \left\{\frac{h}{\pi}\sqrt{\frac{1}{k_f \mu}}\left(v+\frac{1}{2}\right)\right\}^{1/2}$$

$$x_{max} = \left\{\frac{6.626\times10^{-34}}{(3.14)} \times \sqrt{\frac{1}{(4.781\times10^2)\times(1.615\times10^{-27})}}\cdot\left(0+\frac{1}{2}\right)\right\}^{1/2}$$

$$x_{max} = \left\{\frac{6.626\times10^{-34}}{2\times(3.14)} \times \sqrt{\frac{1}{(4.781\times10^2)\times(1.615\times10^{-27})}}\right\}^{1/2}$$

$$x_{max} = 1.096\times10^{-11}\,\text{m}$$

4.14 Anharmonic Oscillator

According to linear harmonic oscillator concept, for HCl molecule we would expect only one frequency ($\bar{\nu}$) at 2990 cm^{-1} (for v = 0 → v = 1). But there is another line with lower intensity at 6000 cm^{-1} and another at 9000 cm^{-1} of much lower intensity. These intensities shows that molecules are not only going from v = 0 → v = 1 but also from v = 0 → v = 2 and v = 0 → v = 3.

This happens because, the above model of linear harmonic oscillator consider the bond between atoms to be perfectly elastic which is not correct. This means, the real bonds although elastic, do not obey exactly the laws of simple harmonic motion. These molecules (like HCl) are not harmonic oscillator but anharmonic oscillator. Thus, if the bond between atoms is stretched, the restoring force becomes weaker and for large amplitude of vibration, there comes a point at which it will break and the molecule dissociate into atoms. Such an oscillator is said to be **Anharmonic oscillator.** Thus, although for small compressions and extensions the bond may be taken as perfectly elastic but for large amplitudes, a more complicated behaviour is assumed.

4.14.1 Potential Energy Curve

We have seen that the vibrational quantum number in the energy expression of a harmonic oscillator can have any value starting from the minimum value of zero to any larger integral value. Thus, a diatomic molecule which is considered to execute harmonic oscillations can have any higher allowed energy without showing any sign of breaking into atoms. Whereas when this classical model is compared with the real diatomic molecule, the real molecule breaks into atoms if it is stretched beyond a certain limit.

A large difference is seen when the classical potential energy curve of a harmonic oscillator is compared with the real/actual potential energy curve of a diatomic molecule:

(1) **In the lower internuclear distance**, the actual potential energy curve rises more steeply than the harmonic parabola energy curve.

(2) **In the higher internuclear distance**, the energy of harmonic oscillator goes on increasing without any limit whereas that in actual energy diagram approaches the limiting dissociation energy at some finite internuclear distance.

(3) **At the equilibrium internuclear distance**, the two curves do not show appreciable departure from each other.

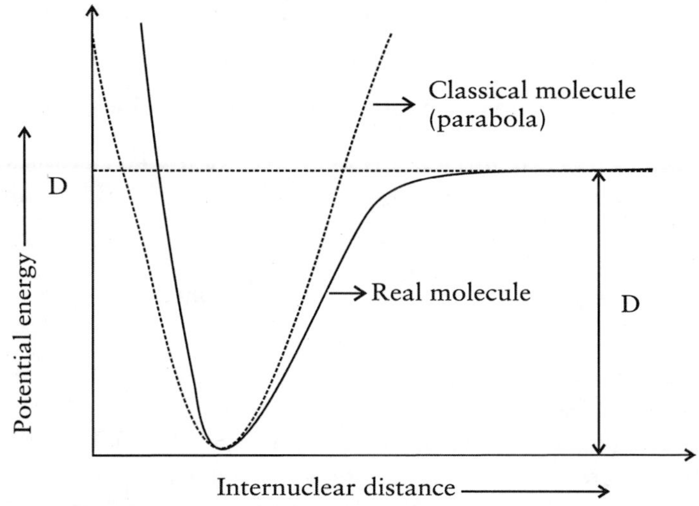

Fig. 4.10: Potential energy curve for classical and real diatomic molecule

4.14.2 Morse Potential Energy

There were many empirical expressions that have been suggested for the correlation of energy of a real diatomic molecule with the internuclear distance. One of the expressions which fits best to the potential energy curve (for a real molecule) is given by P.M. Morse and is known as **Morse Potential.** The energy expression is given by

$$V = D\left[1-\exp\left\{a\left(r_{eq}-r\right)\right\}\right]^2 \quad ...(4.13)$$

where 'a' is a constant for a particular molecule and D is the dissociation energy.

4.14.3 Vibrational Energy of a Diatomic Molecule using Morse Potential Energy

Substituting the Morse potential energy (equation 4.13) expression in Schrödinger equation and the approximate solution of the resultant equation gives the energy expression for the various allowed vibrational potential energy levels.

P.M. Morse in 1929 suggested that the above equation of energy (equation 4.12) has to be modified for an anharmonic oscillator by including an additional term as follows:

$$E_v = h\nu_e\left(v+\frac{1}{2}\right) - h\nu_e x_e\left(v+\frac{1}{2}\right)^2 \quad \text{(in Joules)} \quad ...(4.14)$$

$$\epsilon_v = \left(v+\frac{1}{2}\right)\bar{\nu}_e - \left(v+\frac{1}{2}\right)^2 \bar{\nu}_e x_e \,(\text{in cm}^{-1}) \qquad (v = 0, 1, 2, 3.....)$$

where, $\bar{\nu}_e$ = equilibrium oscillation frequency of an anharmonic oscillator (in cm^{-1})

x_e = anharmonicity constant

For bond stretching, the anharmonicity constant has a very small and positive value (≈ 0.01) that is why on increasing 'v' the vibrational level becomes more closer.

The equation for energy equation (4.14) is only an approximate solution. To solve the energy expression precisely, it requires cubic, quadratic, etc. terms in $\left(v+\frac{1}{2}\right)$ with anharmonicity constant y_e, z_e, etc. These terms are only important when we use large values of v, and we shall ignore them. Therefore, the energy expression given in equation (4.14) for anharmonic oscillator can be rewritten as (equation 4.15)

$$\epsilon_v = \bar{\nu}_e\left[1-\left(v+\frac{1}{2}\right)x_e\right]\left(v+\frac{1}{2}\right) \quad \left(\text{cm}^{-1}\right) \quad ...(4.15)$$

Putting v = 0, 1, 2, 3 ... in equation (4.15), the energy levels of the anharmonic oscillator are obtained. It is seen that whereas the energy levels of a harmonic oscillator are equally spaced, those of anharmonic oscillator are not equally spaced but fall more closely together as the quantum number increases. Further, whereas the potential energy curve for a harmonic oscillator is parabola (shown in Fig. 4.8), that of anharmonic oscillator is shown in Fig. 4.11 with vibrational levels crowding more closely together with increasing value of v.

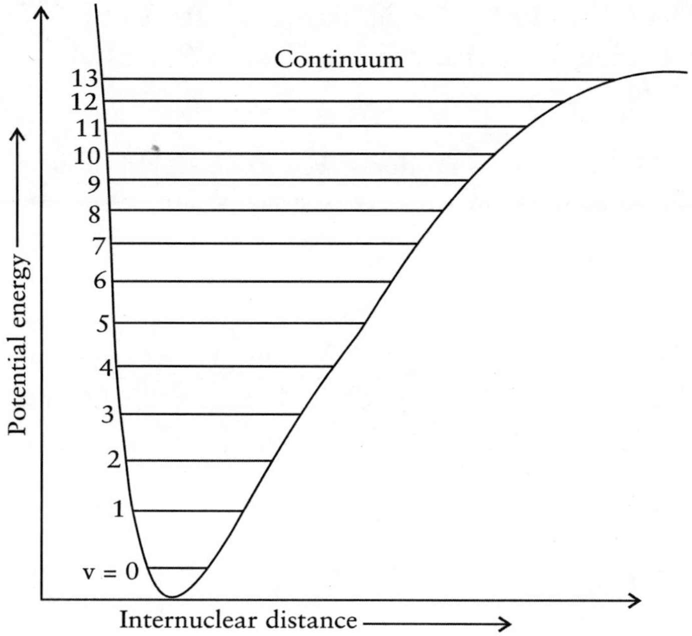

Fig. 4.11: Morse potential energy curve for anharmonic oscillator

4.14.4 Zero Point Energy

For the lowest vibrational level, v = 0, the energy for the level will be

$$\epsilon_0 = \frac{1}{2}\bar{\nu}_e\left(1-\frac{1}{2}x_e\right) \quad \left(\text{in cm}^{-1}\right) \qquad ...(4.16)$$

This is called zero point energy. The zero point energy of anharmonic oscillator differs slightly from that of zero point energy of harmonic oscillator.

4.14.5 Equilibrium Oscillation Frequency

On comparing the energy of harmonic oscillator with that of anharmonic oscillator, i.e. comparing equation (4.12) and equation (4.15)

$$\epsilon_v = \bar{\nu}\left(v+\frac{1}{2}\right) \quad \left(\text{in cm}^{-1}\right) \qquad ...(4.12)$$

$$\epsilon_v = \bar{\nu}_e\left[1-\left(v+\frac{1}{2}\right)x_e\right]\left(v+\frac{1}{2}\right) \quad \left(\text{in cm}^{-1}\right) \qquad ...(4.15)$$

We get,

$$\bar{\nu} = \bar{\nu}_e\left[1-\left(v+\frac{1}{2}\right)x_e\right] \qquad ...(4.17)$$

Thus, anharmonic oscillator behaves like the harmonic oscillator but with an oscillation frequency which decreases steadily with increasing value of 'v'.

4.14.6 Vibrational Selection Rules for Anharmonic Oscillator

Whereas the selection rule for the harmonic oscillator is $\Delta v = \pm 1$, the selection rules for the anharmonic oscillator are

$$\Delta v = \pm 1, \pm 2, \pm 3 \ldots$$

Thus, all types of vibrational transitions are possible. However, the transition from v = 0 → v = 1 is found to be most intense because usually most of the molecules are in v = 0 level and is called the **fundamental transition.**

Also, $$\frac{N_{v=1}}{N_{v=0}} = 0.01$$

i.e. the population of v = 1 state is nearly 0.01 or some 1% of the ground state population. Thus, we may ignore all the transitions originating from v = 1 and more.

Thus, we restrict ourselves to the three transitions only:

1. v = 0 → v = 1, Δv = +1

$$\Delta \epsilon_v = \epsilon_{v=1} - \epsilon_{v=0}$$

$$\Delta \epsilon_v = \left[\frac{3}{2}\bar{\nu}_e - \frac{9}{4}\bar{\nu}_e x_e\right] - \left[\frac{1}{2}\bar{\nu}_e - \frac{1}{4}\bar{\nu}_e x_e\right]$$

$$\Delta \epsilon_v = \bar{\nu}_e - 2\bar{\nu}_e x_e$$

$$\Delta \epsilon_v = \bar{\nu}_e (1 - 2x_e)$$

$$\Delta \epsilon_v \approx \bar{\nu}_e \text{ cm}^{-1}$$

This transition from v = 0 → v = 1 is considerably intense and this absorption line is called **fundamental transition.**

2. v = 0 → v = 1, Δv = +2

$$\Delta \epsilon_v = \epsilon_{v=2} - \epsilon_{v=0}$$

$$\Delta \epsilon_v = \left[\frac{5}{2}\bar{\nu}_e - \frac{25}{4}\bar{\nu}_e x_e\right] - \left[\frac{1}{2}\bar{\nu}_e - \frac{1}{4}\bar{\nu}_e x_e\right]$$

$$\Delta \epsilon_v = 2\bar{\nu}_e - 6\bar{\nu}_e x_e$$

$$\Delta \epsilon_v = 2\bar{\nu}_e (1 - 3x_e)$$

$$\Delta \epsilon_v \approx 2\bar{\nu}_e \text{ cm}^{-1}$$

This transition from v = 0 → v = 2 is of very weak intensity and this absorption line is called **first overtone.**

3. v = 0 → v = 3, Δv = +3

$$\Delta \epsilon_v = \epsilon_{v=3} - \epsilon_{v=0}$$

$$\Delta \epsilon_v = \left[\frac{7}{2}\bar{\nu}_e - \frac{49}{4}\bar{\nu}_e x_e\right] - \left[\frac{1}{2}\bar{\nu}_e - \frac{1}{4}\bar{\nu}_e x_e\right]$$

$$\Delta \epsilon_v = 3\bar{\nu}_e - 12\bar{\nu}_e x_e$$

$$\Delta \epsilon_v = 3\bar{\nu}_e (1 - 4x_e)$$

$$\Delta \epsilon_v \approx 3\bar{\nu}_e \text{ cm}^{-1}$$

This transition from v = 0 → v = 3 has negligible intensity and this absorption line is called **second overtone.**

Since $x_e \approx 0.01$ is very small, so the three spectral lines lies very close to $\overline{\nu}_e$, $2\overline{\nu}_e$ and $3\overline{\nu}_e$.

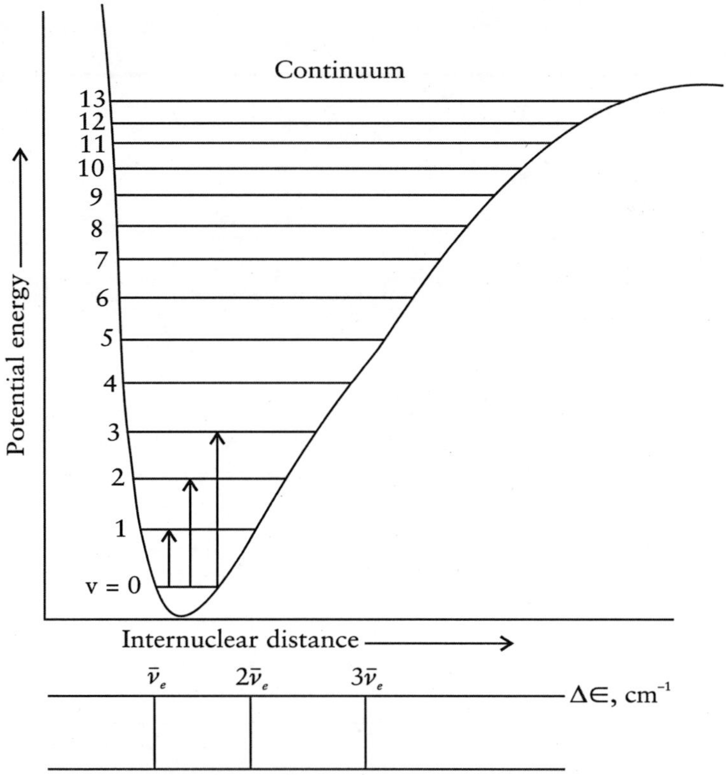

Fig. 4.12: Vibrational spectra of anharmonic oscillator

Like: Spectrum of HCl molecule shows three absorption peaks

Fundamental transition at 2886 cm^{-1}

First Overtone at 5668 cm^{-1}

Second Overtone at 8347 cm^{-1}

Thus, $\overline{\nu}_e (1 - 2x_e) = 2886\ cm^{-1}$

$2\overline{\nu}_e (1 - 3x_e) = 5668\ cm^{-1}$

$3\overline{\nu}_e (1 - 4x_e) = 8347\ cm^{-1}$

Solving any two above equations, we get

$\overline{\nu}_e = 2990\ cm^{-1}$ and $x_e = 0.0174$

In anharmonic oscillator, usually most of the molecules are in v = 0 level. But some molecules are also seen in v = 1 level. If there are some molecules in v = 1 level, then they might go to v = 2 level. Thus, v = 1 → v = 2 transition is also seen and this would be again fundamental transition originating from v = 1 level.

The intensity of this transition would be small.

v = 1 → v = 2

$$\Delta \in_v = \in_{v=2} - E_{v=1}$$

$$\Delta \in_v = \left[\frac{5}{2}\bar{\nu}_e - \frac{25}{4}\bar{\nu}_e x_e\right] - \left[\frac{3}{2}\bar{\nu}_e - \frac{9}{4}\bar{\nu}_e x_e\right]$$

$$\Delta \in_v = \bar{\nu}_e - 4\bar{\nu}_e x_e$$

$$\Delta \in_v = \bar{\nu}_e (1 - 4x_e)\text{cm}^{-1}$$

This is of weaker intensity than fundamental transition (v = 0 → v = 1).

Hence, all types of transitions from any lower vibrational levels to any of the higher vibrational levels are allowed. Since only the lowest level is thickly populated and because the population decreases very rapidly with increasing v, only a few transitions from the lowest vibrational level (v = 0) to the next nearest vibrational levels are actually observed at room temperature. The transition v = 0 → v = 1 is found to be the most intense and is known as the **fundamental absorption.** The transition v = 0 → v = 2, is found to have a poor intensity and is known as the **first overtone.** The transition v = 0 → v = 3 is found to have negligible intensity and is known as the **second overtone.** The higher overtones are too weak to be realized experimentally.

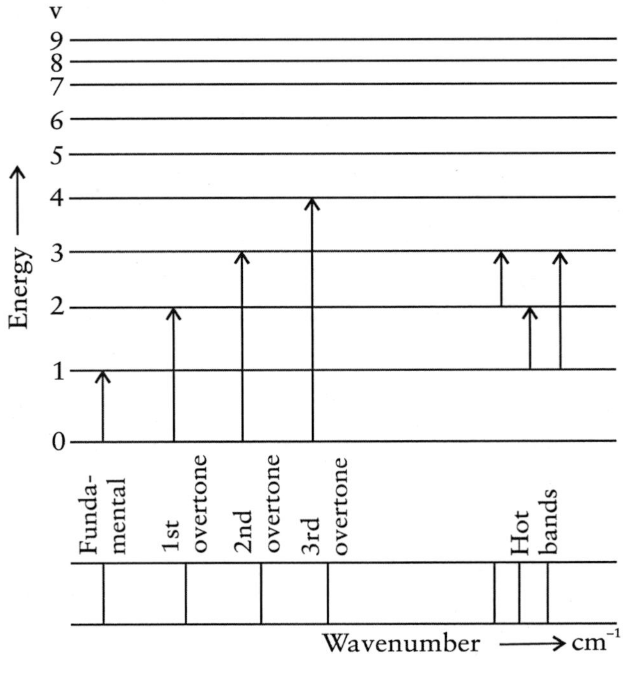

Fig. 4.13: Every levels showing fundamental, overtones and hot bands

Hot Bands: The probability of transition from a level can be increased by increasing the population in that level. The population can be increased by increasing the temperature. When the intensity is increased by raising the temperature, then called as **hot bands**. Thus, the absorption lines generated from the excited states are very weak and are generally known as 'hot bands' as the high temperature is one of the requirements for their generation.

Like: Transition from v = 1 → v = 2 is called hot band because v = 1 has less population and thus less transition. But on heating, the transition becomes quite intense, since as temperature increases, greater number of molecules will occupy v = 1 level.

PROBLEMS BASED ON ANHARMONIC OSCILLATOR

Q. 1. What are hot bands?

Q. 2. Explain briefly the appearance of hot bonds in the IR spectroscopy. In IR spectra, the intensities of hot bands increase with increase in the temperature. Explain.

Q. 3. The fundamental and first overtone transitions of NO molecule are centered at 1876.06 cm^{-1} and 3724.20 cm^{-1}, respectively. Calculate the equilibrium vibration frequency, anharmonicity constant and zero-point energy of the molecule.

Ans. 3. For the fundamental band, we have

$$\overline{\nu}_{0\to 1} = \overline{\nu}_e(1-2x_e) = 1876.06 \text{ cm}^{-1}$$

and for the first overtone, we have

$$\overline{\nu}_{0\to 2} = 2\overline{\nu}_e(1-3x_e) = 3724.20 \text{ cm}^{-1}$$

From these two equations, we have

Equilibrium vibration frequency $= \overline{\nu}_e$ = 1903.98 cm^{-1}

Anharmonicity constant = x_e = 7.332 × 10^{-3}

The zero-point energy is (using equation 4.16)

$$\epsilon_0 = \frac{1}{2}\overline{\nu}_e\left(1-\frac{x_e}{2}\right) = \frac{1}{2}(1903.98)(1-3.666\times 10^{-3}) = 948.5 \text{ cm}^{-1}$$

Q. 4. The fundamental and first overtone transitions of HBr are centred at 2560 cm^{-1} and 5028 cm^{-1} respectively. Calculate the an anharmonicity constant (x_e) and the force constant of HBr.

Ans. Fundamental vibrational = $\overline{\nu}_e$ (1 – 2 x_e) = 2560 cm^{-1} ...(1)

First overtone = 2 $\overline{\nu}_e$ (1 – 3x_e) = 5028 cm^{-1} ...(2)

(i) Divide (2) by (1)

$$\frac{2\cancel{\overline{\nu}_e}(1-3x_e)}{\cancel{\overline{\nu}_e}(1-2x_e)} = \frac{5028}{2560}$$

$$x_e = 0.017$$

(ii) $\overline{\nu}_e (1 - 2x_e) = 2560 \text{ cm}^{-1}$

Putting, $x_e = 0.017$

$\overline{\nu}_e = 2650.1 \text{ cm}^{-1}$

(iii) $\mu_{HBr} = 1.64 \times 10^{-27}$ kg

$k_f = 4\pi^2 c^2 \mu \overline{\nu}_e^2$

$k_f = 4 \times (3.14)^2 \times (3 \times 10^8 \text{ ms}^{-1})^2 (1.64 \times 10^{-27} \text{ kg}) \times (2650.1 \times 10^2 \text{ m}^{-1})^2$

$= 4.08 \times 10^2 \text{ kg s}^{-2} = 4.08 \times 10^2 \text{ Nm}^{-1}$

Q. 5. The fundamental and first overtone of HCl are centred at 2886 cm^{-1} and 5668 cm^{-1} respectively. Evaluate the equilibrium vibrational frequency, the anharmonicity constant and zero point energy.

Ans. Fundamental vibrational = $\overline{\nu}_e (1 - 2x_e) = 2886 \text{ cm}^{-1}$ ...(1)

First overtone = $2\overline{\nu}_e (1 - 3x_e) = 5668 \text{ cm}^{-1}$ ...(2)

(i) Divide (2) by (1)

$$\frac{2\overline{\nu}_e (1-3x_e)}{\overline{\nu}_e (1-2x_e)} = \frac{5668}{2886}$$

$$x_e = 0.017$$

(ii) $\overline{\nu}_e (1 - 2x_e) = 2886 \text{ cm}^{-1}$

Putting, $x_e = 0.017$

$\overline{\nu}_e = 2987.5 \text{ cm}^{-1}$

(iii) $$\epsilon_v = \left(v + \frac{1}{2}\right)\overline{\nu}_e - \left(v + \frac{1}{2}\right)^2 \overline{\nu}_e x_e \text{ cm}^{-1}$$

For zero point energy, v = 0

$$\epsilon_v = \frac{1}{2}\overline{\nu}_e - \frac{1}{4}\overline{\nu}_e x_e$$

$$\epsilon_v = \frac{1}{2} \times (2987.5) - \frac{1}{4}(2987.5)(0.017)$$

$$\epsilon_v = 1493.75 - 12.70$$

$$\epsilon_v = 1481.05 \text{ cm}^{-1}$$

Q. 6. The fundamental line for $^{12}C^{16}O$ occurs at 2143.0 cm^{-1} and the first overtone occurs at 4260 cm^{-1}. Calculate the equilibrium oscillation frequency, the anharmonicity constant for this molecule and the force constant for the bond (in N/m). Reduced mass CO = 1.138×10^{-26} kg.

Ans. (same as Q. 3 and Q. 4)

Q. 7. For a diatomic molecule, the Morse potential is given by

$$V = D\left[1 - \exp\left\{a\left(r_{eq} - r\right)\right\}\right]^2$$

Show that molecule oscillates simple harmonically provided the displacement from mean position is small and force constant $k_f = 2Da^2$

or

Show that the Morse potential:

$$V = D\left[1 - \exp\left\{a\left(r_{eq} - r\right)\right\}\right]^2$$

is reducible to harmonic potential for the lower vibrational levels. Also show that the force constant is given as $k = 2Da^2$

or

The vibrational potential energy of a diatomic molecule can be represented by function $V = b\,[1 - \exp\,\{a(r_{eq} - r)\}]^2$

where, a and b are constants characteristics of molecule, r is the internuclear distance and r_{eq} is the equilibrium bond length. Show that:

(i) the potential has a minimum value at $r = r_{eq}$

(ii) for a small displacement from the equilibrium position, the molecule oscillates simple harmonically.

Ans. (a) For minimum potential $\frac{dV}{dr} = 0$ and proceed to show $\frac{d^2V}{dr^2} > 0$

$$V = b\left[1 - \exp\left\{a\left(r_{eq} - r\right)\right\}\right]^2$$

$$\frac{dV}{dr} = 2b\left[1 - \exp\left\{a\left(r_{eq} - r\right)\right\}\right]\left[a\exp\left\{a\left(r_{eq} - r\right)\right\}\right]$$

$$\frac{dV}{dr} = 2ba\exp\left\{a\left(r_{eq} - r\right)\right\} - 2ba\exp\left\{2a\left(r_{eq} - r\right)\right\}$$

At $r = r_{eq}$, we have

$$\left(\frac{dV}{dr}\right) = 0$$

$$\frac{d^2V}{dr^2} = -2ba^2\exp\left\{a\left(r_{eq} - r\right)\right\} + 4ba^2\exp\left\{2a\left(r_{eq} - r\right)\right\}$$

$$\frac{d^2V}{dr^2} = 2ba^2 > 0$$

(b) $V = b\left[1 - \exp\left\{a\left(r_{eq} - r\right)\right\}\right]^2$

$$\exp(x) = 1 + x + \frac{x^2}{2!} + \ldots$$

Neglecting the higher terms for smaller values of x
exp (x) = 1 + x

$$\therefore \quad V = b\left[1 - \left\{1 + a\left(r_{eq} - r\right)\right\}\right]^2$$

$$V = b\left[1 - 1 - a\left(r_{eq} - r\right)\right]^2$$

$$V = ba^2\left(r_{eq} - r\right)^2 \quad \text{...(1)}$$

For harmonic oscillator

$$V = \frac{1}{2}k_f\,(r_{eq} - r)^2 \quad \text{...(2)}$$

Comparing (1) and (2):

$$\frac{1}{2}k_f = ba^2$$

$$k_f = 2ba^2$$

Q. 8. An equilibrium oscillation frequency of iodine molecule I_2 is 215 cm^{-1} and anharmonicity constant is 0.003. At 300 K, what is the intensity of first overtone relative to that of the fundamental transition?

$\bar{\nu}_e = 215\ cm^{-1}$, $x_e = 0.003$

Ans. Fundamental transition $= \Delta \epsilon_1 = \Delta \epsilon_{v=0\to1} = \bar{\nu}_e\left(1 - 2x_e\right)$

$$= 215\ (1 - 2 \times 0.003) = 213.71\ cm^{-1}$$

$$= 213.71 \times 10^2\ m^{-1}$$

First overtone $= \Delta \epsilon_2 = \Delta \epsilon_{v=0\to2} = 2\bar{\nu}_e\,(1 - 3x_e)$

$$= 426.13\ cm^{-1} = 426.13 \times 10^2\ m^{-1}$$

$$\Delta E_1 = \Delta\epsilon_1 \times hc$$

$$= (213.71 \times 10^2\ m^{-1}) \times (6.626 \times 10^{-34}\ Js) \times (3 \times 10^8\ ms^{-1})$$

$$= 4245.295 \times 10^{-24}\ J$$

$$\frac{N_1}{N_0} = \exp\left(-\Delta E_1 / kT\right) = \exp\left(\frac{-4245.295 \times 10^{-24}\ J}{1.38 \times 10^{-23}\ JK^{-1} \times 300\,K}\right)$$

$$\frac{N_1}{N_0} = \exp\left(-1.024\right) = 0.3586$$

Similarly,

$$\Delta E_2 = \Delta\epsilon_2 \times hc$$

$$= (426.13 \times 10^2\ m^{-1})\ (6.626 \times 10^{-34}\ Js)\ (3 \times 10^8\ ms^{-1})$$

$$= 8464.97 \times 10^{-24}\ J$$

$$\frac{N_2}{N_0} = \exp(-\Delta E_2 / kT) = \exp\left(\frac{-8464.97\times10^{-24}\,J}{1.38\times10^{-23}\,JK^{-1}\times300K}\right)$$

$$\frac{N_2}{N_0} = \exp(-2.0446) = 0.1295$$

$$\text{Intensity ratio} = \frac{\begin{array}{c}\text{Difference in population of v = 0 level and}\\ \text{difference in population of v = 2 level}\end{array}}{\begin{array}{c}\text{Difference in population of v = 0 level and}\\ \text{difference in populaion of v = 1 level}\end{array}}$$

$$= \frac{1-0.1295}{1-0.3526} = 1.36$$

Q. 9. The IR spectrum of HCl shows the fundamental, the first and second overtone at 2886, 5668 and 8547 cm^{-1} respectively. Calculate the following:

(i) The equilibrium vibration frequency in m^{-1}

(ii) The anharmonicity constant

(iii) The force constant in N/m, if the reduced mass is 1.62×10^{-27} kg

Ans. $\bar{v}_{0\to1} = \bar{\nu}_e(1-2x_e) = 2886\ cm^{-1}$ ...(1)

$\bar{v}_{0\to2} = 2\bar{\nu}_e(1-3x_e) = 5668\ cm^{-1}$ ...(2)

Multiply equation (1) by 3 and subtracting equation (2) from it:

$3\bar{\nu}_e - 6x_e = 8658$

$2\bar{\nu}_e - 6x_e = 5668$

$\bar{\nu}_e = 2990\ cm^{-1}$

Equilibrium vibration frequency = $\bar{\nu}_e$ = 299000 m^{-1}

Substitute the value of $\bar{\nu}_e$ in equation (1)

$$\bar{\nu}_e - 2\bar{\nu}_e x_e = 2886$$

$$x_e = \frac{\bar{\nu}_e - 2886}{2\bar{\nu}_e}$$

$$\text{Anharmonicity constant} = x_e = \frac{2990-2886}{2\times2886}$$

$$x_e = 0.0174$$

Force constant $= k_f = 4\pi^2c^2\bar{\nu}_e^2\mu$

$k_f = 4 \times (3.14)^2 \times (3 \times 10^8\ ms^{-1})^2 \times (2990 \times 10^2\ m^{-1})^2 \times (1.62 \times 10^{-2}\ kg)$

$k_f = 514\ kg\ s^{-2} = 514\ Nm^{-1}$

Q. 10. The vibrational wavenumber of the following molecules in their v = 0 states are:

$HCl = 2885\ cm^{-1}$

$DCl = 1990\ cm^{-1}$

$D_2 = 2990\ cm^{-1}$

$HD = 3627\ cm^{-1}$

Using the concept of zero point energy, calculate the energy change in KJ/mole of the reaction

$$HCl + D_2 \rightarrow DCl + HD$$

Ans. Zero Point Energy $= \frac{1}{2} h\nu$ (Joule)

$$= \frac{\bar{\nu}}{2}\ (cm^{-1})$$

$$HCl + D_2 \rightarrow DCl + HD$$

ZPE for $(HCl + D_2)$ (Zero Point Energy) $= \frac{2885}{2} + \frac{2990}{2} = 2937.5\ cm^{-1}$

ZPE for $(DCl + HD)$ (Zero Point Energy) $= \frac{1990}{2} + \frac{3637}{2} = 2808.5\ cm^{-1}$

Energy change $(\Delta \in) = 2937.5 - 2808.5 = 129\,cm^{-1}$

$$\Delta \in = (129\,cm^{-1}) \times \left(6.626 \times 10^{-34}\ Js^{-1}\right)\left(3 \times 10^{10}\ cms^{-1}\right) \times \left(6.023 \times 10^{23}\ mol^{-1}\right)$$

$$\Delta \in = 1.542 \times 10^{3}\ J/mol = 1.542\,kJ/mol$$

Q. 11. What are the selection rules for observing IR spectra of an anharmonic oscillator? Derive the expression for energy required for fundamental transition and first overtone.

Ans. (Refer section 4.14.6)

4.14.7 Dissociation Energy of a Diatomic Molecule

The vibrational energy of a molecule increases with increase in the vibrational quantum number. Eventually, a stage is reached where vibrational energy in quite sufficient to break the molecules into atoms. This energy is known as the **dissociation energy of the diatomic molecule.** Thus at the dissociation stage of the molecule, the vibration energy has a maximum value.

The value of vibrational quantum number at this stage may be determined by applying the mathematical criterion of maximum, i.e. $\partial E/\partial v = 0$.

$$E_v = h\nu_e \left[1 - \left(v + \frac{1}{2}\right) x_e\right]\left(v + \frac{1}{2}\right) \quad \text{...(4.14)}$$

For maxima, $\partial E/\partial v = 0$:

$$\frac{\partial E}{\partial v} = h\nu_e\left[\left\{1-\left(v_{max}+\frac{1}{2}\right)x_e\right\}+\left\{\left(v_{max}+\frac{1}{2}\right)(-x_e)\right\}\right]=0$$

$$h\nu_e\left[1-2\left(v_{max}+\frac{1}{2}\right)x_e\right]=0$$

$$1-2\left(v_{max}+\frac{1}{2}\right)x_e=0$$

$$v_{max}=\frac{1}{2x_e}-\frac{1}{2} \quad \text{...(4.18)}$$

Substituting the value of v_{max} equation (4.18) in equation (4.14); we get value of E_{max}:

$$E_{max}=\frac{h\nu_e}{4x_e} \text{ (in Joules)} \quad \text{...(4.19)}$$

Thus, expression in equation (4.19) represents the dissociation energy of the molecule.

However, the true dissociation energy will be given by:

$$D = E_{max} - E_0 \quad \text{(using equation 4.16 and equation 4.19)}$$

$$D=\frac{h\nu_e}{4x_e}-\frac{1}{2}h\nu_e\left(1-\frac{1}{2}x_e\right) \text{ (in Joules)} \quad \text{...(4.20)}$$

Dividing by hc, the energy will convert into cm^{-1}.

$$D = \epsilon_{max} - \epsilon_0$$

$$D=\frac{\overline{\nu}_e}{4x_e}-\frac{1}{2}\overline{\nu}_e\left(1-\frac{1}{2}x_e\right) \text{ (in } cm^{-1}) \quad \text{...(4.21)}$$

PROBLEMS BASED ON DISSOCIATION ENERGY

Q. 1. Calculate the dissociation energy of HCl (J/mole) when the values of equilibrium vibrational frequency and anharmonicity constant are given as

$$\overline{\nu}_e = 2990\,cm^{-1} \text{ and } x_e = 0.0174$$

Ans.

$$D = \epsilon_{max} - \epsilon_0$$

$$D=\frac{\overline{\nu}_e}{4x_e}-\frac{1}{2}\overline{\nu}_e\left(1-\frac{1}{2}x_e\right)$$

$$D=\frac{2990}{4(0.0174)}-\frac{1}{2}\times 2990\left(1-\frac{1}{2}\times 0.0174\right)$$

$$D=\frac{2990}{4(0.0174)}-\frac{2090}{2}+\frac{2990\times 0.0174}{4}$$

$$D = \frac{2990}{0.0696} - \frac{2090}{2} + \frac{52.026}{4}$$

$$D = 42959.7701 - 1495 + 13.0065$$

$$D = 41477.7766\ \text{cm}^{-1}$$

$$D = 41477.7766 \times \text{hc Joules}$$

$$D = 41477.7766 \times (6.6 \times 10^{-34}) \times (3 \times 10^{8})\ \text{Joules}$$

$$D = 8.23 \times 10^{-19}\ \text{J}$$

$$D = (8.23 \times 10^{-19}) \times (6.023 \times 10^{23})\text{J/mole}$$

$$D = 4.956 \times 10^{5}\ \text{J/mole}$$

Q. 2. The value of $\bar{\nu}_e$ and x_e for NO molecule are found to be 1904 cm^{-1} and 14.187 respectively. Calculate the dissociation energy of NO molecule.

Ans. [same as Q. 1]

Q. 3. The energy expression of an anharmonically vibrating HBr molecule is given by

$$\epsilon_v = 2649\left(v + \frac{1}{2}\right) - 45.217\left(v + \frac{1}{2}\right)^2\ \text{cm}^{-1}$$

Derive the following numerical values from this relation:

(i) Fundamental vibration

(ii) First overtone

(iii) First hot band

(iv) Calculate the maximum value of 'v' for which HBr will dissociate and thus calculate its dissociation energy.

Ans. The energy of anharmonic oscillation is given by:

$$\epsilon_v = \bar{\nu}_e\left(v + \frac{1}{2}\right) - \bar{\nu}_e x_e\left(v + \frac{1}{2}\right)^2\ \text{cm}^{-1} \quad \text{...(1)}$$

$$\epsilon_v = 2649\left(v + \frac{1}{2}\right) - 45.217\left(v + \frac{1}{2}\right)^2\ \text{cm}^{-1} \quad \text{...(2)}$$

Comparing the two equations:

$$\bar{\nu}_e = 2649\ \text{cm}^{-1}$$

and, $\bar{\nu}_e x_e = 45.217$

or, $x_e = 0.017$

(i) Fundamental vibration $= \bar{v}_{0\to1} = \bar{\nu}_e(1 - 2x_e)\ \text{cm}^{-1}$

$$\bar{v}_{0\to1} = 2649(1 - 2 \times 0.017)$$

$$= 2558.9\ \text{cm}^{-1}$$

(ii) First overtone $\bar{v}_{0\to2} = 2\bar{\nu}_e(1-3x_e)\ cm^{-1}$

$$= 2\times2649(1-3\times0.017)$$

$$= 5027.8\ cm^{-1}$$

(iii) First hot band $\bar{v}_{1\to2} = 2\bar{\nu}_e(1-4x_e)\ cm^{-1}$

$$= 2649(1-4\times0.017)$$

$$= 2468.8\ cm^{-1}$$

(iv) $$v_{max} = \frac{1}{2x_e}-\frac{1}{2}$$

$$= \frac{1}{2\times0.017}-\frac{1}{2}$$

$$= 29.41-0.5$$

$$v_{max} = 28.91 \approx 29$$

Substituting v_{max} in equaton (2), we get $\in_{max}$

$$\in_{max} = 2649\left(29+\frac{1}{2}\right)-45.217\left(29+\frac{1}{2}\right)^2$$

$$= 78145.5-39350.1$$

$$\in_{max} = 38795.4\ cm^{-1}$$

Substituting $v = 0$ in equation (2), we get $\in_0$

$$\in_0 = 2649\left(\frac{1}{2}\right)-45.217\left(\frac{1}{2}\right)^2$$

$$\in_0 = 1324.5-11.30 = 1313.19\ cm^{-1}$$

Dissociation energy $= D = \in_{max} - \in_0$

$$D = 38795.4-1313.19$$

$$D = 38482.21\ cm^{-1}$$

Q. 4. A heteronuclear molecule AB is undergoing vibrational motion, derive the expression for the highest vibrational quantum number, v_{max}.

Q. 5. The spectrum of HCl gas shows a very intense absorption at 2886 cm^{-1}, a weaker one at 5668 cm^{-1} and a very weak at 8347 cm^{-1} at 300 K. Calculate vibration frequency and dissociation energy of HCl gas.

Ans. $\bar{v}_{0\to1} = \bar{\nu}_e(1-2x_e) = 2886\,cm^{-1}$ ...(1)

$\bar{v}_{0\to2} = 2\bar{\nu}_e(1-3x_e) = 5668\,cm^{-1}$ ...(2)

Multiply equation (1) by 3 and subtracting equation (2) from it:

$$3\overline{\nu}_e - 6x_e = 8658$$
$$2\overline{\nu}_e - 6x_e = 5668$$
$$\overline{\nu}_e = 2990 \text{ cm}^{-1}$$

Substitute the value of $\overline{\nu}_e$ in equation (1):

$$\overline{\nu}_e(1-2x_e) = 2886$$
$$2990(1-2x_e) = 2886$$
$$x_e = 0.0174$$

Dissociation energy = D = $\in_{max} - \in_0$

$$D = \frac{\overline{\nu}_e}{4x_e} - \frac{1}{2}\overline{\nu}_e\left(1-\frac{1}{2}x_e\right)$$
$$D = \frac{2990}{4(0.0174)} - \frac{1}{2}\times 2990\left(1-\frac{1}{2}\times 0.0174\right)$$
$$D = \frac{2990}{4(0.0696)} - \frac{2990}{2} + \frac{2990\times 0.0174}{4}$$
$$D = \frac{2990}{0.0696} - \frac{2990}{2} + \frac{52.026}{4}$$
$$D = 42959.7701 - 1495 + 13.0065$$
$$D = 41477.7766 \text{ cm}^{-1}$$

Q. 6. Dissociation energy of $^{12}C^{16}O$ is 1.107×10^6 J/mole. The anharmoncity constant of the molecule is 5.680×10^{-3}. Find

(a) equilibrium frequency of vibration

(b) force constant of the molecule

4.15 Diatomic Vibrating Rotator

The energy required for rotation is much less than required for vibration, thus the vibrational motion is always accompanied by rotational motion and hence we do not have a pure vibrational spectra, instead we have vibrational-rotational spectra lying in the infrared region.

4.15.1 Energy of Diatomic Vibrating Rotator

The energy of vibration is larger than energy of rotation by 1000 times, thus rotational motion and vibrational motion takes place independent of each other, and they do not interact. Since no interaction, thus the total energy of the molecule may be taken as the sum of the rotational energy and the vibrational energy. Thus,

$$\in_{Total} = \in_{rot} + \in_{vib}$$

For non-rigid rotator:

$$\in_J = \overline{B}J(J+1) - \overline{D}J^2(J+1)^2 \text{ cm}^{-1}$$

For anharmonic oscillator:

$$\in_v = \left(v+\frac{1}{2}\right)\overline{\nu}_e - \left(v+\frac{1}{2}\right)\overline{\nu}_e x_e \text{ cm}^{-1}$$

Thus, total energy will be given as:

$$\in_{J,v} = \in_J + \in_v$$

$$\in_{J,v} = \left[\overline{B}J(J+1) - \overline{D}J^2(J+1)^2\right] + \left[\left(v+\frac{1}{2}\right)\overline{\nu}_e - \left(v+\frac{1}{2}\right)^2\overline{\nu}_e x_e\right]\text{cm}^{-1}$$

Now, $\overline{\nu}_e \sim 1000 \text{ cm}^{-1}$

$\overline{B} \sim 1-10 \text{ cm}^{-1}$

and, $\overline{D} \sim 10^{-3} \text{ cm}^{-1}$

Ignoring centrifugal distortion ($\overline{D}$), as it has very low value, thus

$$\in_{J,v} = \overline{B}J(J+1) + \left(v+\frac{1}{2}\right)\overline{\nu}_e - \left(v+\frac{1}{2}\right)^2\overline{\nu}_e x_e \quad ...(4.22)$$

From equation (4.22), it is clear that in the same vibrational level 'v' there will be number of rotational levels corresponding to J = 0, 1, 2, 3 ...

4.15.2 Selection Rules for Diatomic Vibrating Rotator

$$\Delta J = \pm 1$$

and $\Delta v = \pm 1, \pm 2...$

Also, $\Delta J = 0$ is not possible, since a vibrational change is always accompanied by a simultaneous rotational change.

$\Delta v = 0$ is also not possible, since this corresponds to the pure rotational transition.

4.15.3 Vibrational-Rotational Spectra for Linear Diatomic Molecule

When a transition takes place from a level with quantum numbers 'v' and 'J' to a level with quantum number v′ and J′, the energy change expressed in terms of wavenumbers will be given by

$$\Delta \in_{J,v} = \left[\overline{B}J'(J'+1) + \left(v'+\frac{1}{2}\right)\overline{\nu}_e - \left(v'+\frac{1}{2}\right)^2\overline{\nu}_e x_e\right]$$
$$- \left[\overline{B}J(J+1) + \left(v+\frac{1}{2}\right)\overline{\nu}_e - \left(v+\frac{1}{2}\right)^2\overline{\nu}_e x_e\right]$$

Restricting to the transition from $v = 0 \rightarrow v'=1$, we get

$$\Delta \in_{J,v} = \in_{J',v'=1} - \in_{J,v=0}$$

$$\Delta \in_{J,v} = \left[\overline{B}J'(J'+1) + \frac{3}{2}\overline{\nu}_e - \frac{9}{4}\overline{\nu}_e x_e\right] - \left[\overline{B}J(J+1) + \frac{1}{2}\overline{\nu}_e - \frac{1}{4}\overline{\nu}_e x_e\right]$$

$$\Delta \in_{J,v} = \left[\overline{\nu}_e - 2\overline{\nu}_e x_e\right] + \overline{B}(J'-J)(J'+J+1)$$

Putting, $\bar{\nu}_e - 2\bar{\nu}_e x_e = \bar{\nu}_0$, we have

$$\bar{v} = \Delta \in_{J,v} = \bar{\nu}_0 + \bar{B}(J' - J)(J' + J + 1) \quad ...(4.23)$$

Here, $\bar{\nu}_0$ represents the **fundamental frequency**. This wavenumber is called **band origin** or **band centre.**

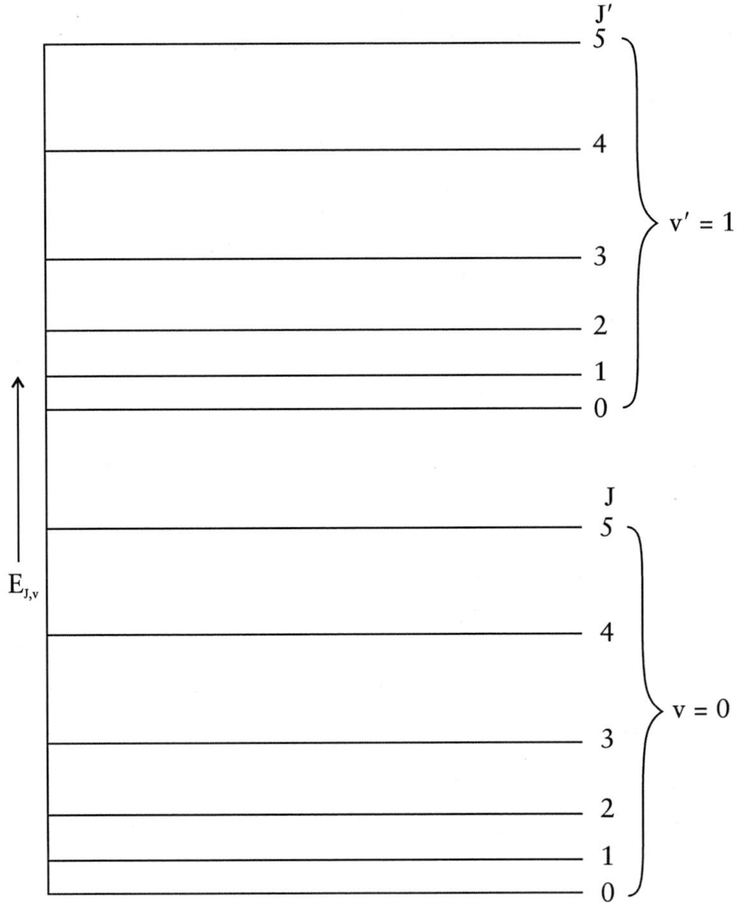

Fig. 4.14: Energy levels for vibrating rotator

(i) For transition with ΔJ = +1

J′ – J = + 1 or J′ = J +1

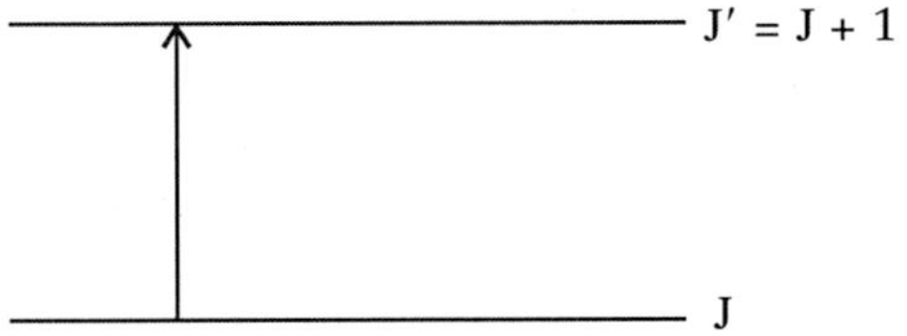

$$\Delta \in_{J,v} = \bar{\nu}_0 \oplus 2\bar{B}(J+1); \; J = 0,1,2,3... \quad ...(4.24)$$

The (+ve) sign shows that rotational energy has increased.

Thus, for $\Delta J = +1$, lines with frequency greater than the fundamental frequency are obtained. These lines are called **R-branch** of the spectrum.

(ii) For transition with $\Delta J = -1$

$J' - J = -1$ or $J = J' + 1$

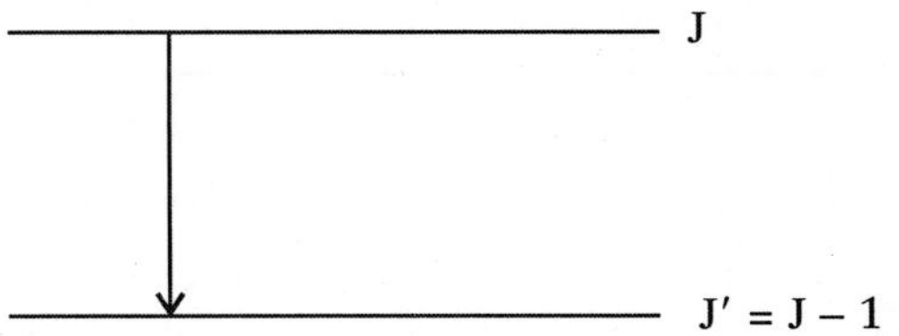

$$\Delta \in_{J,v} = \bar{\nu}_0 \ominus 2\bar{B}\,(J'+1);\; J' = 0,1,2,3... \qquad ...(4.25)$$

The (–ve) sign shows that rotational energy has decreased.

Thus, for $\Delta J = -1$, lines with frequency lower than the fundamental frequency are obtained. These lines are called **P-branch** of the spectrum.

(iii) For $\Delta J = 0$, i.e. when a vibrational transition occurs without accompanying by rotational transition (e.g. in case of NO)

$$\Delta \in_{J,v} = \bar{\nu}_0$$

i.e. a single line should be obtained at the centre of the spectrum. In actual practice, as the rotational constants of two vibrational levels are slightly different, a cluster of closely spaced lines appears at the centre. This group of lines is called the **Q-branch** of the spectrum. Usually, this type of transition is forbidden, a gap appears at the centre (Fig. 4.15).

Now, combining equations (4.24) and (4.25):

$$\Delta \in_J = \bar{\nu}_0 \pm 2\bar{B}m \text{ cm}^{-1} \qquad ...(4.26)$$

where, $m = (J + 1)$ in equation (4.24)

and $m = (J' + 1)$ in equation (4.25)

(a) As $J = 0, 1, 2, 3$ ..., this means that in equation (4.26), $m = \pm 1, \pm 2, \pm 3$ ...

(+ve sign for $\Delta J = +1$ and –ve sign for $\Delta J = -1$).

(b) Also $m \neq 0$, because it means J or J′ is negative, which is not possible. Therefore, line corresponding to the frequency $\bar{\nu}_0$ will not appear in the spectrum. This wavenumber is called **band centre** or **band origin.**

(c) As $m = \pm 1, \pm 2, \pm 3$ ..., it follows that the spectrum will consist of equally spaced lines on each side of the band centre, with spacing equal to $2\bar{B}$ between any two adjacent lines.

Keeping in view the selection rule, $\Delta J = \pm 1$, the absorption transitions then take place from level $v = 0 \rightarrow v = 1$ and the corresponding lines in the spectrum may be represented in Fig. 4.15.

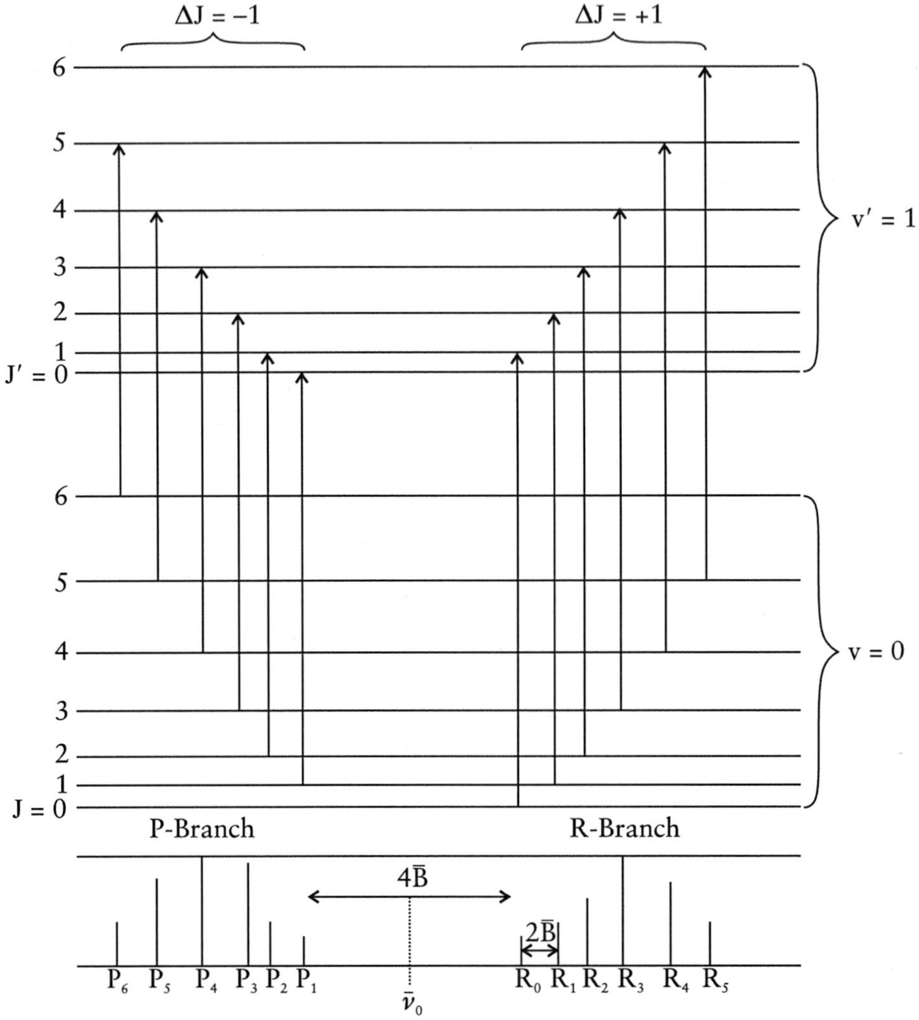

Fig. 4.15: Combined vibrational-rotational spectra for linear diatomic molecule (PR structure)

Lines arising from Δ J	−2	−1	0	+1	+2
Branch	O	P	Q	R	S

The separation of lines in either P branch or R branch provide the value of $2\overline{B}$ and hence the moment of inertia and the bond length.

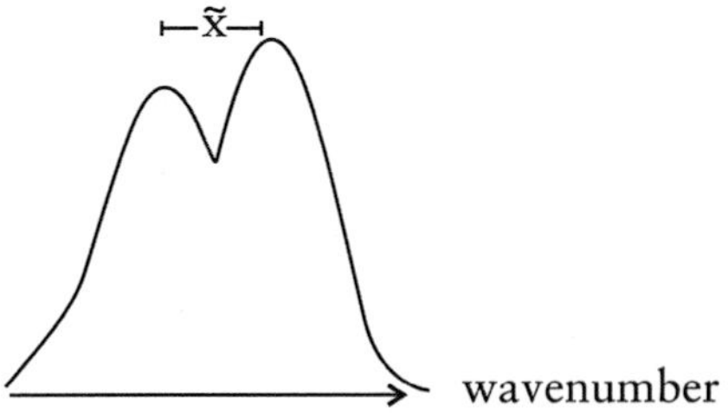

Fig. 4.16: Vibration-rotation spectrum

PROBLEMS BASED ON VIBRATING-ROTATOR

Q. 1. Show that the separation between the maxima in P and R branches of a vibration-rotation spectra of a hetro-nuclear diatomic molecule is given by

$$\sim \left(\frac{8kT\bar{B}}{hc}\right)^{1/2}$$

where the various symbols have usual meaning.

Ans. The separation between P and R branches of vibration-rotation spectrum is given by

$$\Delta \in = 2\bar{B}m + \bar{\nu}_e(1 - 2x_e)$$

where m = ±1, ±2....

m represents (J + 1) in case of ΔJ = + 1 and (J′ + 1) in case of ΔJ = –1 respectively.

Now, $$J_{max} = \sqrt{\frac{kT}{2\bar{B}hc}} - \frac{1}{2}$$

Since m = ± (J + 1), we have

$m_{max} = \pm\left(\sqrt{\frac{kT}{2\bar{B}hc}} + \frac{1}{2}\right)$ and hence the maxima of P and R branches will be located at

$$\Delta \in = \pm 2\bar{B}\left(\sqrt{\frac{kT}{2\bar{B}hc}} + \frac{1}{2}\right) + \bar{\nu}_e(1 - 2x_e)$$

For P - branch :

$$\Delta \in_P = \bar{\nu}_e(1 - 2x_e) - 2\bar{B}\left(\sqrt{\frac{kT}{2\bar{B}hc}} + \frac{1}{2}\right)$$

For R - branch :

$$\Delta \in_R = \bar{\nu}_e(1 - 2x_e) + 2\bar{B}\left(\sqrt{\frac{kT}{2\bar{B}hc}} + \frac{1}{2}\right)$$

Hence, the separation between them is

$$\tilde{x} = \Delta \in_R - \Delta \in_P = 4\bar{B}\left(\sqrt{\frac{kT}{2\bar{B}hc}} + \frac{1}{2}\right)$$

$$\tilde{x} = \left(\sqrt{\frac{8kT\bar{B}}{hc}} + 2\bar{B}\right)$$

The factor $2\bar{B}$ is much smaller than the factor $\sqrt{8kT\bar{B}/hc}$ and hence the above relationship can be approximated to

$$\tilde{x} \approx \left(\sqrt{\frac{8kT\bar{B}}{hc}}\right)$$

(or) $$\bar{B} \approx \frac{hc\tilde{x}^2}{8kT}$$

Q. 2. Sketch rotational-vibrational spectrum of a diatomic molecule. Derive expressions for energy levels of P and R branches of spectra.

Q. 3. What are P, Q and R branches in vibrational-rotational spectrum?

4.16 Influence of Nuclear Spin on Vibrational-Rotational Spectra

In pure rotational and pure vibrational spectroscopy, we did not talk about O_2, H_2, $HC \equiv CH$ molecules because they do not have dipole moment. Although they are IR and microwave inactive, they have vibrational-rotational spectra.

For molecules having a centre of symmetry, the nuclear spin also affect the population in the rotational levels. If a molecule has centre of symmetry, then its total wavefunction (ψ_{total}) should be symmetric (S) or antisymmetric (A).

The total wavefunction is symmetric or antisymmetric depends on whether the molecule is **fermion** or **boson.**

Fermions: $I = \frac{1}{2}, 1;$ E.g.: $H_2, N_2, HC \equiv CH$

Boson: $I = 0;$ E.g.: CO_2, O_2

where, I is the **nuclear spin quantum number.**

1. Bosons:

$$\psi_{Total} = \psi_{electronic} \cdot \psi_{vib} \cdot \psi_{rot} \cdot \psi_{nuclear\ spin\ (ns)}$$

Since $\psi_{electronic}$ is independent of change of nuclear spin, therefore it is not considered.

The CO_2 and O_2 molecule are boson with I = 0, thus 2I + 1 = 1 and hence $\psi_{ns} = 1$

So, ψ_{Total} depends only on ψ_{vib} and ψ_{rot},

$\therefore \ \psi_{Total} = \psi_{vib} \cdot \psi_{rot}$

(a) When v = 0, $\psi_{vib} = ⓢ$ (S = Symmetric)

For a boson, ψ_{Total} must by symmetric and to make ψ_{Total} to be symmetric, ψ_{rot} has to be symmetric

$$\psi_{Total} = \psi_{vib} \cdot \psi_{rot}$$

$$\psi_{Total} = S \times S = ⓢ$$

Thus, only symmetrical rotational levels are allowed, i.e. J = 0, 2, 4, 6...

So the alternate rotation bands are missing in PR band.

Thus, the spacing between the rotational lines is not $2\overline{B}$ but it is $4\overline{B}$ in the spectra.

(b) When v = 1, ψ_{vib} = Ⓐ (A = Asymmetric)

For ψ_{Total} to be symmetric, ψ_{rot} has to be antisymmetric

$$\psi_{Total} = \psi_{vib} \cdot \psi_{rot}$$

$$\psi_{Total} = A \times A = Ⓢ$$

Thus, only asymmetrical rotational levels are allowed, i.e. J = 1, 3, 5...

Note: In case of Bosons, every alternative rotational level is unoccupied.

By removing the centre of symmetry, $^{16}O—C—O^{18}$, the separation between alternative rotational lines is $2\overline{B}$ only, i.e. there is no effect on spectrum due to nuclear spin.

2. Fermions:

$$\psi_{Total} = \psi_{electronic} \cdot \psi_{vib} \cdot \psi_{rot} \cdot \psi_{nuclear\ spin\ (ns)}$$

Since $\psi_{electronic}$ is independent of change of nuclear spin, therefore it is not considered. The HC ≡ CH molecule is a fermion with I = 1/2, thus 2I + 1 = 2 and hence $\psi_{ns} \neq 1$. So, ψ_{Total} depends on ψ_{vib}, ψ_{rot} and ψ_{ns}

$\therefore \; \psi_{Total} = \psi_{vib} \cdot \psi_{rot} \cdot \psi_{ns}$

(a) When v = 0, and J = 0, 2, 4, 6...

$$\psi_{vib} = Ⓢ \text{ and } \psi_{rot} = Ⓢ$$

For a fermion, ψ_{Total} must be antisymmetric and to make ψ_{Total} to be antisymmetric, ψ_{ns} should be antisymmetric (A) in this case.

Possible nuclear spins (ns) wavefunction are:

1. αα
2. αβ ⎤ αβ and βα are indistinguishable
3. βα ⎦
4. ββ

Since αβ and βα are indistinguishable, therefore combinations of these are taken.

Thus, the four nuclear spin wavefunction are:

1. αα → Ⓢ
2. αβ + βα → Ⓢ
3. αβ – βα → Ⓐ
4. ββ → Ⓢ

This nature of symmetric and asymmetric is because on exchange of electrons we observe that

1. $\alpha(1)\alpha(2) \xrightarrow[\text{electrons}]{\text{exchange of}} \alpha(2)\alpha(1) \Rightarrow$ Ⓢ
2. $\alpha(1)\beta(2) + \beta(1)\alpha(2) \rightarrow \alpha(2)\beta(1) + \beta(2)\alpha(1) \Rightarrow$ Ⓢ
3. $\alpha(1)\beta(2) - \beta(1)\alpha(2) \rightarrow \alpha(2)\beta(1) - \beta(2)\alpha(1) \Rightarrow$ Ⓐ
4. $\beta(1)\beta(2) \rightarrow \beta(2)\beta(1) \Rightarrow$ Ⓢ

$$\psi_{Total} = \psi_{vib} \times \psi_{rot} \times \psi_{ns}$$
$$\downarrow \quad \downarrow \quad \downarrow$$
$$\psi_{Total} = \textcircled{S} \times \textcircled{S} \times \textcircled{A} = \textcircled{A}$$

The only one nuclear spin wavefunction, which is antisymmetric is

$$\psi_{ns} = \alpha\beta - \beta\alpha$$

Thus, only one degenerate level possible.

(b) When v = 0 and J = 1, 3, 5, 7 ...

$\psi_{vib} = \textcircled{S}$ and $\psi_{rot} = \textcircled{A}$

For a fermion, ψ_{Total} must be antisymmetric and to make ψ_{Total} to be antisymmetric, ψ_{ns} should be symmetric (S) in this case.

$$\psi_{Total} = \psi_{vib} \times \psi_{rot} \times \psi_{ns}$$
$$\downarrow \quad \downarrow \quad \downarrow$$
$$\psi_{Total} = \textcircled{S} \times \textcircled{A} \times \textcircled{S} = \textcircled{A}$$

There are three nuclear wavefunction, which are symmetric, i.e.

$$\alpha\alpha,\ \beta\beta \text{ and } (\alpha\beta + \beta\alpha)$$

So degeneracy = 3, as there are three wavefunctions with same energy. Thus, we can combine either of these nuclear wavefunction to make Ψ_{Total} to be antisymmetric.

Value of J	Degeneracy (g_J)
0	1
1	3
2	1
3	3

Thus in fermions (like HC ≡ CH) alternate rotational lines have intensity in the ratio 1:3.

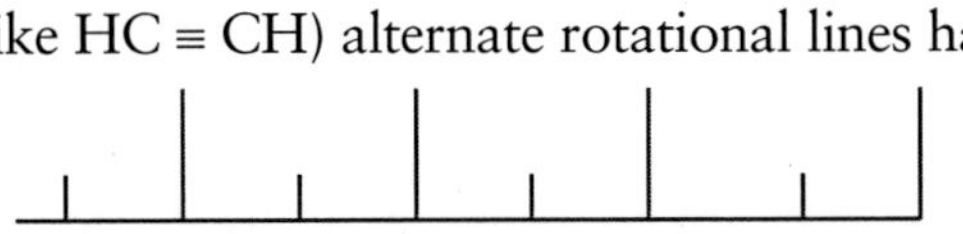

NUMERICALS BASED ON EFFECT OF NUCLEAR SPIN ON VIBRATIONAL-ROTATIONAL SPECTRA

Q. 1. Why do alternate lines appear in the vibrational-rotational spectra of CO_2?

Ans. Alternate lines is due to the existence of nuclear spin.

Q. 2. The line spacing (on each side of the band origin $\overline{\nu}_0$) in PR spectrum of CO_2 is $4\overline{B}$ instead of the usual $2\overline{B}$ as observed in case of HCN. Explain briefly.

Ans. Nuclear spin effects the population of rotational levels, if the molecule has centre of symmetry. As CO_2 molecule has centre of symmetry, every alternate J level is unoccupied. The line spacing in PR branches is generally $2\overline{B}$. Since alternate rotational levels are missing in CO_2, the line spacing becomes $4\overline{B}$. Since HCN does not have centre of symmetry, so no such rotational levels are missing in HCN.

4.17 Application of Infrared Spectroscopy to Organic Molecules

Discussion: IR spectroscopy is the most common technique used by chemists to determine the functional group present in the organic compounds. IR spectroscopy deals with the recording of the absorption of radiations in the infrared region by a molecule. The region of interest for IR spectroscopy is the mid-IR region, i.e. wavelength from 2.5 to 15 μ (4000–667 cm^{-1}), since most diagnostically useful vibration occurs in this region.

The IR region is commonly divided in three regions

	Region	Wavenumber or absorption frequency (cm^{-1})
1.	Near IR	12500–4000
2.	Mid IR	4000–667
3.	Far IR	667–50

IR spectra originates from transitions in vibrational and rotational energy levels within a molecule on absorption of IR radiations. When a molecule absorbs IR radiations below 100 cm^{-1}, transition in its rotational energy levels occurs and discrete lines are formed in the spectrum due to molecular rotation. But when a molecule absorbs radiation in the range 100–10,000 cm^{-1}, the transitions in its vibrational energy level occurs. The energy difference between rotational energy levels of a molecule are far less than that between its vibrational energy levels. Hence, a single transition in a vibrational energy level is accompanied by a large number of transitions in rotational energy levels, resulting in appearance of spectra as vibration-rotational bands instead of discrete lines, i.e. IR spectra of a molecule results due to the transitions between different vibrational energy levels. The absorption of IR radiations causes the various bonds in a molecule to stretch and bend with respect to one another.

The absorption, IR radiation is expressed in terms of wavenumber ($\bar{v}$) or absorption frequency (ν) of wavelength (λ). IR spectra of organic compounds are plotted as % transmittance against wavenumber ($\bar{v}$).

4.17.1 Finger Print Region

One of the important functions of infra-red spectroscopy is to determine the identity of two compounds. Two identical compounds have exactly the same spectra when run in the same medium under similar conditions. The region below 1500 cm^{-1} is rich in many absorptions which are caused by bending vibrations and those resulting from the stretching vibrations of C—C, C—O and C—N bonds. In a spectrum, the number of bending vibrations is usually more than the number of stretching vibrations. The said region is usually rich in absorption bands and shoulders. It is called **Finger print region**. Some molecules containing the same functional group show similar absorption above 1500 cm^{-1} but their spectra differ in finger print region. The identity of an unknown compound can also be revealed by comparing its infra-red spectrum with a set of spectra of known compounds under identical conditions. The identity of infra-red spectra of two compounds is much more characteristic than the comparison of their many physical properties. It is not possible to distinguish between two enantiomers even if their spectra are run with the same machine under exactly identical conditions such as sampling, scan speed, etc. Also it is not possible to distinguish the infra-red spectra of straight chain alkanes containing 30 carbon atoms or more and it becomes necessary to make distinction with the help of their mass spectra. Finger print region can be subdivided into three regions as follows:

(i) 1500–1350 cm^{-1}
(ii) 1350–1000 cm^{-1}
(iii) Below 1000 cm^{-1}

4.17.2 Stretching and Bending Vibrations

Atoms in a molecule can be represented as balls of different masses and covalent bonds between them an tiny springs attached to the balls (Fig. 4.17). Atoms in the molecule are not still, but vibrate and the two types of fundamental molecular vibrations are:

(1) Stretching vibration, (2) Bending vibration

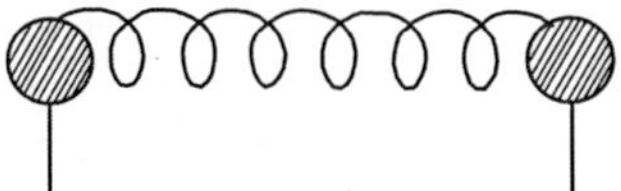

Fig. 4.17: Representation of Molecule

1. **Stretching Vibration:** In this mode of vibration, the distance between the two atoms increases or decreases, but the atoms remain in the same bond axis. These are of two types:
   (a) **Symmetric Stretching:** In symmetric stretching, the movement of atoms with respect to a particular atom is in the same direction [Fig. 4.18(a)]

ν_s (Symmetric stretching) (a) ν_{as} (Asymmetric stretching) (b)

Fig. 4.18: Types of stretching vibrations of $-CH_2-$ group

   (b) **Asymmetric stretching:** In asymmetric stretching, one atom approaches the central atom while the other departs from it [Fig. 4.18(b)].

2. **Bending Vibrations:** In this mode of vibration, the position of atoms change with respect to the original bond axis. These are of four types:
   (a) **Scissoring:** In this, the movement of atoms takes place in opposite direction with change in their bond axis and bond angle they form with the central atom and thus the two atoms approach each other [Fig. 4.19(a)].
   (b) **Rocking:** In this, the movement of atoms is in the same direction with change in their bond axis [Fig. 4.19(b)].
   Scissoring and Rocking are in-plane bending.
   (c) **Twisting:** In this, one of the atoms moves up the plane while other moves down the plane with respect to the central atom [Fig. 4.19(c)].

(d) **Wagging:** In this, both atoms move up and below the plane with respect to the central atom [Fig. 4.19(d)].

Twisting and Wagging are out-of-plane bending.

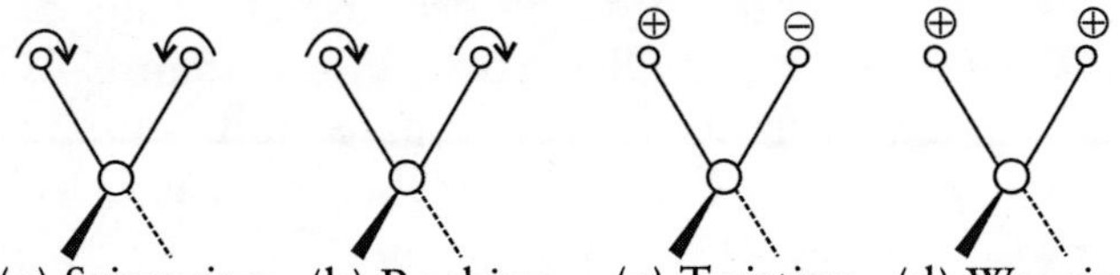

(a) Scissoring (b) Rocking (c) Twisting (d) Wagging

Fig. 4.19: In-plane bending of —CH_2 group and out-of-plane bending of —CH_2 group

Bending vibrations of a bond requires lesser energy than stretching vibration, hence absorption bands for bending vibration appear at lower wavenumber/absorption frequency than stretching of the same bond.

Absorption frequency required for various vibration :	Asymmetric vibration	>	Symmetric vibration	>	Bending vibration

IR radiations are absorbed when the oscillating dipole moment caused due to molecular vibration interacts with the oscillating electric field of the IR beam. Absorption bands in IR spectra appears only when molecular vibrations produces change in the dipole moment of the molecule. It is not necessary for a molecule to have permanent dipole moment for IR absorption. Larger the change in the dipole moment, higher is the intensity of absorption.

4.17.3 Hook's Law (Calculation of Stretching Vibrational Frequencies)

If two bonded atoms are considered to be two vibrating masses connected by a spring (Fig. 4.17) then the stretching vibrational frequency of a bond can be calculated by the application of Hooke's law.

$$\nu = \frac{1}{2\pi c}\sqrt{\frac{k_f}{\mu}}$$

where ν is the wavenumber or vibrational frequency (cm^{-1})
c is the velocity of radiation (3×10^8 cm/sec.)
μ is the reduced mass (g)

$$\mu = \frac{m_1 \times m_2}{m_1 + m_2}$$

where m_1 and m_2 are the masses of atoms linked to a particular bond.
k_f is the force constant of the bond (dynes/cm). Value of k_f increases with the strength of bond.

From Hooke's law, its clear that value of ν depends on two factors:

1. **Bond Strength:** As the force constant increases, i.e. bond strength increases, value of vibrational frequency increases or wavenumber increases or wavelength decreases,

e.g. —C ≡ C—bond stretching absorbs at higher frequencies than $-\overset{|}{C}=\overset{|}{C}-$ bond stretching.

2. **Reduced Mass:** As the value of reduced mass decreases, the value of vibrational frequency increases, e.g. O—H bond has smaller reduced mass than C—C bond, hence O—H stretching absorbs at higher frequency as compared to C—C stretching vibrations and same is true for C—H and C—C stretching vibrations, as C—H has smaller reduced mass than C—C, former appear at higher frequency.

4.17.4 Instrumentation

Double-beam IR spectrophotometer consists of following main parts (Fig. 4.20):

1. **Source of Radiation:** Coil of nichrome wire or a rod of silicon carbide or a Nernst filament upon electrical heating upto 1000–1800°C, emits radiations in the infrared region. The radiation from the source is divided into reference and sample beam of equal intensity by a beam divider.
2. **Reference and Sample Cells:** Reference and sample beam pass through the reference and sample cells respectively. Alkali metal halides such as NaCl, NaBr, KCl and KBr are used as windows of reference and sample cells as these are transparent to the IR radiations, whereas glass and quartz cannot be used as windows of cells because they absorb strongly in most of the IR region.
3. **Photometer:** Reference beam passes through the attenuator and sample beam through the comb. The photometer combines and reference and sample beam into a single beam of alternating segments.
4. **Monochromator:** The single beam from photometer passes through the monochromator which disperses it into various frequencies and sends individual frequency bands to the detector.
5. **Detector and Amplifier:** The common detector devices are photocell, photomultiplier, photographic plane and thermocouple. Detector measures the radiant energy by means of its heating effect that produces current. From detector, an alternating current flows to the amplifier where it is amplified and relayed to the recorder.
6. **Recorder:** Recorder records the IR spectra as a plot of wavenumber of absorbed radiations against intensity of absorbance A or transmittance T.

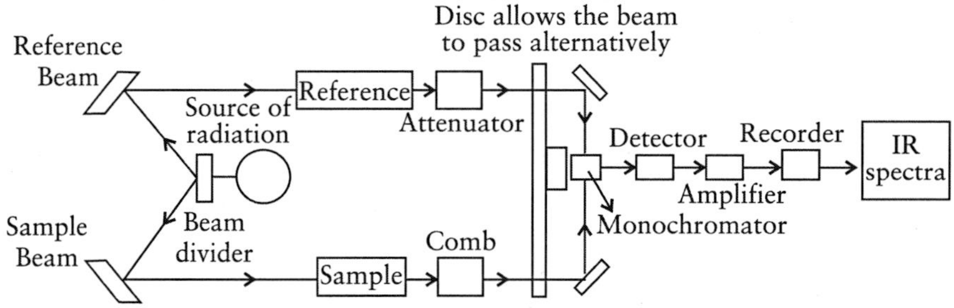

Fig. 4.20: Schematic diagram of a double-beam IR spectrophotometer

Sample Handling: Infrared spectra of compounds can be recorded in a pure liquid form, in the solution form, in the vapour phase or in the solid state. As water absorbs in IR region (~3710 cm^{-1} and ~1630 cm^{-1}), so the sample should be dry.

1. **In liquid form (as a liquid film):** This is the simplest sample handling technique. A drop of pure liquid is placed between the two flat plates of NaCl to give a thin film (plates of NaCl are used as NaCl is IR transparent).
2. **In solution form:** In this technique, two cells of 1 mm thickness made up of NaCl are taken. In one cell (the reference cell), pure solvent is taken and placed in the reference beam. In the another cell 1–5% solution of the sample (in the same solvent). The solvent taken must not absorb in the region of interest, e.g. CCl_4, CS_2 and $CHCl_3$.
3. **In vapour phase:** The technique is limited because of very low vapour pressure of organic compounds. The vapour of sample is introduced in a cell whose walls are made up of NaCl.
4. **In solid state:**
   (a) **As a mull or paste:** About 3–5 mg of solid sample is ground in a mortar with two drops of mulling agent. The mull or paste is examined as a thin film between two NaCl flat plates. The most common mulling agent is nujol (a mixture of saturated long chain hydrocarbons). To avoid interference of C—H bands with the spectrum, another mulling agent, i.e. hexachlorobutadiene may be used.
   (b) **As a pressed disc:** The solid sample is mixed with 50 mg to KBr in a dry mortar and pestle (KBr is IR transparent). The mixture is pressed with special dies under high pressure and made into pellets or discs. KBr gives better spectra than nujol mull as it is throughout transparent in the IR region.

4.17.5 Factors Affecting Vibrational Frequency

Any factor which affects the force constant (k_f) of a bond will affect its stretching vibrational frequency (or absorption frequency). Usually, the force constant of a bond changes due to steric effects, electronic effects, hydrogen bonding in the molecule and hence the absorption frequencies are also shifted from their normal values (calculated using Hook's law) and are not exactly equal to the experimental observed values.

Some of the factors responsible for shifting the vibrational frequencies from their normal values are:

1. **Physical state of the compound:** When the IR spectra of the same compound is taken in different states (solid, liquid, or gas), then a compound absorbs at higher absorption frequency in the vapour state than in the solid or liquid state.
2. **Electronic effects:** It includes inductive effect, mesomeric effect (resonance), conjugation, etc.

   (a) **Inductive effect:**
   (i) **(+I) effect:** Presence of electron donating group (like alkyl group), results in the lengthening or weakening of bond, force constant of bond decreases and absorption frequency decreases. Like:

Absorption frequency for C=O bond, $\nu_{C=O}$ (cm^{-1}) :	$CH_3 \rightarrow \underset{\underset{O}{\|}}{C} \leftarrow CH_3$	<	$CH_3 \rightarrow \underset{\underset{O}{\|}}{C} - H$	<	$H - \underset{\underset{O}{\|}}{C} - H$
	1715		1745		1750

It is seen from the above case, as the number of —CH_3 group (alkyl group is electron donating) attached to $>C=O$ group from increases, and the C = O absorption frequency decreases.

Aldehydes absorb at higher frequencies than ketones

(ii) **(–I) effect:** Presence of electron withdrawing group (like halogen atom), result in the strengthening of the bond, force constant of bond increases and absorption frequency increases. Like:

Absorption frequency for C=O bond, $\nu_{C=O}$ (cm^{-1}) :	$CH_3 - \underset{\underset{O}{\|}}{C} - CH_3$	<	$CH_3 - \underset{\underset{O}{\|}}{C} - CH_2 \rightarrow Cl$	<	$CH_3 - \underset{\underset{O}{\|}}{C} - HC \begin{smallmatrix} \nearrow Cl \\ \searrow Cl \end{smallmatrix}$
	1715		1725		1740

(b) **Conjugation:** In general, the conjugation cause weakening of bond and absorption frequency decreases.

(i) **Conjugation due to aromatic ring (Mesometric effect or Resonance effect):** Conjugation with aromatic ring, result in the lengthening or weakening of bond, and absorption frequency decreases. Like:

Absorption frequency for C=O bond, $\nu_{C=O}$ (cm^{-1}) :	$C_6H_5 \rightarrow \underset{\underset{O}{\|}}{C} - CH_3$	<	$CH_3 - \underset{\underset{O}{\|}}{C} - CH_3$
	1693		1715

(ii) **Conjugation due to double bond (or α, β-unsaturation):** Conjugation with a double bond, results in the weakening of bond, and absorption frequency decreases. Like:

Absorption frequency for C=O bond, $\nu_{C=O}$ (cm^{-1}) :	$H_2C{=}CH - \underset{\underset{O}{\|}}{C} - CH_3$	<	$CH_3 - \underset{\underset{O}{\|}}{C} - CH_3$
	1706		1715

Conjugation due to aromatic ring is more than the conjugation due to a double bond. Thus, the lowering in absorption frequency is more in case of conjugation with aromatic ring.

Absorption frequency for C=O bond, $\nu_{C=O}$ (cm^{-1}) :	$C_6H_5 - \underset{\underset{O}{\|}}{C} - CH_3$	<	$\boxed{CH_2{=}CH} - \underset{\underset{O}{\|}}{C} - CH_3$
	1693		1706

(iii) **Conjugation due to lone pair of electrons:** The conjugation due to lone pair of electrons on an atom depends on the electronegativity of an atom. Lesser the electronegativity of an atom, more easily the lone pair of electron of the atom can participate in conjugation, and cause decrease in absorption frequency of bond with which it is in conjugation. Like:

Absorption frequency for C═O bond, : $H_2\ddot{N}-C(C_6H_5)=\ddot{O}$ (Amide) < $CH_3-\ddot{O}-C(C_6H_5)=\ddot{O}$ (Ester)

$\nu_{C=O}$ (cm^{-1}): 1693 (Amide), 1730 (Ester)

In this case, nitrogen atom is less electronegative than oxygen atom, so lone pair present on nitrogen atom participates in conjugation with C = O group more easily than oxygen atom and hence absorption frequency is much less in amides than in ester.

3. **Hydrogen Bonding:** In general, the H-bonding, causes weakening of bond and absorption frequency decreases. IR spectroscopy is a useful technique to distinguish between Intermolecular hydrogen bonding (IMHB) and Intramolecular hydrogen bonding.

(a) **For Intermolecular Hydrogen Bonding:** Broad bands seen in IR spectra. These broad bands are seen only at high concentration (or condensed form) of the compound because in condensed phases molecules are associated as dimers or polymers through IMHB. Whereas, in dilute solution or in vapour state or in inert solutions, molecules exist in free state and no IMHB and high absorption frequency. The absorption frequencies difference between free and associated molecule is larger in case of IMHB. Like:

(i) Absorption frequency for N — H bond, : $R-NH_2$ (Concentrated solution) (IMHB present) < $R-NH_2$ (Dilute solution) (No IMHB)

ν_{N-H} (cm^{-1}): 3300, 3500

Due to IMHB in amines in condensed form, the N—H bond becomes weaker and hence absorption frequency decreases.

(ii) Absorption frequency for O — H bond, : $R-OH$ (Concentrated solution) (IMHB present) < $R-OH$ (Dilute solution) (No IMHB)

ν_{O-H} (cm^{-1}): 3350, 3650

Also nitrogen is less electronegative than oxygen atom, thus IMHB is more stronger in case of alcohols than in amines. Thus, the decrease in absorption frequencies from dilute to concentrated solution is more in case of alcohol.

(iii) Carboxylic acids in condensed phase, exists as dimers and there is formation of IMHB between the molecules. Due to formation of dimer, there is weakening of C = O and O — H bonds and absorption frequency of both bonds decreases.

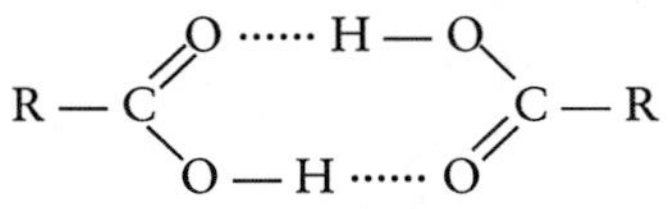

Dimer of carboxylic acid

(b) **For Intramolecular Hydrogen Bonding:** Sharp and well-defined bands seen in IR spectra. The intramolecular H-bonding is seen in dilute solution. The absorption frequency difference between free and associated molecule is smaller in case of intramolecular hydrogen bonding than that in IMHB, which indicates that intramolecular H-bonding is weaker than IMHB.

E.g. Staggered syn conformation of ethylene glycol in dilute solution, shows two ν_{O-H} bands at 3644 cm^{-1} and 3612 cm^{-1}. The band at 3644 cm^{-1} is due to free O—H bond and the band at 3612 cm^{-1} is due to O—H bond which is involved in intramolecular hydrogen bonding. The same molecule at higher concentration shows a third strong and broad band due to O—H at 3477 cm^{-1} which is involved in IMHB.

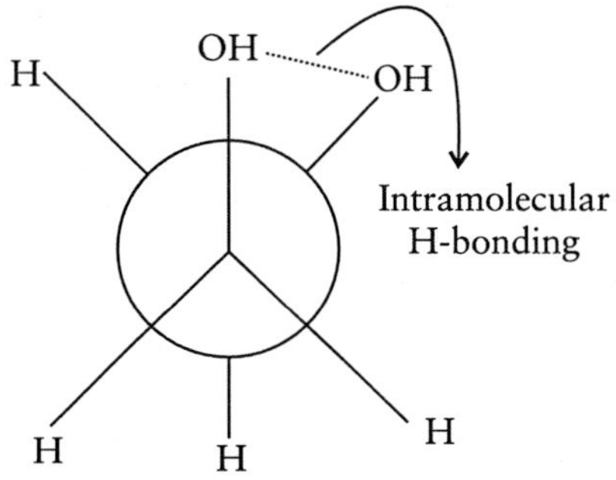

ν_{O-H} (cm^{-1}) : O — H involved in Intermolecular H-bonding < O — H involved in Intramolecular H-bonding < Free O — H

Usually, non-associating solvents like CS_2, $CHCl_3$, CCl_4, etc., are used to prepare the solution of the compound whose IR spectra is to be recorded because these solvents does not absorb any radiation in the region of interest and does not influence O—H and N—H bands. Whereas the solvents like benzene, acetone, etc. influence it to a considerable amount.

4. **Angle Strain in Cyclic Compounds:** In cycloalkanes, the value of C—H stretching absorption frequency increases with the increase in angle of strain in the ring:
   (a) The C—H stretching absorption frequency for cycloproane occurs at higher absorption frequency as compared to that in cyclohexane.

Absorption frequency for C — H bond, : 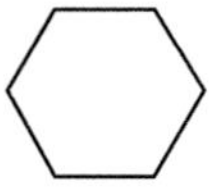 < 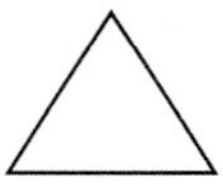

	Cyclohexane	Cyclopropane
ν_{C-H} (cm^{-1})	2840	3040

   (b) In cyclic ketones, the C = O absorption frequency increases as the size of the ring decreases and ring strain increases.

Absorption frequency for C=O bond, : 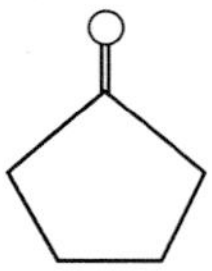 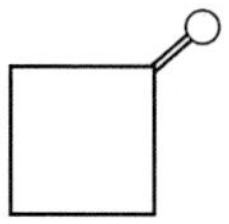 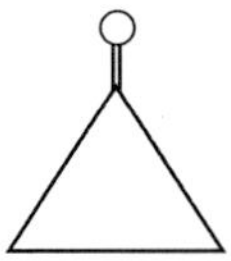

	Cyclopentanone		Cyclobutanone		Cyclopropanone
$\nu_{C=O}$ (cm^{-1})	1746	<	1788	<	3040

(c) α, β– unsaturation in cyclic ketones causes the decrease in absorption frequency of C = O bond.

Absorption frequency for C=O bond, : $\nu_{C=O}$ (cm^{-1})

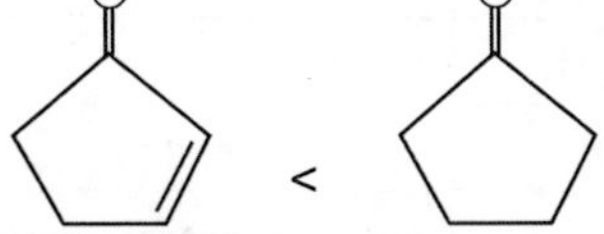

5. Accidental Degeneracy and Fermi Resonance

In infrared spectrum, absorption bands are spread over a wide range of frequencies. The complete IR spectrum of a polyatomic molecule, therefore, consists of the highly intense fundamental vibrational frequencies and the less intense overtones as well combination bands. Sometimes it may happen that a fundamental mode of vibration interact with an overtone or a combination band due to same symmetry and comparable energy, i.e. **accidental degeneracy**. As a result, a type of resonance takes place and during this the molecule transfers energy from fundamental to overtone. Thus, intensity of overtone or the combination band which is otherwise low is enhanced by a resonance phenomenon. This type of resonance is known as **Fermi Resonance.**

This phenomenon was first observed by **E. Fermi**. The classical example of Fermi Resonance is seen in CO_2 molecule.

(1)

IR Spectra of CO_2 molecule					
ν_3	667 cm^{-1}	Bending	π_u	IR active	—
ν_1	1330 cm^{-1}	Symmetric stretching	Σ_g^+	IR inactive	Raman active
ν_2	2350 cm^{-1}	Asymmetric stretching	Σ_g^+	IR active	—

Fundamental frequency = ν_1 = 1330 cm^{-1}; Σ_g^+

Overtone of $\nu_3 = 2\nu_3 = 2 \times 667 = 1334$ cm^{-1}; $\pi_u \times \pi_u = \Sigma_g^+$

Since ν_1 and $2\nu_3$, both have the same symmetry as Σ_g^+ and have almost same energy, thus there will be interaction between them. Thus, 1330 cm^{-1} and 1334 cm^{-1} interacts with each other and appears as two lines of equal intensity at 1285 cm^{-1} and 1385 cm^{-1} because of distribution of intensity of 1330 cm^{-1}.

Also the two bands of CO_2 (1330 cm^{-1} and 1334 cm^{-1}) are not observed in IR spectra but they are observed in the Raman spectrum at 1285 cm^{-1} and 1385 cm^{-1} respectively (i.e. 50 cm^{-1} below and 50 cm^{-1} above). Their mean is near 1330 cm^{-1}.

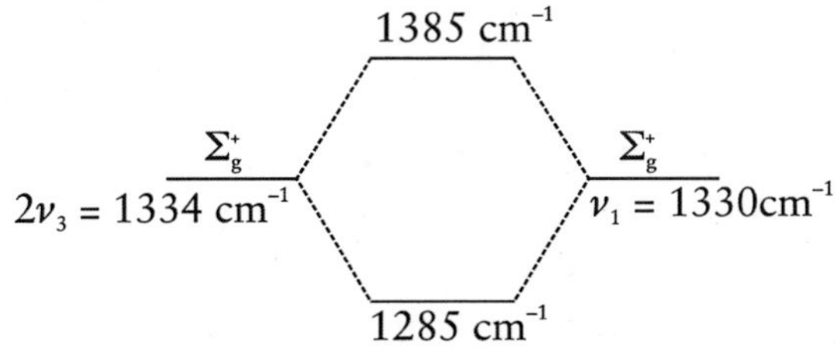

(2)

IR Spectra of H_2O molecule				
ν_2	1595 cm^{-1}	Bending	A_1	IR active
ν_1	3600 cm^{-1}	Symmetric	A_1	IR inactive
ν_3	3700 cm^{-1}	Asymmetric	B_2	IR active

As the energy of ν_1 and ν_3 are almost equal, so it may be mislead to the conclusion that Fermi resonance occurs between them. But this does not happen as ν_1 and ν_3 have different symmetry, i.e. A_1 and B_2, thus no interaction.

But the interaction between ν_1 and overtone of ν_2 ($2\nu_2$) might be possible but the interaction is less.

Fundamental frequency = ν_1 = 3600 cm^{-1}; A_1

Overtone of $\nu_2 = 2\nu_2 = 2 \times 1595 = 3190$ cm^{-1}, $A_1 \times A_1 = A_1$.

The difference between, 3600 – 3190 = 410 cm^{-1}, is quite large and hence less interaction.

Although they have the same symmetry as A_1, but energy is not comparable, thus less interaction.

PROBLEMS ON FACTORS AFFECTING VIBRATIONAL FREQUENCY

Q. 1. What is the effect of hydrogen bonding in vibrational frequencies?

Q. 2. Explain briefly Fermi Resonance, with examples.

Q. 3. Which region of the IR spectrum is referred to the finger print region? Why it is called finger print region and what information can be obtained from it?

Q. 4. What are "fingerprint bands" in reference to IR spectra of organic molecules?

Q. 5. What is accidental degeneracy? Explain.

Q. 6. How can IR be helpful in distinguishing between isomeric:

(i) aldehyde and ketone

(ii) alcohol and ketone

4.18 Modes of Vibrations of Atoms in Polyatomic Molecules

The mass of the atom is concentrated in its nucleus and the nucleus is very small size, thus the atom may be considered as **mass points**. To represent each mass point, we require three coordinates. If there are N atoms in a molecule, then to represent the position of N mass points in space, we require 3N co-ordinates. The number of coordinates required to specify the position of all the mass points, i.e. atoms in a molecule is called the number of **degrees of freedom.**

In a polyatomic molecule, each atom is having three degrees of freedom in three directions, which are perpendicular to one another. Consequently, a polyatomic molecule requires three times as many degrees of freedom as the number of its atoms. Thus, a molecule of N atoms has 3N degrees of freedom.

According to kinetic theory of matter, when thermal energy is absorbed by a molecule, it is stored within the molecule in the form of:

(i) Translational motion of the molecule.

(ii) Internal movement of the atoms within the molecule, i.e. **rotational motion** and **vibrational motion.**

The translational motion of the molecule means the motion of the centre of mass of the molecule as a whole. The centre of mass of the molecule can be represented by three coordinates. Hence, there are three translational degrees of freedom. The remaining (3N–3) coordinates represent the internal degree of freedom, i.e. rotational and vibrational degree of freedom.

In a **linear molecule**, only two degrees of freedom are required to describe rotation and thus remaining degrees of freedom for the linear molecule are equal to (3N–5) which describe the motion of nuclei with respect to one another thus (3N–5) represent the vibrational degrees of freedom. The vibrational degrees of freedom represent the number of independent vibrational modes (stretching and bending) that can occur in a molecule. In a linear molecule, out of (3N–5) vibrational degrees of freedom, (N–1) of the vibrations are bond stretching vibrations and (2N–4) are bending vibrations.

In a **non-linear molecule,** three degrees of freedom are required to describe rotation and thus remaining degrees of freedom for the non-linear molecule are equal to (3N–6) which describe the motion of nuclei with respect to one another and thus (3N–6) represent the vibrational degrees of freedom. Out of these (3N – 6) vibrational degrees of freedom, (N–1) of the vibrations are bond stretching vibrations and (2N–5) are bending vibrations.

Examples on Calculation of Vibrational Degrees of Freedom

Number of atoms in a molecule = N

Total degree of freedom = 3N

Linear Molecule	Translational	Rotational	Vibrational
	3	2	3N–5
			Stretching = N – 1
			Bending = 2N –4
Non-linear molecule	3	3	3N–6
			Stretching = N – 1
			Bending = 2N –5

1. **Linear Diatomic Molecule**

   HCl; Linear; N = 2

   Vibrational degrees of freedom = 3N – 5 = 3(2) – 5 = 1

   Thus, there is one fundamental vibration in HCl.

2. **Linear Polyatomic Molecule**

(a) CO_2; Linear; N = 3

   Vibrational degrees of freedom = 3N – 5 = 3(3) – 5 = 4

   Thus, there are four fundamental vibrations in CO_2 molecule.

(i) Symmetric stretching

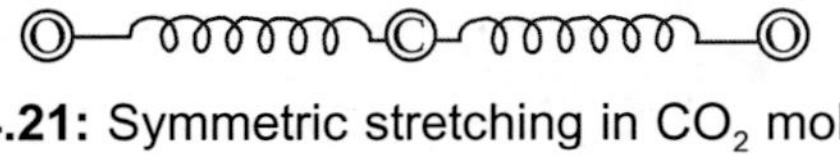

Fig. 4.21: Symmetric stretching in CO_2 molecule

This vibration of CO_2 is IR inactive since there is no dipole moment change during vibration. Since during stretching the dipole moment is changing along bond axis (axis which lies on the bond), thus it is called **parallel vibration.** It occurs at 1330 cm^{-1} in the IR spectra.

(ii) **Asymmetric stretching**

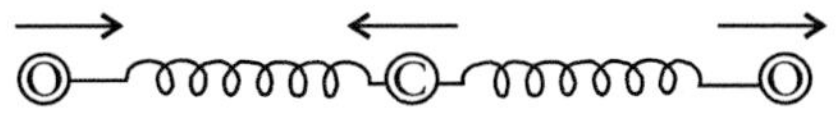

Fig. 4.22: Asymmetric stretching in CO_2 molecule

This vibration in CO_2 is IR active, since there is dipole moment change during vibration along the bond axis. Since during the stretching the dipole moment is changing along bond axis, thus it is called **parallel vibration.** It occurs at 2349.3 cm^{-1} in the vibrational spectra.

(iii) **Bending**

This involves two vibrations, i.e. one in the plane of the paper and the other in which the oxygen atoms moves simultaneously into and out of the plane. These two vibrations are degenerate. Thus, in CO_2 molecule, bending mode is doubly degenerate mode and both the bending vibration occurs at 667 cm^{-1}. It is a **perpendicular vibration.**

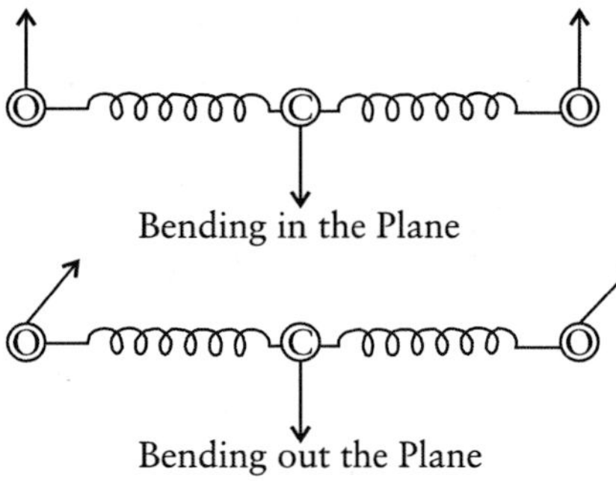

Fig. 4.23: Bending vibration in CO_2 molecule

(b) $HC \equiv CH$; Linear, N = 4

Vibrational degrees of freedom = $3N - 5 = 3(4) - 5 = 7$

There are seven fundamental vibrations in acetylene molecule.

3. **Non-Linear Polyatomic Molecule**

(a) H_2O; Non-linear; N = 3

Vibrational degrees of freedom = $3N - 6 = 3(3) - 6 = 3$

Thus, there are three fundamental vibrations or there are three normal modes of vibration in H_2O molecule.

(i) **Symmetric Stretching**

This vibration in H_2O is IR active, since there is dipole moment change during vibration. It is a **parallel vibration** and occurs at 3651.7 cm^{-1} in the IR spectra.

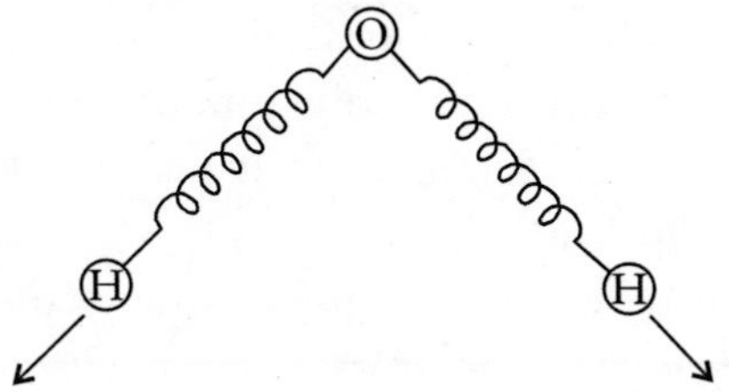

Fig. 4.24: Symmetric stretching in H_2O molecule

(ii) **Asymmetric Stretching**

This vibration in H_2O is also IR active. It is a **parallel vibration** and occurs at 3755.8 cm^{-1} in the IR spectra.

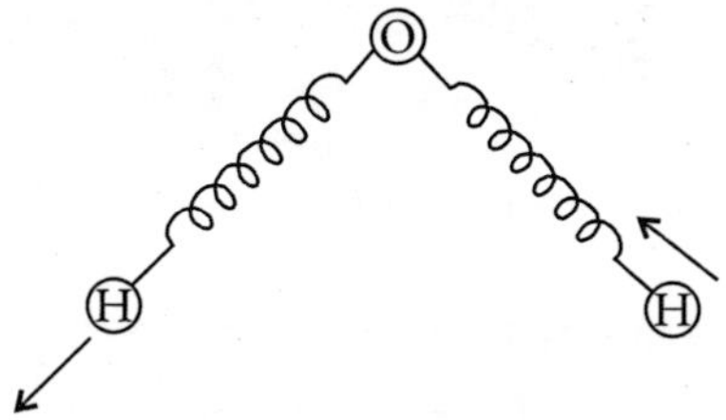

Fig. 4.25: Asymmetric stretching in H_2O molecule

(iii) **Symmetric Bending**

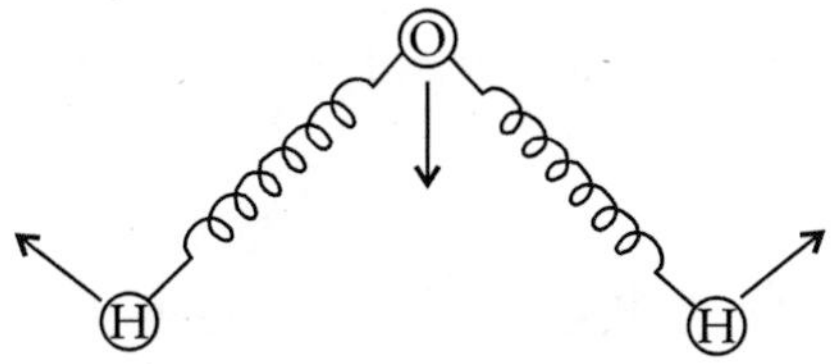

Fig. 4.26: Symmetric bending in H_2O molecule

This vibration in H_2O is also IR active. It is a **parallel vibration** and occurs at 1595.0 cm^{-1} in the IR spectra.

(b) COS; Non-linear; N = 3
Vibrational degrees of freedom = 3N – 6 = 9 – 6 = 3

(c) SO_2; Non-linear; N = 3
Vibrational degrees of freedom = 3N – 6 = 9 – 6 = 3

(d) C_6H_6; Non-linear; N = 12
Vibrational degrees of freedom = 3N – 6 = 3(12) – 6 = 30

PROBLEMS BASED ON VIBRATION IN POLYATOMIC MOLECULES

Q. 1. Roughly sketch the fundamental vibrations of CO_2 molecule. Show which of them will be IR active.

Q. 2. What are symmetric and asymmetric stretching and bending modes? Illustrate these using N_2O as an example.

Q. 3. Draw all the fundamental modes of vibration of H_2O and show that they are IR active.

Q. 4. Short note on modes of vibration in polyatomic molecules.

Q. 5. Calculate the degrees of freedom for CO_2, Benzene, Ne, HCHO and C_2H_6.

Q. 6. The linear N_2O molecule shows three fundamental vibrational frequencies in the IR spectrum. Suggest molecular structure for this molecule. Justify briefly.

Or

The linear N_2O molecule shows three fundamental vibrational frequencies in the spectrum. Is the structure N—N—O or N—O—N? Explain.

Ans. 6. NO_2 is a linear molecule, so the total vibrational modes will be 3N – 5 = 4.

We see only three vibrational modes, this means either two of them are degenerate or one is IR inactive. The probable structure may be N–N–O or N–O–N.

(a) **First we take N–N–O structure**

It will give four vibrational modes: symmetric stretching, asymmetric stretching and two bending vibrations. The two bending modes are degenerate. Thus, we get three IR active modes.

(b) **If we consider N–N–O structure**

The two stretching vibration in this case are degenerate as the their is no dipole moment change. Thus, we get a set of two IR active vibrational modes. Therefore, the structure of the molecule is N–N–O.

4.19 Selection Rules for Parallel and Perpendicular Vibration in Linear Molecule

1. Parallel Vibrations

The selection rule for parallel vibration is identical with that for diatomic molecule, since a diatomic molecule is linear and can undergo only parallel vibrations.

$$\Delta J = \pm 1,\ \Delta v = \pm 1 \text{ (Simple harmonic motion)}$$

$$\Delta J = \pm 1,\ \Delta v = \pm 1, \pm 2, \pm 3 \ldots \text{ (Anharmonic motion)}$$

The spectra will thus be similar in appearance consisting of 'P' and 'R' branches with lines equally spaced on each side and no line occurring at the band centre (shown in Fig. 4.15).

2. Perpendicular Vibrations

$\Delta J = 0, \pm 1$ and $\Delta v = \pm 1$ (Simple harmonic motion)

Now for the first time, a vibrational change can take place with no simultaneous rotational transition ($\Delta v = \pm 1$ and $\Delta J = 0$). The transition with $\Delta J = 0$ corresponds to 'Q' branch, whose lines may be derived from the equation

$$\Delta \epsilon_{J,v=1} = \epsilon_{J,v=1} - \epsilon_{J,v=0}$$

$$\Delta \epsilon_{J,v} = \left[\bar{B}J(J+1) + \frac{3}{2}\bar{\nu}_e - \frac{9}{4}\bar{\nu}_e x_e\right] - \left[\bar{B}J(J+1) + \frac{1}{2}\bar{\nu}_e - \frac{1}{4}\bar{\nu}_e x_e\right]$$

$$\Delta \epsilon_{J,V} = \bar{\nu}_e - 2x_e\bar{\nu}_e$$

$$\Delta \epsilon_{J,V} = \bar{\nu}_0 \text{ cm}^{-1}$$

Here, $\overline{\nu}_0$ represents the **fundamental frequency**. This wavenumber is called **band origin** or **band centre**. This corresponds to 'Q' branch. The 'Q' branch consists of lines superimposed upon each other at band centre, thus the resultant line is very intense. Thus, the spectrum consists of P, Q and R branches as shown in Fig. 4.28.

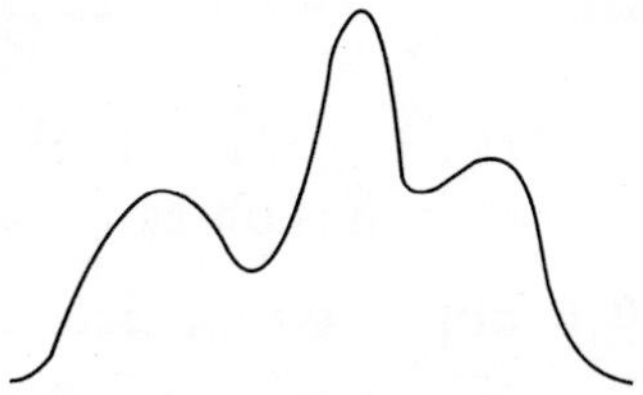

Fig. 4.27: Contour of PQR band

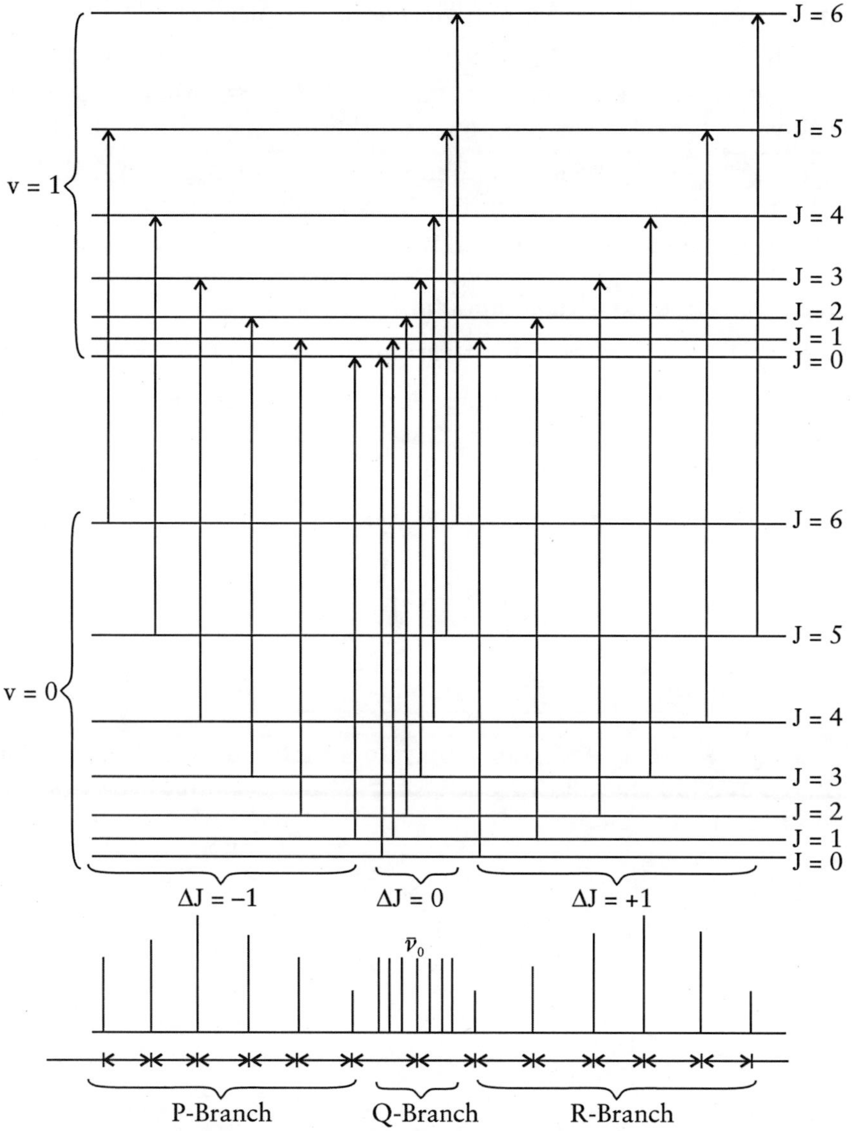

Fig. 4.28: Combined vibrational-rotational spectra for perpendicular vibrations in linear diatomic molecule (PQR structure)

PROBLEM BASED ON PQR CONTOUR

Q. 1. Derive the mathematical expression to evaluate the rotational constant $\overline{B}$, for a spectrum which is observed as a PQR contour.

4.20 Group Frequencies

Polyatomic molecule exhibits 3N—5 (linear) or 3N—6 (non-linear) fundamental vibrational modes of frequencies in their spectra. These vibrations can be further divided into two categories as **skeletal vibrational frequencies** and **group vibrational frequencies** or **group frequencies.**

Skeletal vibration modes involve more or less all atoms of the molecule, whereas group vibrational modes involve a small group of atoms and their absorption bands appear at higher wavelength in comparison to skeletal vibration absorption.

Skeletal vibration absorption bands are found between 400–1500 cm^{-1}. They are helpful in knowing about the general structure of the molecule such as linear, branch or benzenoid structure. These vibrational absorption bands are characteristics of a compound and hence 400–1500 cm^{-1} region is termed as **finger print region.** This region helps in establishing the identity of compound by comparing the spectrum of known compounds.

Group frequencies involve transitions of various functional groups such as —OH, >C=O, etc. They are quite independent of the full molecular structure. E.g. >C=O stretching vibrational absorption band occur almost at same place in spectrum (nearly 1700 cm^{-1}) irrespective of chain length.

Group frequencies follow approximately the equation

$$\overline{\nu} = \frac{1}{2\pi c}\sqrt{\frac{k_f}{\mu}}$$

PROBLEM BASED ON GROUP FREQUENCY

Q. 1. Under group frequency concept, in a polyatomic molecule the C–H stretching frequency is observed at 3030 cm^{-1}. If hydrogen is replaced by isotopic deuterium, at what value will this absorption be observed?

5

Raman Spectroscopy

5.1 Introduction

In 1923, A. Smekal predicted that in the case of light scattered by transparent medium, the scattered radiation might contain frequencies other than those present in the original incident radiation. In 1928, Sir C.V. Raman discovered that when a beam of monochromatic light was allowed to pass through some organic substances like benzene, toluene, etc. the scattered light contained some additional frequencies over and above that of incident frequencies. This is known as '**Raman Scattering**' or '**Raman Effect**' and is a beautiful confirmation of the Smekal's prediction.

There are number of ways through which photons can be scattered by the molecules. The scattering may be **elastic**, in which there is no exchange between the photon and molecule takes place. But a few undergo inelastic scattering. During inelastic scattering, there is exchange of energy between the photon and molecule. Thus, the inelastically scattered photons may have frequencies lower or higher than the incident frequency. This scattering is known as the '**Raman Scattering**' or the '**Raman Effect**'. For this brilliant discovery, Raman was awarded Nobel Prize in 1930.

The elastic scattering gives rise to Rayleigh scattering and the line at the incident frequency is called **Rayleigh line.** The lines on the low and high frequency sides of the Rayleigh line are known as the '**Stokes lines**' and '**Anti-Stokes lines**', respectively or collectively known and **Raman lines.**

If ν_0 represents the frequency of the incident radiation and ν_r is the frequency of the light scattered by a given molecular species. The difference between the frequencies (or wavenumbers) of the Rayleigh line and a Raman line is known as **Raman Shift,** $\Delta\nu$ is defined as:

$$\Delta\nu = \nu_0 - \nu_r$$

where subscripts 0 $\rightarrow$ incident and r $\rightarrow$ raman

For Stokes line, $\Delta\nu$ is positive and for Anti-Stokes line $\Delta\nu$ is negative (shown in Fig. 5.1)

The Raman shift does not depend upon the frequency of the incident light but it is regarded as a characteristic of substance causing Raman effect.

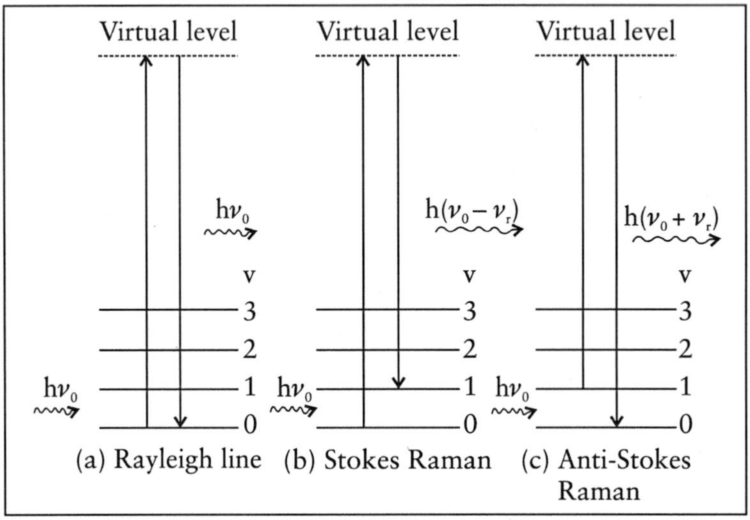

Fig. 5.1: Energy level diagram showing Rayleigh line, Raman Stokes line and Raman Anti-Stokes line

In terms of wavenumber,

(a) $\bar{\nu}_0$ is the wavenumber of central (exciting) line, i.e. Rayleigh line

(b) $\bar{\nu}_s = \bar{\nu}_0 - \Delta\bar{\nu}$ is the wavenumber of the series of lines on the low frequency side of the exciting line. These are **Stokes lines**

(c) $\bar{\nu}_{as} = \bar{\nu}_0 - \Delta\bar{\nu}$ is the wavenumber of the series of lines on the high frequency side of the exciting line. These are **Anti-Stokes lines.**

PROBLEMS BASED ON STOKES LINE, ANTI-STOKES LINE AND RAMAN SHIFT

Q. 1. What is Raman effect? Explain the origin of Stokes and Anti-Stokes lines.

Q. 2. When the photons of wavelength 4358 Å incident on CCl_4, we get Raman lines at 4400, 4419 and 4447 Å. Find out Raman shift for each of them in cm^{-1}.

Ans. Wavenumber of incident radiation $\bar{\nu}_0 = \dfrac{1}{\lambda_0} = \dfrac{10^8}{4358}\ \text{cm}^{-1}$

Raman shift of 4400 Å line $= \dfrac{10^8}{4358} - \dfrac{10^8}{4400} = 219.0\ \text{cm}^{-1}$

Raman shift of 4419 Å line $= \dfrac{10^8}{4358} - \dfrac{10^8}{4419} = 316.8\ \text{cm}^{-1}$

Raman shift of 4447 Å line $= \dfrac{10^8}{4358} - \dfrac{10^8}{4447} = 459.2\ \text{cm}^{-1}$

Q. 3. The exciting line in an experiment is 5650 Å and the Stokes line is 5810 Å. Calculate the wavelength of the Anti-Stokes line.

Ans. The Stokes and Anti-Stokes lines have the same wavenumber (energy) displacement with respect to the exciting line. The wavenumber of the exciting line is

$$\overline{v}_0 = \frac{1}{5650 \times 10^{-8}} = 17699.1 \text{ cm}^{-1}$$

The wavenumber of Stokes line is

$$\overline{v}_s = \frac{1}{5810 \times 10^{-8}} = 17211.7 \text{cm}^{-1}$$

Hence, the wavenumber displacement is

$$\Delta\overline{v} = 17699.1 - 17211.7 = 487.4 \text{ cm}^{-1}$$

Thus, the wavenumber of Anti-Stokes line is

$$\overline{v}_{as} = \overline{v}_0 + \Delta\overline{v} = 17699.1 + 487.4 = 18186.5 \text{ cm}^{-1}$$

Hence, the wavelength of Anti-Stokes line is

$$\lambda = \frac{1}{18186.5} = 5498.5 \times 10^{-8} \text{ cm} = 5498.5 \text{ Å}$$

Q. 4. A substance shows a Raman line at 4568 Å when the exciting line 4332 Å is used. Calculate the wavelengths of Stokes and Anti-Stokes lines for the same substance when the exciting line 4036 Å is used.

Ans. The wavenumber of the exciting line 4332 Å is

$$\frac{1}{(4332 \times 10^{-8})} = 23084.0 \text{ cm}^{-1}$$

The wavenumber of the Raman line at 4568 Å is

$$\frac{1}{(4568 \times 10^{-8})} = 21891.4 \text{ cm}^{-1}$$

Hence, the wavenumber displacement is

$$\Delta\overline{v} = 23084.0 - 21891.4 = 1192.6 \text{ cm}^{-1}$$

The Raman displacement is the characteristic of the substance, whatever be the exciting line. Now, the wavenumber of the exciting line is

$$\overline{v}_0 = \frac{1}{(4036 \times 10^{-8})} = 24777.0 \text{ cm}^{-1}$$

Thus, the wavenumber of the Stokes line is

$$\overline{v}_s = \overline{v}_0 - \Delta\overline{v} = 24777.0 - 1192.6 = 23584.4 \text{ cm}^{-1}$$

The corresponding wavelength is

$$\lambda_s = \frac{1}{23584.4} = 4240.1 \times 10^{-8} \text{cm} = 4240.1 \text{Å}$$

and the wavenumber of the Anti-Stokes line is

$$\bar{\nu}_{as} = \bar{\nu}_0 + \Delta\bar{\nu} = 24777.0 + 1192.6 = 25969.6 \text{ cm}^{-1}$$

The corresponding wavelength is

$$\lambda_{as} = \frac{1}{25969.6} = 3850.7 \times 10^{-8} \text{ cm} = 3850.7 \text{ Å}$$

Q. 5. When a substance is irradiated with 4365 Å line, a Raman line attributed to the symmetrical stretching vibration is found at 4814 Å. Calculate the fundamental frequency for this vibration.

Ans. The wavenumber of the exciting line 4365 Å is

$$\frac{1}{(4365 \times 10^{-8})} = 22909.5 \text{cm}^{-1}$$

The wavenumber of the exciting line 4814 Å is

$$\frac{1}{(4814 \times 10^{-8})} = 20772.7 \text{cm}^{-1}$$

Thus, the vibrational Raman displacement is

$$\Delta\bar{\nu} = 22909.5 - 20772.7 = 2136.8 \text{ cm}^{-1}$$

The corresponding frequency is

$$\nu = \bar{\nu}c = 2136.8 \times (3 \times 10^{10}) = 6.41 \times 10^{13} \text{ Hz}$$

5.2 Theory of Raman Scattering

Raman effect can be understood with the help of classical as well as quantum considerations.

5.2.1 Classical Theory of Raman Effect

When a static electric field is applied to a molecule, the molecule is polarised, i.e. its electrons and nuclei are displaced. The electrons are attracted to the positive pole while the nuclei to the negative pole. An induced electric dipole moment is thus produced in the molecule due to the displacement of the electrons and nuclei and the molecule is said to be **polarised.**

Suppose E is the strength of the electronic field, μ is the magnitude of electric dipole moment then we can write

$$\mu = \alpha E$$

Thus, the size of the induced electric dipole moment (μ) depends on two factors:

(a) magnitude of applied electric field (E) and

(b) ease with which molecule can be distorted (α), here α is the polarizability of the molecule.

So, Raman spectroscopy is observed only when there occurs a periodic change in the component of molecular polarizability of molecule due to its vibrational or rotational oscillation.

Polarizability change can be caused due to change in either the direction or magnitude of polarizability ellipsoid. Since the polarizability has the dimension of volume in Gaussian units, thus it may be concluded that only those vibrations which are associated with the change in volume will be Raman active.

5.2.2 Quantum Theory of Raman Effect

This treats the radiation of frequency 'ν_0' as consisting of a stream of particles (photons) having energy '$h\nu_0$'. Photons can be imagined to undergo collisions with molecules. If the collision is perfectly elastic, then there is no exchange of energy between photon and molecule, the photon will be deflected unchanged, this gives rise to **Rayleigh Scattering** and thus **Rayleigh line** is observed.

However, it may happen that energy is exchanged between photon and molecule during the collision, such collisions are called **inelastic collision.** Thus, small portion of the scattered radiation has lower frequency and some have higher frequency than the original frequency. The lines which have same frequency as that of the frequency of incident radiation is called **Rayleigh line** and if it is different, then the lines are called **Raman lines.**

The exchange of energy or the Raman lines is of two types, i.e. **Stokes line and Anti-Stokes lines.** If $\Delta\in$ is the change in vibrational or rotational energy of molecule then:

(a) **Stokes Lines:** If the molecule gains energy, $\Delta\in$, the photon will be scattered with energy ($h\nu_0 - \Delta\in$). Thus, the radiation is scattered with a frequency lower than that of incident beam and thus a Raman line on the low frequency side is observed. Also, the molecule before the collision occupies a lower vibrational or rotational energy level and after the emission of the photon, it may be present in any of the higher energy level.

(b) **Anti-Stokes Lines:** If the molecule loses energy, $\Delta\in$, the photon will be scattered with energy ($h\nu_0 + \Delta\in$). Thus the radiation is scattered with a frequency higher than that of incident radiation and thus a Raman line on the high frequency side is observed. Also, the molecule before the collision occupies a higher vibrational or rotational energy level and after the emission of the photon, it may be present in any of the lower energy level.

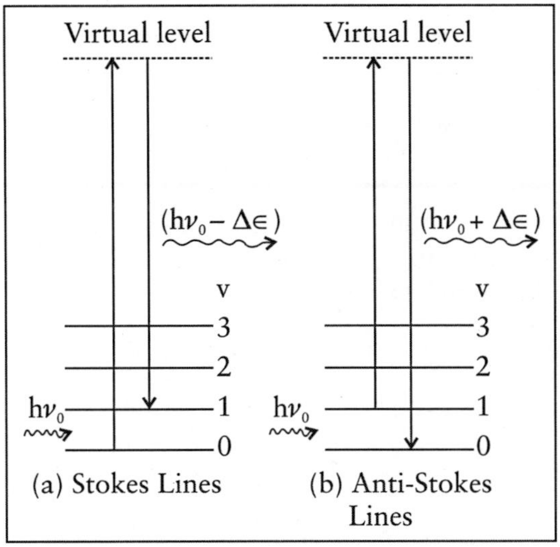

Fig. 5.2: Energy level diagram for Stokes and Anti-Stokes lines

Raman lines are, in general, of very weak intensities as compared to the Rayleigh line.

5.3 Intensity of Stoke's and Anti-Stoke's Lines

The intensity of a line depends on many factors but the most important being the population of the initial state from which the transition originates. The Stokes line originate from v = 0 level and the Anti-Stokes line originates from v = 1 level. Since most of the population is present in v = 0 level, thus intensity of Stoke line are more than the corresponding Anti-Stoke's line.

5.4 Effect of Temperature on Stoke's Lines and Anti-Stoke's Line

In case of **Stoke radiation,** the molecule before collision is present in lower vibrational level (v = 0) and after the collision with a photon, the molecule is present in any higher vibrational level (v = 1). Thus, molecular energy increases in case of Stoke's radiation.

Whereas in case of **Anti-Stoke's radiation,** the molecule before collision is present in higher vibrational level (v = 1) and after the collision with a photon it is present in any lower vibrational level (v = 0). Thus, molecular energy decreases in case of Anti-Stoke's radiation.

Thus, the **intensity of Anti-Stokes lines can be increased by increase in temperature.**

On heating the sample, the temperature of the sample increases and the population of v = 1 (ground state of Anti-Stokes) increases and hence more molecules would go to the excited state (v = 0 of excited state of Anti-Stokes). Therefore, intensity of Anti-Stokes radiation increases.

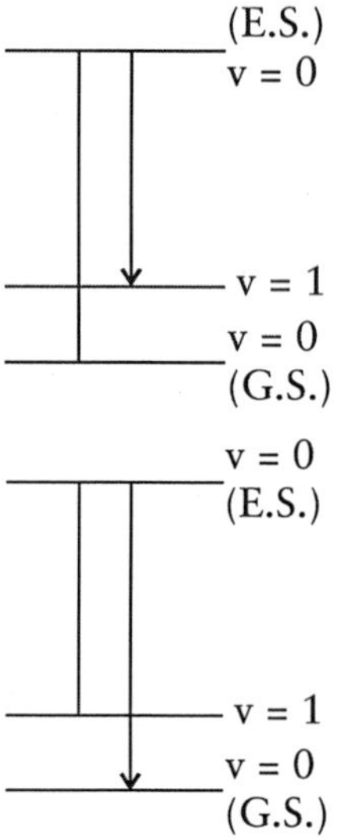

PROBLEM BASED ON EFFECT OF TEMPERATURE ON RAMAN LINES

Q. 1. To obtain Raman Anti-Stokes lines, should we heat or cool the molecules?

Ans. To obtain Anti-Stoke line, we have to heat the molecule because on heating the molecule, population in the excited state increases. Thus, the intensity of Anti-Stoke line increases on heating.

5.5 Raman Spectrum

The scattered photons have wavenumber shifts (10–4000 cm^{-1}) which are due to the rotational or vibrational-rotational transitions in the molecule. These weak lines of modified wavenumber are generally known as the **Raman spectrum.**

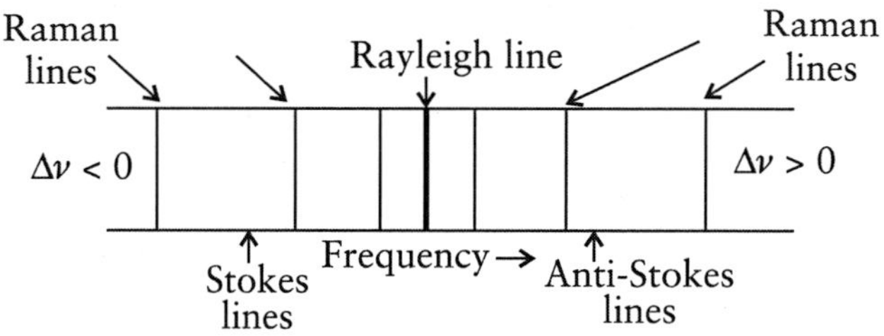

Fig. 5.3: Raman Spectrum

As Raman lines are of weaker intensity, concentrated solution must be utilised to increase the intensity of Raman lines. Raman lines are symmetrically displaced about the parent line. The frequency difference between the modified and parent line represents the frequency of the absorption band of the material.

PROBLEM BASED ON RAMAN SPECTRA

Q. 1. How is Fluorescence spectra different from Raman spectra?

Ans. Raman scattering and Fluorescence emission are two competing phenomena. Generally, a laser photon bounces off a molecule and looses a certain amount of energy that allows a molecule to vibrate (Stokes process). The scattered photon is therefore less energetic and the associated light exhibits a frequency shift. Sometimes if molecule is vibrationally excited, it can give this extra energy to the photon (Anti-Stokes process). So Raman scattering is inelastic scattering of photons and there is no excited state in Raman scattering, therefore the process is almost instantaneous. The energy of the photon is not resonant with some molecular transition, i.e. energy of photon is not equal to the energy difference between ground energy state and one of the excited states of the molecule. Thus, Raman scattering is a **non-resonant** process, so the energy of the photon is not important.

However fluorescence is a type of luminescence caused by photons exciting a molecule, raising it to an electronic excited state. It is brought about by absorption of photons in the singlet ground state promoted to a singlet-excited state. As the excited molecule returns to ground state, emits a photon of lower energy, which corresponds to a longer wavelength, than the absorbed photon. Fluorescence has characteristic lifetime of nanoseconds and it is a resonant process.

5.6 Raman Spectrometer

The Stokes line have intensity of 1/1000 times that of the original intensity. So in Raman spectroscopy, we require high energy source like lasers are used for Raman's experiment otherwise intensity will be very low. Thus for good detection of Raman scattering, one requires an intense source with small size. A laser beam may be focused to produce a beam of much smaller diameter with the help of a lens.

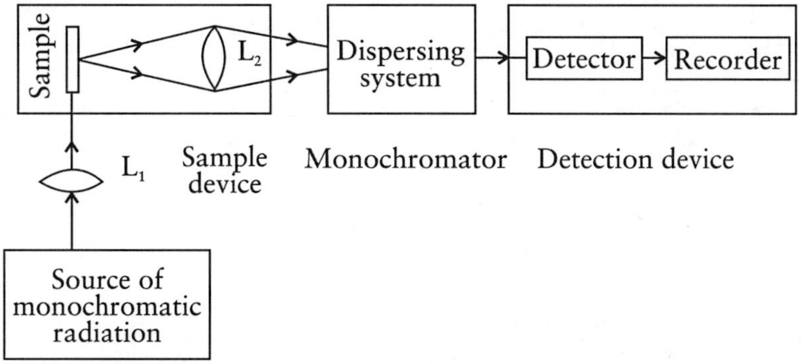

Fig. 5.4: Raman spectrometer with a laser source

The Raman effect was first observed in an organic compound where the sun was used as the source. Although the sun appears very intense, but when Raman tried to use a particular colour, the intensity was found to be very weak and consequently the intensities of Raman lines were also very weak. However, Raman recorded them with big efforts. Before the discovery of lasers, the commonly used sources have been the 435.8 nm (blue) and the 253.6 nm (UV) emission lines of mercury vapour. These sources had limitations and more often the detection of Raman scattering was not feasible. After the discovery of lasers, the scenario has changed drastically. Now, highly directional and intense lasers allow us to record the spectrum down up to 10 cm^{-1} produced by the highly exciting levels. The commonly used sources are He-Ne laser (632.8 nm), Argon ion laser (488 and 514.5 nm), Krypton laser (647.1, 568.2, 530.8, 520.8, 482.5, 476.2 nm), Argon-Krypton mixed laser (488.0, 514.5, 647.1 nm).

5.7 Pure Rotational Raman Spectra for Diatomic Molecules

Raman scattering can be used for the study of rotational spectrum in the linear molecules. The selection rule for pure Rotational-Raman spectra for linear diatomic molecules are

$$\Delta J = 0, \pm 2$$

which is different from the selection rule in Rotational spectroscopy ($\Delta J = \pm 1$).

In Raman spectra, the change in rotational quantum number by two units is due to symmetry of polarizability ellipsoid. In rotation around the bond axis in linear molecule, there is no change in the polarizability ellipsoid. However, in end-over-end rotation, the ellipsoid has same appearance to observer twice for a complete rotation. Hence in Rotational-Raman spectra, only end-over-end rotation is considered.

The **ΔJ = 0 corresponds to the Rayleigh scattering**, i.e. the scattered Raman radiation will be of the same frequency as that of incident light.

The **ΔJ = ± 2 gives rise to Raman lines.** Also, $\Delta J = -2$ may be ignored as for a pure rotational change, the upper state J must necessary be greater than the lower state J. Thus, we restrict overselves only to $\Delta J = +2$. The rotational energy expression is:

$$\epsilon_J = \bar{B}J(J+1)\,cm^{-1}; \quad (J = 0, 1, 2, 3.....)$$

$$\Delta \epsilon_J = \epsilon_{J+2} - \epsilon_J$$

$$\Delta \epsilon_J = \bar{B}(J+2)(J+3) - \bar{B}J(J+1)$$

$$\Delta \epsilon_J = \bar{B}(4J+6)\,cm^{-1}$$

During collision, the molecule gains energy from the photon, thus producing a series of lines on the low frequency side of the exciting line (Rayleigh line). These are Stokes lines and these corresponds to **S-branch**. Their wavenumber are:

$$\bar{v}_s = \bar{v}_0 - \Delta \epsilon_J$$

$$\bar{v}_s = \bar{v}_0 - \bar{B}(4J+6)]\,cm^{-1}$$

where, $\bar{v}_0$ is the wavenumber of the central (exciting line) Rayleigh line.

Similarly, if the molecule gives energy to the photon, produces a series of lines on the high frequency side of the exciting line (Rayleigh line). These are Anti-Stokes lines and these corresponds to **O-branch**.

Their wavenumber are

$$\bar{\nu}_{as} = \bar{\nu}_0 + \Delta \in_J$$

$$\bar{\nu}_{as} = \bar{\nu}_0 + \bar{B}(4J+6)]\ \text{cm}^{-1}$$

Thus, the Raman lines will appear at wavenumbers given by the expression

$$\bar{\nu}_r = \bar{\nu}_0 \pm \Delta \in_J$$

$$\bar{\nu}_r = \bar{\nu}_0 \pm \bar{B}(4J+6)\ \text{cm}^{-1}$$

where, o stands for incident and r stands for raman.

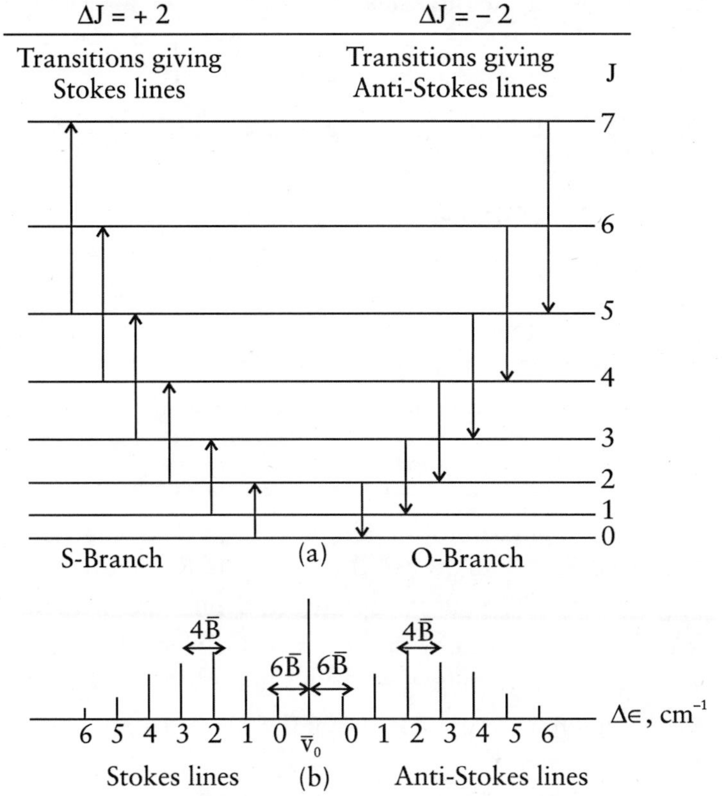

Fig. 5.5: Pure Rotational-Raman Spectra

- It is seen that the Raman shift of the first Stokes and Anti-Stokes lines from the exciting (Rayleigh) line is $6\bar{B}$. The separation between successive lines on either side of the exciting line is $4\bar{B}$.
- Just as the presence of a permanent dipole moment in a molecule is regarded as a criterion of IR spectrum. Similarly, change in polarizability of the molecule will determine whether the Raman spectrum will be observed or not. Thus, the essential

criterion for a molecule to be Raman active is that a molecular vibration or rotation must cause a change in the component of molecular polarizability.

- Rotation about bond axis produces no change in polarizability, hence the microwave concern only with end-over-end rotation.
- In Raman spectroscopy, the rotational quantum number changes by two units ($\Delta J = \pm 2$) because of the symmetry of the polarizability ellipsoid. For a linear molecule, during end-over-end rotation, the ellipsoid presents the same appearance to an observer twice in every complete rotation.

5.7.1 Application of Pure Rotational-Raman Spectra

Homonuclear diatomic molecules (O_2, H_2 etc.) gives no pure IR or microwave spectra, since they possess no dipole moment. Whereas they do give a Rotational-Raman spectra as there is a change in polarizability ellipsoid of the molecule.

For diatomic and other light molecules, the Rotational-Raman spectrum is generally resolved. Thus by measuring the wavenumbers of the spectral lines, one can calculate the rotational constant ($\overline{B}$) and have the moment of inertia (I) and in turn the internuclear separation for such molecules can be calculated.

5.7.2 Intensities of Pure Rotational-Raman Spectra

The intensities of Raman lines depend on the population of the initial level from where the molecules are promoted or demoted to the final level. Since all the rotational levels belong to v = 0 vibrational state and therefore all the rotational levels are of reasonable population. Thus, the Stokes and Anti-Stokes lines appear with considerable intensity.

For molecules having a **centre of symmetry**, the nuclear spins also effect the population in the rotational levels.

(a) For molecules having nuclear spin zero (I = 0), every alternate rotational level is unoccupied.

Like: The levels, J = 0, 2, 4 ... in case of O_2 and J = 1_1 3, 5, ... in case of CO_2 are unoccupied. Thus, the first Stokes and Anti-Stokes absorptions in Raman-Rotational spectrum of O_2 occur at $(\overline{\nu}_0 - 10\overline{B})$ and $(\overline{\nu}_0 + 10\overline{B})$ respectively. While, the first Stokes and Anti-Stokes absorptions in Raman-Rotational spectrum of CO_2 occurs at $(\overline{\nu}_0 - 6\overline{B})$ and $(\overline{\nu}_0 + 6\overline{B})$ respectively. While the remaining absorptions in both cases occur at the interval of $8\overline{B}$.

(b) For molecules having non-zero nuclear spin ($I \neq 0$), the rotational levels show alteration in population.

Like: (1) For H_2 (I = 1/2), the alteration in intensities is 1 : 3.

$$\text{Half-integral spin: } \frac{\text{Even J}}{\text{Odd J}} = \frac{I(2I+1)}{(I+1)(2I+1)} = \frac{I}{(I+1)}$$

(2) For N_2 (I = 1), the alteration in intensities is 2 : 1.

$$\text{Integral spin: } \frac{\text{Even J}}{\text{Odd J}} = \frac{(I+1)(2I+1)}{I(2I+1)} = \frac{(I+1)}{I}$$

PROBLEMS BASED ON ROTATIONAL-RAMAN SPECTRA

Q. 1. Bond length of H_2 molecule is 0.7417 Å. Determine the position of the first three Rotational-Raman lines in the spectrum. Given, mass of $^1H = 1.673 \times 10^{-24}$ g.

Ans. Reduce mass μ of H_2 molecule is

$$\mu = \frac{1.673\times10^{-24}}{2} = 0.8365\times10^{-24} \text{ g}$$

Hence, the rotational constant is

$$\bar{B} = \frac{h}{8\pi^2\mu r^2 c}$$

$$\bar{B} = \frac{(6.626\times10^{-27})}{8\times(3.14)^2\times(0.8365\times10^{-24})\times(0.7417\times10^{-8})^2\times(3\times10^{10})} = 60.73 \text{ cm}^{-1}$$

The first three lines are at the wavenumbers $\bar{\nu} = 6\bar{B}, 10\bar{B}, 14\bar{B} = 364.4, 607.3, 850.2$ cm^{-1}.

Q. 2. In the Raman spectrum of a linear triatomic molecule, the first three lines are at 4.86, 8.14 and 11.36 cm^{-1}. Determine the rotational constant $\bar{B}$ and the moment of inertia of the molecule.

Ans. In the Raman spectrum of a linear molecule, the first three lines occur at $6\bar{B}$, $10\bar{B}$ and $14\bar{B}$. Thus, we have

$6\bar{B} = 4.86 \text{ cm}^{-1}$ $\quad 10\bar{B} = 8.14 \text{ cm}^{-1}$ $\quad 14\bar{B} = 11.36 \text{ cm}^{-1}$

Average $\bar{B} = 0.812 \text{ cm}^{-1} = 81.2 \text{ m}^{-1}$.

The moment of inertia is

$$I = \frac{h}{8\pi^2\bar{B}c} = \frac{6.626\times10^{-27}}{8\times(3.14)^2\times81.2\times(3\times10^{10})} = 34.45\times10^{-40} \text{ g cm}^2$$

Q. 3. Wavenumbers of rotational levels are expressed as

$$\in_J = \bar{B}J(J+1) - \bar{D}J^2(J+1)^2$$

where $\bar{B}$ and $\bar{D}$ are, respectively, the rotational and centrifugal distortional constants for the ground vibrational state. Derive an expression for wavenumbers of Rotational-Raman lines.

Ans. We have

$$\in_J = \bar{B}J(J+1) - \bar{D}J^2(J+1)^2$$

and

$$\in_{J+2} = \bar{B}(J+2)(J+3) - \bar{D}(J+2)^2(J+3)^2$$

Thus, wavenumber of Rotational-Raman line is

$$\begin{aligned}\bar{\nu} &= \in_{J+2} - \in_J \\ &= [\bar{B}(J+2)(J+3) - \bar{D}(J+2)^2(J+3)^2] - [\bar{B}J(J+1) - \bar{D}J^2(J+1)^2] \\ &= 2\bar{B}(2J+3) - 4\bar{D}(J^2+3J+3)(2J+3)\end{aligned}$$

Q. 4. In the Raman spectrum of $^{16}O_2$ molecule, the first three Stokes lines are separated by 14.4 cm^{-1}, 25.8 cm^{-1} and 37.4 cm^{-1} from the exciting radiation. Using the rigid rotator model, calculate the bond length. Given, Avogadro number = 6.023 × 10^{23}.

Ans. Because of zero value of nuclear spin of oxygen atom, for O_2 molecule, the alternate lines in Rotational spectrum will be missing, thus the first three Stokes lines are at $\bar{\nu} = 10\bar{B}, 18\bar{B}, 26\bar{B}$. Thus, we have

$10\bar{B} = 14.4$ cm^{-1} $\quad 18\bar{B} = 25.8$ cm^{-1} $\quad 26\bar{B} = 37.4$ cm^{-1}

Average $\bar{B} = 1.43704$ cm^{-1}.

The reduced mass is

$$\mu = \frac{16}{2 \times (6.023 \times 10^{23})} = 1.328 \times 10^{-23} \text{ g}$$

The bond length r_0 is

$$r_0 = \left(\frac{h}{8\pi^2 c\mu\bar{B}}\right)^{1/2}$$

$$r_0 = \left(\frac{6.26 \times 10^{-27}}{8 \times (3.14)^2 \times (3 \times 10^{10}) \times (1.328 \times 10^{-23}) \times 1.43704}\right)^{1/2}$$

$r_0 = 1.211 \times 10^{-8}$ cm = 1.211 Å

Q. 5. A diatomic molecule, AB has a rotational constant $\bar{B} = 10$ cm^{-1}. When a sample of this gas was irradiated with the Argon gas laser for recording its Rotational-Raman spectra, a Rayleigh line was observed at 19430 cm^{-1}.

(a) Draw the energy level diagram for the molecule and show the Stokes and Anti-Stokes lines on it.

(b) Calculate the frequency at which the first Stokes and Anti-Stokes lines will be observed.

Ans. (a) (Same as Fig. 5.5)

(b) $\bar{B} = 10$ cm^{-1} (given).

The first Anti-Stoke line will appear at,

$$\bar{\nu}_{as} = \bar{\nu}_0 + 6\bar{B}$$

$$\bar{\nu}_{as} = 19430 + 60 = 19490 \text{ cm}^{-1}$$

The frequency of first Anti-Stoke line $= \nu_{as} = \frac{c}{\lambda} = c\bar{\nu}_{as}$

$\nu_{as} = 3 \times 10^{10}$ cm/s × 19490 cm^{-1}

$\nu_{as} = 5.84 \times 10^{14}$ Hz

The first Stoke line will appear at $\bar{\nu}_s = \bar{\nu}_0 - 6\bar{B}$

$\bar{\nu}_s = 19430 - 60 = 19370$ cm^{-1}

The frequency of first Stoke line $= \nu_s = \frac{c}{\lambda} = c\bar{\nu}_s$

$\nu_s = 3 \times 10^{10}$ cm/s $\times$ 19370 cm^{-1}

$\nu_s = 5.8 \times 10^{14}$ Hz

Q. 6. Selection rule for pure rotational spectra in microwave region is $\Delta J = \pm 1$ whereas for Rotational-Raman spectra is $\Delta J = 0, \pm 2$. Give reason.

Ans. The rotational quantum number is Rotational-Raman spectroscopy changes by two units is due to symmetry of the polarizability ellipsoid. For a linear molecule during end-over-end rotation, the ellipsoid present the same appearance twice to an observer in every rotation. However, rotation around the bond axis will not cause any change in polarizability ellipsoid. Hence, only end-over-end rotation is considered. The transition with $\Delta J = 0$ does not correspond to a molecular transition but rather to Rayleigh scattering. Hence, the selection rule for Rotational-Raman spectra is $\Delta J = 0, \pm 2$.

Q. 7. H_2 is microwave inactive but it is Rotation-Raman active.

Ans. As H_2 does not have a permanent dipole moment, hence it is microwave inactive. Whereas it is Raman active due to change in polarizability ellipsoid.

Q. 8. The pure Rotational-Raman spectrum of N_2 was obtained in a specially constructed cell and excited by mercury light of 22946 cm^{-1}, when the following lines were observed 22910, 22918, 22926, 22934, 22946, 22958, 22966, 22974, 22982 cm^{-1}.

(i) Draw the spectrogram to show line position and thus assign each line to transition.

(ii) Calculate the rotational constant of N_2 molecule.

Ans. Wavenumber for Stokes line ($\bar{\nu}_s$) is less than $\bar{\nu}_0$ (22946 cm^{-1}) and wavenumber of Anti-Stokes line ($\bar{\nu}_{as}$) is greater than $\bar{\nu}_0$ (22946 cm^{-1}).

(a) 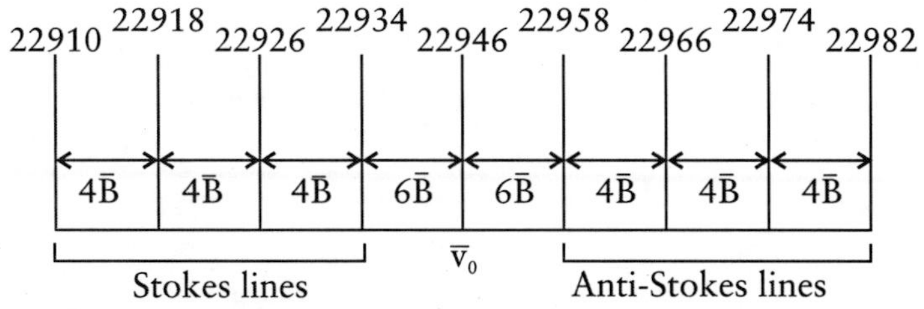

(b) $\bar{\nu}_0 - \bar{\nu}_s = 22946 - 22934 = 6\bar{B}$

$$12 = 6\bar{B} \quad \Rightarrow \bar{B} = 6 \text{ cm}^{-1}$$

Q. 9. What would be the spacing of the lines in the Raman-Rotational spectrum of H_2, if its bond length is 0.0757 nm?

Ans. Mass of hydrogen atom = 1.673×10^{-27} kg

$$\mu_{H_2} = \frac{m_H \times m_H}{m_H + m_H} = \frac{m_H}{2} = 8.365 \times 10^{-28} \text{ kg}$$

$r_0 = 0.0757 \times 10^{-9}\,m = 7.57 \times 10^{-11}m.$

$I = \mu r_0^2 = 4.79 \times 10^{-48}\ kgm^2$

Now $\bar{B} = \dfrac{\lambda}{8\pi^2 Ic} = \dfrac{6.626\times10^{-34}}{8\times(3.14)^2\times(4.79\times10^{-48})\times(3\times10^8)} = 5.84\times10^3\,m^{-1}$

The spacing between lines in Rotational-Raman spectrum of H_2 is $4\bar{B}$

$4\bar{B} = 4 \times (5.84 \times 10^3)\ m^{-1}$

$4\bar{B} = 22.16 \times 10^3\ m^{-1}$

$4\bar{B} = 221.6\ cm^{-1}$

Q. 10. Homonuclear diatomic molecule, Br_2, is microwave inactive but is Rotational-Raman active. Why?

Ans. For Br_2 molecule, its dipole moment is zero, so it is microwave inactive. But rotation brings about a change in molecular polarizability, therefore, it is Rotational-Raman active.

Q. 11. For H_2 molecule, the value of spacing between the S-branch line was found to be 237.3 cm^{-1}. Calculate the bond length of hydrogen molecule. Reduced mass of H_2 is 8.365×10^{-28} kg.

Ans. For hydrogen molecule, the spacing between lines is $4\bar{B}$.

$\therefore\ 4\bar{B} = 237.3\ cm^{-1} \qquad \Rightarrow \bar{B} = 59.325\ cm^{-1} = 5.93 \times 10^3\ m^{-1}$

$$I = \frac{\lambda}{8\pi^2\bar{B}c} = \frac{(6.626\times10^{-34})}{8\times(3.14)^2\times(5.93\times10^3)\times(3\times10^8)}$$

$I = 4.72 \times 10^{-48}\ kgm^2$

Now, $I = \mu r_0^2$

$$r_0 = \left(\frac{I}{\mu}\right)^{1/2} = \left(\frac{4.72\times10^{-48}}{8.365\times10^{-28}}\right)^{1/2}$$

$r_0 = 7.50 \times 10^{-11}\ m$

$r_0 = 75\ pm$

Q. 12. The wavenumber of the incident radiation in a Raman spectrometer is 20490 cm^{-1}. What is the wavenumber of the scattered Anti-Stokes line for $J = 0 \rightarrow 2$ transition. ($\bar{B} = 1.99\ cm^{-1}$)

Ans. For $J = 0 \rightarrow 2$

$$\Delta \in_J = \Delta\bar{v} = \bar{B}(4J+6) = \bar{B}(0+6) = 6\bar{B}$$

$$\bar{v}_0 = 20490\,cm^{-1}$$

Also, $\bar{v}_{as} = \bar{v}_0 + \Delta\bar{v} = 20490 + 6\bar{B}$

$= 20490 + 6(1.99)$

$= 20501.94\ cm^{-1}$

Q. 13. The separation between lines in the Rotation-Raman spectra is 4.12 cm^{-1} in the S-branch. Find the value of rotational constant for $CO_2(g)$.

Ans. As CO_2 has centre of symmetry, so because of zero value of nuclear spin, alternate lines in Rotational-Raman spectra will be missing

$$\Delta\bar{v} = 8\bar{B}$$

$$4.12 = 8\bar{B}$$

$$\bar{B} = 4.12/8$$

$$\bar{B} = 0.515 \text{ cm}^{-1}$$

5.8 Pure Vibration-Raman Spectra for Diatomic Molecules

The vibrational energy of an anharmonic oscillator is

$$\epsilon_v = \left(v + \frac{1}{2}\right)\bar{\nu}_e - \left(v + \frac{1}{2}\right)^2 x_e\bar{\nu}_e$$

where, v = 0, 1, 2, 3 ......

$\bar{\nu}_e$ = oscillation frequency

x_e = anharmonicity constant

Similar to those in IR, the selection rule for Vibrational-Raman spectra is

$$\Delta v = \pm 1, \pm 2, \pm 3, \ldots.$$

At normal temperature, we consider only v = 0 → 1 transition, as the population of vibrationally excited states v = 1, 2, 3... is negligibly small. Also, as the Raman scattering is very weak, overtones and combinations generally do not appear in the spectra. Thus, we restrict ourselves only to the fundamental transition v = 0 → 1, we have

$$\Delta\epsilon_v = \epsilon_1 - \epsilon_0$$

$$\Delta\epsilon_v = \left(\frac{3}{2}\bar{\nu}_e - \frac{9}{4}x_e\bar{\nu}_e\right) - \left(\frac{1}{2}\bar{\nu}_e - \frac{1}{4}x_e\bar{\nu}_e\right)$$

$$\Delta\epsilon_v = \bar{\nu}_e(1 - 2x_e) \text{ cm}^{-1}$$

During collision, the molecule gains energy from the photon, thus producing a series of lines on the low frequency side of the exciting line (Rayleigh line). **These are Stokes lines.** The wavenumber of these Stokes lines are:

$$\bar{v}_s = \bar{v}_0 - \Delta\epsilon_v$$

$$\bar{v}_s = \bar{v}_0 - \bar{\nu}_e(1 - 2x_e) \text{ cm}^{-1}$$

where, $\bar{v}_0$ is the wavenumber of the central (exciting line).

Similarly, if the molecule gives energy to the photon, produces a series of lines on the high frequency side of the exciting line. **These are Anti-Stokes lines.** The wavenumber of these Anti-Stokes lines are:

$$\overline{v}_{as} = \overline{v}_0 + \Delta \in_v$$

$$\overline{v}_{as} = \overline{v}_0 + \overline{\nu}_e(1 - 2x_e)\ \text{cm}^{-1}$$

Thus, we observe the Raman lines at the wavenumbers given by the expression

$$\overline{v}_r = \overline{v}_0 \pm \Delta \in_v$$

$$\overline{v}_r = \overline{v}_0 \pm \overline{\ }_e(1 - 2x_e)\ \text{cm}$$

The transition v = 1 → v = 0 is expected to be very weak as very few molecules exist in the v = 1 at normal temperature. On increasing the temperature, the population of v = 0 level is decreased, whereas that of v = 1 level is increased. It is thus expected that the intensity of Stokes lines will be decreased, whereas that of Anti-Stokes will be increased on increasing the temperature of the sample.

PROBLEMS BASED ON VIBRATION-RAMAN SPECTRA

Q. 1. In Vibration-Raman spectra at room temperature, the Anti-Stokes lines are quite weak as compared to Stokes lines.

Ans. At room temperature, the population in the excited state is very less as more molecules are present in the ground state. For Anti-Stokes lines excitation takes place from excited state. Hence, Anti-Stokes lines are quite weak in comparison to Stokes lines.

Q. 2. The IR spectra of deuterated alkane show sharp C–D bending lines at 1032 cm^{-1} and 971 cm^{-1}. Predict the position of Stokes and Anti-Stokes lines in its Raman spectrum, if alkane is irradiated with an Argon laser of wavelength 514.53 nm (19430 cm^{-1}).

Ans. For Stokes lines,

$$\overline{v}_s = \overline{v}_0 - \Delta\in_v$$

Hence, Stokes lines will lie between

$\overline{v}_s$ = (19430–1032) cm^{-1} and $\overline{v}_s$ = (19430–971) cm^{-1} i.e. 18398 cm^{-1} and 18459 cm^{-1}.

Similarly for Anti-Stokes lines,

$$\overline{v}_{as} = \overline{v}_0 + \Delta\in_v$$

Hence, Anti-Stokes line will lie between

$\overline{v}_{as}$ = (19430 + 971) cm^{-1} and $\overline{v}_{as}$ = (19430+1032) cm^{-1}, i.e. 20401 cm^{-1} and 20462 cm^{-1}.

Q. 3. Spectral analysis shows that molecule has a fundamental vibrational frequency equal to 2960 cm^{-1}. Predict the position of the Stokes and Anti-Stokes lines for this vibrational mode if the Raman spectrometer is using mercury light of wavelength 435.8 nm.

Ans. $\lambda = 435.8\ \text{nm} = 435.8 \times 10\ \ \ \text{cm}$

$$\overline{v}_0 = \frac{1}{\lambda_0} = \frac{1}{435.8 \times 10^{-7}} = 22946.30\ \text{cm}^{-1}$$

Also, $\Delta\in_v$ = 2690 cm^{-1} (fundamental transition)

For Stokes lines:

$$\bar{v}_s = \bar{v}_0 - \Delta\epsilon_v$$

$$\bar{v}_s = 22946.30 - 2960$$

$$\bar{v}_s = 19986.3 \text{ cm}^{-1}$$

For Anti-Stokes lines

$$\bar{v}_{as} = \bar{v}_0 + \Delta\epsilon_v$$

$$\bar{v}_{as} = 22946.30 + 2960$$

$$\bar{v}_{as} = 25906.3 \text{ cm}^{-1}$$

Hence, Stokes and Anti-Stokes lines appear at 19986.3 cm^{-1} and 25906.3 cm^{-1} respectively.

Q. 4. A substance shows a Vibration-Raman line at 456.8 nm when the incident radiation is at 433.2 nm. Calculate the wavelength for the Stokes and Anti-Stokes line for this substance when the incident radiation is at 403.6 nm.

Ans. Wavelength of incident radiation = λ_0

$$\lambda_0 = 433.2 \text{ nm} = 433.2 \times 10^{-7} \text{ cm}$$

$$\text{Wavenumber of incident radiation, } \bar{v}_0 = \frac{1}{\lambda_0} = \frac{1}{433.2\times10^{-7}} = 23084.02 \text{ cm}^{-1}$$

$$\text{Wavenumber of Raman line} = \bar{v}_r = \frac{1}{\lambda_r} = \frac{1}{456.8\times10^{-7}} = 2189.41 \text{ cm}^{-1}$$

Since the Vibration-Raman line occurs at 456.8 nm, which is higher wavelength than 433.2 nm, thus this Vibration-Raman line occurs on left of the Rayleigh line.

$$\bar{v}_r = \bar{v}_0 - \Delta\epsilon_v \quad \Rightarrow \quad \Delta\epsilon_v = \bar{v}_0 - \bar{v}_r$$

$$\Delta\epsilon_v = 23084.02 - 2189.41 \quad \Rightarrow \quad \Delta\epsilon_v = 1192.61 \text{ cm}^{-1}$$

Now, Wavelength of new incident radiation = λ_0 = 403.6 nm = 403.6 × 10^{-7} cm

$$\text{Wavenumber of incident radiation} = \bar{v}_0 = \frac{1}{\lambda_0} = \frac{1}{403.6\times10^{-7}} = 24777 \text{ cm}^{-1}$$

$$\therefore \text{Wavenumber of Stoke's line} = \bar{v}_s = \bar{v}_0 - \Delta\epsilon_v = 24777 - 1192.61 = 23584.39 \text{ cm}^{-1}$$

$$\text{Wavelength of Stoke's line} = \lambda_s = \frac{1}{\bar{v}_s} = \frac{1}{23584.39} = 4.24 \times 10^{-5} \text{ cm}$$

$$\text{Wavenumber of Anti-Stoke's } \bar{v}_{as} = \bar{v}_0 + \Delta\epsilon_v = 24777 + 1192.61 = 25969.61 \text{ cm}^{-1}$$

$$\text{Wavelength of Anti-Stoke's line} = \lambda_{as} = \frac{1}{\bar{v}_{as}} = \frac{1}{25969.61} = 3.85 \times 10^{-5} \text{ cm}$$

Q. 5. Stokes and Anti-Stokes lines in pure Rotational-Raman spectrum have similar intensities while Stokes lines are more intense than Anti-Stokes lines in Vibrational-Raman spectrum. Explain.

Ans. The intensities of Raman lines depend on the population of the initial level from where the molecules are promoted or demoted to the final level.

In case of **Rotational-Raman spectra,** since all the rotational levels belong to v = 0 vibrational state and thus all the rotational levels are of reasonable population. Thus, the Stokes and Anti-Stokes lines appear with considerable intensity.

Whereas in case of **Vibrational-Raman spectra,** at room temperature the population in the excited state (v = 1) is very less as more molecules are present in the ground state (v = 0). For Anti-Stokes lines, excitation takes place from excited state (v = 1). Hence, Anti-Stokes lines are quite weak in comparison to Stokes lines.

Q. 6. The Raman spectra of acetylene using 445.90 mmHg incident radiation shows one of the lines at 542 nm. Determine the wavenumber of the vibration band $(\Delta\overline{v})$.

Ans. Wavelength of incident radiation $= \lambda_0 = 445.90$ mmHg

$= 445.90 \times 10^{-7}$ cm.

Wavenumber of incident radiation $= \overline{v}_0 = \dfrac{1}{\lambda_0} = \dfrac{1}{445.90 \times 10^{-7}\ \text{cm}^{-1}}$

Also, Wavelength of Raman line $= \lambda_r = 542$ nm $= 542 \times 10^{-7}$ cm

Wavenumber of Raman line $= \overline{v}_r = \dfrac{1}{\lambda_r} = \dfrac{1}{542 \times 10^{-7}\ \text{cm}^{-1}}$

Since $\lambda_r > \lambda_0$ ($542\times10^{-7} > 445.90\times10^{-7}$), thus it must be a Stokes line

Wavenumber of Stokes line $\overline{v}_s = \overline{v}_0 - \Delta\overline{v}$

Wavenumber of Vibration band $\Delta\overline{v} = \overline{v}_0 - \overline{v}_s$

$$\Delta\overline{v} = \frac{1}{445.90 \times 10^{-7}} - \frac{1}{542 \times 10^{-7}}$$

$$\Delta\overline{v} = 3976\ \text{cm}^{-1}$$

5.9 Vibration-Rotation Spectra in Raman Spectroscopy

The rotational fine structure (i.e. vibrational-rotational transitions) associated with the Vibrational-Raman spectrum for large molecules is ignored as it cannot be resolved. It can be resolved only in the case of light diatomic gaseous molecules.

For vibration-rotation transitions, the selection rules are

$\Delta v = \pm 1, \pm 2, \pm 3, \ldots$ and $\Delta J = 0, \pm 2$

The vibration-rotation energy is given as

$$\Delta \in_{v,J} = \overline{\nu}_e \left(v + \frac{1}{2}\right) - \left(v + \frac{1}{2}\right)^2 x_e \overline{\nu}_e + \overline{B}J(J+1)$$

$$\Delta \in_{v,J} = \overline{\nu}_e (1 - 2x_e) + \overline{B}(4J + 6)\,\text{cm}^{-1}$$

For ΔJ= +2, the wavenumber of the Stokes lines are:

$$\bar{\nu}_s = \bar{\nu}_0 - \Delta\epsilon_{v,J}$$

$$\bar{\nu}_s = \bar{\nu}_0 - [\bar{\nu}_e(1-2x_e) + \bar{B}(4J+6)] \quad [S-\text{Branch}]$$

For ΔJ= –2, the wavenumber of the Anti-Stokes lines are:

$$\bar{\nu}_{as} = \bar{\nu}_0 + \Delta\epsilon_{v,J}$$

$$\bar{\nu}_{as} = \bar{\nu}_0 + [\bar{\nu}_e(1-2x_e) + \bar{B}(4J+6)] \quad [O-\text{Branch}]$$

For ΔJ= 0

$$\bar{\nu} = \bar{\nu}_0 - \bar{\nu}_e(1-2x_e) \quad [Q-\text{Branch}]$$

5.10 Criterion of Observing Raman Lines

Polarizability is related to how easily the molecular orbital of a molecule can be deformed. An electric field will distort the molecular orbital (electron cloud) of a molecule, creating, an "induced electric dipole moment."

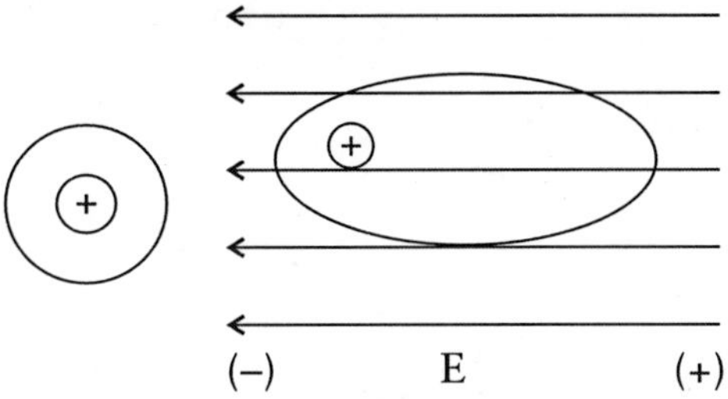

Fig. 5.6: Distortion of electron cloud by electric field

The oscillating electric field of the incident electromagnetic radiation will create an oscillating induced electric dipole moment which will emit electromagnetic radiation (scattered radiation).

Thus, in the presence of a beam of radiation, the molecule acquires an induced dipole moment due to the electrical component of the incident light. This electrical component is not static but varies and consequently the induced dipole moment oscillates.

The induced dipole moment (μ_{ind}) is given by

$$\mu_{ind} = \alpha E$$

where, 'α' is the polarizability of the molecule and 'E' is the electrical component of the incident light and 'E' depends on the frequency of radiation (ν).

Now, an oscillating dipole emits radiation of its own frequency, which results in the occurrence of **Rayleigh line** in the light scattered by the substance. Whereas the occurrence of Raman lines is due to the periodic change in the polarizability of molecule due to its vibrational and rotational motion.

The polarizability of a molecule can be expressed according to the equation:

$$\alpha = \alpha_0 + \left(\frac{d\alpha}{d\xi}\right)\xi_0 \sin(2\pi\nu_i t)$$

where, $\frac{d\alpha}{d\xi}$ is the change in polarizability with the displacement coordinate ξ.

α_0 is the polarizability in the equilibrium position ξ_0.

ν_i is the frequency of internal motion.

This periodic change in polarizability will change periodically the induced dipole moment (as $\mu_{ind} = \alpha E$), i.e. the induced dipole moment of the molecule oscillates and thus emit radiations of frequency, ν (Raleigh line) as well frequencies, $\nu - \nu_i$ and $\nu + \nu_i$ (Raman lines).

Now, if the periodic change in polarizability $\left(\frac{d\alpha}{d\xi}\right)$ is zero, i.e. there occurs no change in the polarizability during the small displacement from the equilibrium position, then the induced dipole moment of the molecule will oscillate with only one frequency corresponding to the frequency of incident light.

In that case, the molecule emits only Rayleigh line and hence Raman lines are not observed and the molecule is said to be **Raman inactive.** Thus, the essential criterion for the molecule to be Raman active is that a molecular vibration or rotation must cause a **change in the component of molecular polarisability.**

Polarisability has the dimension of volume and thus only those vibrations which are associated with the change in volume will be Raman active.

For small molecule, polarisability can be figured like an **ellipsoid.**

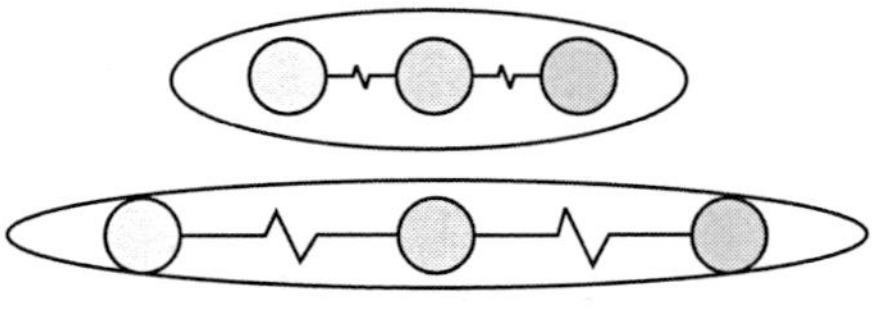

Fig. 5.7: Change in Polarizability Ellipsoid

5.11 Molar Polarisation

When a polar molecule is subjected to an electric field, the positive and negative charge distribution in the molecule is disturbed which is known a **polarisation.**

The polarisation so caused is quantified in terms of **molar polarisation** (P_M) which is the polarisation of one mole of a substance.

Polarisation of the molecule is due to the distortion of the electronic charge cloud in a molecule caused by the applied electric field. Thus, polarisation is independent of temperature. Its value depends on the strength of an applied field in distorting the electron cloud of the molecule.

The ability of a molecule to undergo distortion of its electronic distribution is called polarizability (α_e). If a molecule contains large atoms with electron fairly distant from the nucleus, the nuclear control is less and the polarizability is greater.

PROBLEM BASED ON MOLAR POLARISATION

Q. 1. Molar polarisation of homonuclear molecule is independent of temperature. Explain.

5.12 Depolarisation Ratio in Raman Spectroscopy

The Raman scattered light is emitted by the simulation of the electric field of the incident light. Therefore, the direction of the vibration of the electric field (or polarization direction), of the scattered light might be expected to be the same as that of incident light. However, in reality, some fraction of the Raman scattered light has a polarization direction that is perpendicular to that of the incident light. This component is called the **perpendicular component.** Usually, the component of the Raman scattered light whose polarisation direction is parallel to that of the incident light is called the **parallel component.** Therefore, the Raman scattered light consists of the **parallel component** and the perpendicular component.

The ratio of the peak intensity of the parallel and perpendicular component is known as the **depolarization ratio** (ρ)

$$\rho = \frac{I_{perpendicular}}{I_{parallel}} \text{ or } \frac{I_{\perp}}{I_{\parallel}} \quad \Rightarrow \quad \text{or } \rho = \frac{I_{depolarized}}{I_{polarised}}$$

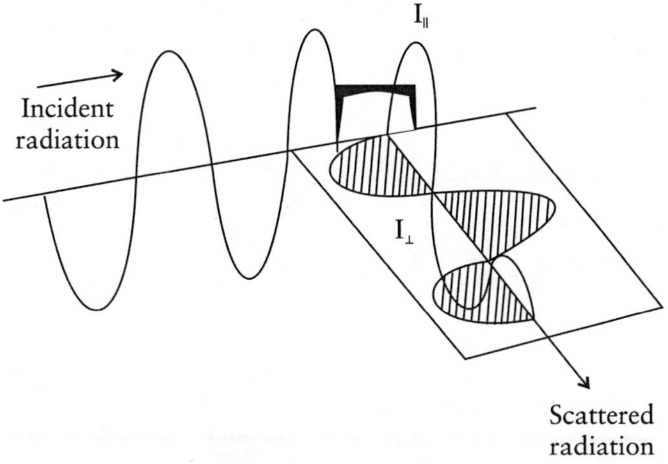

Fig. 5.8: Parallel and perpendicular components of scattered radiation

- A Raman band whose depolarization ratio is less than 0.75 is called **polarized band.** Whereas, a band with a depolarization ratio equal to or greater than 0.75 is called a **depolarized band.**
- The value of depolarization ratio of a Raman band depends on the symmetry of the molecule and on the symmetry of normal vibrational mode.

  Polarised laser radiation = ρ_{max} = 0.75

⇒ $\rho = 0$: total symmetric vibration mode

⇒ $0 < \rho < 0.75$: partial symmetric vibration mode

⇒ $\rho = 0.75$: total asymmetric vibration mode

- For completely polarised light, $I_{\perp} = 0$, thus $\rho = 0$
- For completely unpolarised light, $I_{\perp} = I_{\parallel}$, thus $\rho = 1$

PROBLEM BASED ON DEPOLARISATION RATIO

Q. 1. Briefly explain depolarisation ratio.

5.13 Raman Activity of Vibrations

If a molecule has little or no symmetry, then usually all its vibrational modes are Raman active. However, if the molecule has considerable symmetry it is not easy to determine that which vibration mode is Raman active, i.e. whether or not the polarizability changes during the vibration.

To explain this, few examples are discussed below:

(1) Asymmetric Top Molecule (H_2O):

Number of atoms = N = 3

Total degrees of freedom = 3N = 9.

Translational degree of freedom = 3

Rotational degree of freedom = 3 (Non-Linear)

Vibrational degree of freedom = 3N – 6 = 9 – 6 = 3

All three fundamental vibration modes (ν_1, ν_2 and ν_3) are Raman active.

Since all the three vibration involves change in size, shape or direction of the polarizability ellipsoid of the water molecule during each of its three vibration modes and thus all are Raman active.

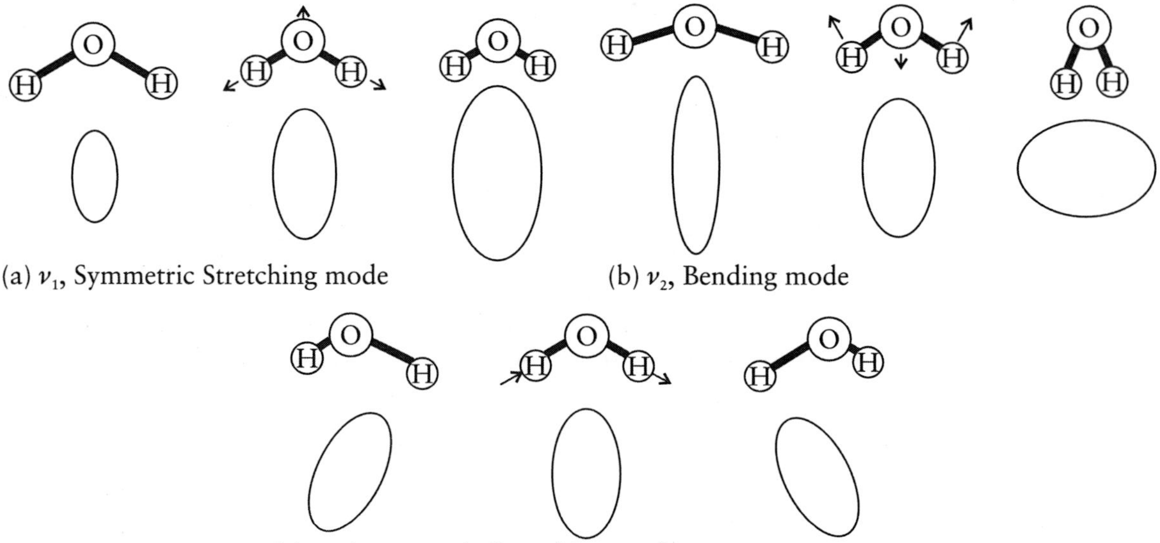

Fig. 5.9: Changes in Polarizability Ellipsoid of water molecule during its three vibrational modes

(2) Linear Triatomic Molecule (CO_2):

Number of atoms = N = 3

Total degrees of freedom = 3N = 9

Translational degree of freedom = 3

Rotational degree of freedom = 2 (Linear molecule)

Vibrational degree of freedom = 3N – 5 = 9 – 5 = 4

Thus there are four fundamental vibrations in CO_2. Out of these four, only one is symmetric stretching vibration (ν_1), one is asymmetric stretching vibration (ν_3) and other two are bending vibration (ν_2). The bending involves two vibration, i.e. one in the plane of the paper and the other in which the oxygen atoms moves simultaneously into and out of the plane. These two vibrations are degenerate. Thus in CO_2 molecule, bending mode is doubly degenerate and both the bending vibration occur at one frequency (ν_2).

(a) **During the Symmetric stretch,**

$\leftarrow O{=}C{=}O \rightarrow$

the dipole moment of the molecule remains zero throughout the vibration i.e. the dipole moment does not alter periodically, and hence this **vibrational mode will be IR inactive.**

Since during the vibration, volume of the molecule changes, thus there occurs a change in the polarizability of the molecule and hence **this vibrational mode is Raman active.** So, there must be small displacement near the equilibrium position of the vibration motion, which produces a change in the polarizability ($d\alpha/d\xi \neq 0$).

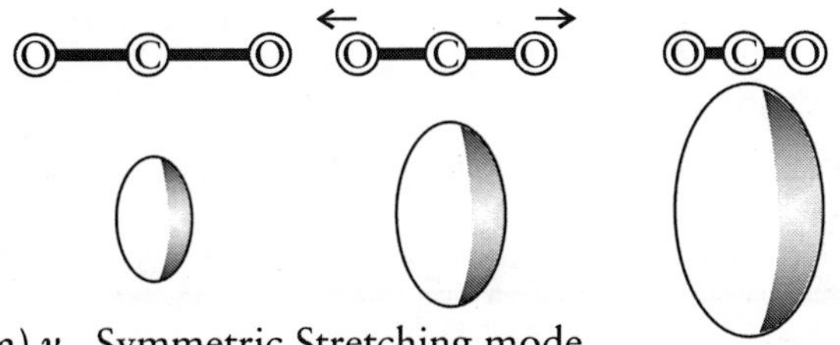

(a) ν_1, Symmetric Stretching mode

Fig. 5.10: Changes in the Polarizability Ellipsoid during Symmetric stretching of CO_2 molecule

Similarly, the ν_2 and ν_3 vibrations of CO_2 molecule are also thought to be Raman active, because the molecule changes shape during each vibration, hence there are changes in the polarizability ellipsoid. However, these modes are observed to be Raman inactive.

(b) **During the Anti-Symmetric stretch,**

$\underset{\rightarrow}{O} = \underset{\leftarrow}{C} = \underset{\rightarrow}{O}$

one of the bonds undergoes lengthening, whereas the other bond undergoes shortening of the bond length. Thus, there is a change in the dipole moment of the molecule and hence **this vibrational mode will be IR active.** But since the lengthening and shortening of bonds occur by the same amounts, the polarizability (volume) of the molecule does not change much during the vibration so $d\alpha/d\xi = 0$ near the equilibrium position of the vibration motion. **Thus, this vibrational mode is Raman inactive.**

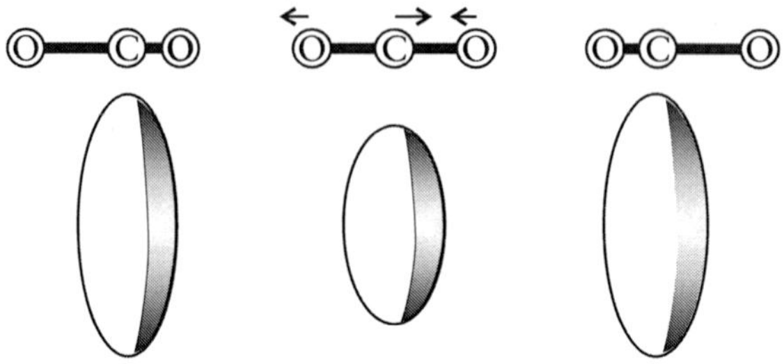

(b) ν_3, Asymmetric Stretching mode

Fig. 5.11: Changes in the Polarizability Ellipsoid during Anti-Symmetric Stretching of CO_2 molecule

(c) **During Bending motion**

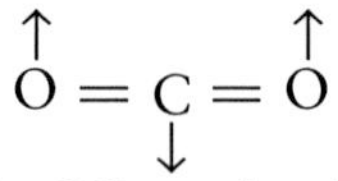

in CO_2 molecule, the dipole moment of the molecule changes and hence this **vibrational mode will be IR active.** But the polarizability (volume) of the molecule does not change much during the vibration ($d\alpha/d\xi = 0$). **This bending in CO_2 molecule is Raman inactive.**

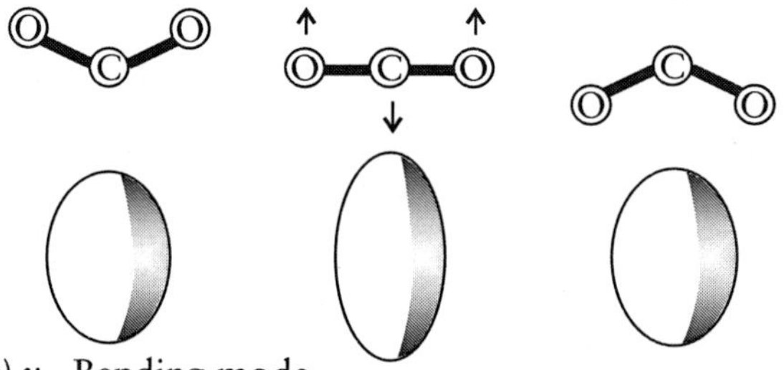

(c) ν_2, Bending mode

Fig. 5.12: Changes in the Polarizability Ellipsoid during Bending Motion of CO_2 molecule

PROBLEMS BASED ON RAMAN ACTIVITY

Q. 1. What is the essential criterion for a molecule to be Raman active. Do homonuclear diatomic molecule exhibit Raman spectrum?

Ans. In order to be Raman active, a molecular rotation or vibration must cause some change in a component of the molecular polarizability. Yes, homonuclear diatomic molecule exhibit Raman spectrum.

Q. 2. The asymmetric stretching and bending modes of vibration of CO_2 molecule are Raman inactive? Why?

Ans. This happens because there is effectively no change in the polarizability in the asymmetric stretching and bending vibration of CO_2 ($d\alpha/d\xi = 0$) and hence they are Raman inactive.

Q. 3. Which of the following molecules will be Raman active and why?

CO_2, H_2, CH_4, N_2O

Ans. CO_2, H_2 and N_2O are Raman active because of change in polarizability ellipsoid.

5.14 Rule of Mutual Exclusion

Certain molecular vibrations interact only with the IR radiation whereas certain others show the Raman effect. A general rule, known as the **rule of mutual exclusion**, for molecules having centre of symmetry was given. This rule states that in the molecules having centre of symmetry (like CO_2), the Raman active vibrations are infrared inactive and vice versa.

Whereas, the molecules which does not have centre of symmetry (like H_2O), all the vibrations are active in both IR and Raman. When a vibration is active in both, it may be easier to observe in one because of intensity considerations.

Mode of vibration	Molecule with centre of symmetry		Molecule with no centre of symmetry	
	CO_2		H_2O	
	Raman	IR	Raman	IR
Symmetric stretching	Active	Inactive	Active	Active
Bending	Inactive	Active	Active	Active
Asymmetric stretching	Inactive	Active	Active	Active

Thus, if some vibrations are observed to give coincident Raman and IR absorption, it is certain that the molecule has no centre of symmetry.

Thus, by comparing Raman and IR spectra of a substance, one can establish whether the molecule has a centre of symmetry or not and thus the structure of a molecule can be determined. If the two spectra (IR and Raman) include common lines, then the molecule has no centre of symmetry. Whereas on the other hand, if the two spectra do not include the common lines, it implies that the molecule has centre of symmetry.

PROBLEMS BASED MUTUAL EXCLUSION PRINCIPLE

Q. 1. Short note on mutual exclusion.

Q. 2. What is the rule of mutual exclusion and its use in structure elucidation?

5.15 Application of IR and Raman Spectra

IR and Raman spectra both are needed to determine the structure of the molecule.

Various rules are helpful in determining the structure of the molecule:

(1) **Rule of mutual exclusion** which states that if a molecule has centre of symmetry, then Raman active vibrational modes are IR inactive and vice versa.

(2) If the molecule is **linear,** the vibrational modes in which linearity is preserved will exhibit PR contours. Whereas the vibrational modes in which linearity is destroyed will exhibit PQR contours.

(3) Symmetric vibration gives polarized or partially polarized Raman line. Whereas an antisymmetric vibration gives depolarized Raman line. Bending also gives polarised Raman line.

(4) The absorption lines follow the order $\nu_{\text{Bending}} < \nu_{\text{symm.stretch}} < \nu_{\text{antisymm.stretch}}$

This is because the bond length deformation is more difficult than bond angle deformation, thus more energy (higher frequency) is required for stretching than bending.

PROBLEMS BASED ON STRUCTURE ANALYSIS

Q. 1. The spectroscopic data for CO_2 molecule is given below. Predict the geometry of the CO_2 molecule.

Wavenumber (cm^{-1})	IR	Raman
1330	Inactive	Active
2349	Active (‖) (PR)	Inactive
667	Active (⊥) (PQR)	Inactive

Ans. The IR and Raman spectra do not include any common line and hence the molecule contains the centre of symmetry. The structure is O=C=O with C in centre.

Wavenumber (cm^{-1})	Mode of Vibration	Explanation (CO_2 is a Linear Molecule)
1330	Symmetric stretching ←O=C=O→	This vibration is IR inactive, i.e. dipole moment does not change. Also this vibration is Raman active as polarizability changes during vibration. Thus, this must be symmetric stretching.
2349	Anti-Symmetric stretching O = C = O → ← → (parallel vibration)	Linearity of molecule is maintained as it includes PR contour in IR spectrum and also it is a parallel vibration. Since during this mode the linearity of molecule is not changed, thus this must be antisymmetric stretching. Antisymmetric stretching has higher frequency than symmetric stretching.
667	Bending ↑ ↑ O = C = O ↓ perpendicular vibration	This transition includes PQR contour, thus linearity is destroyed. Also it is a perpendicular vibration, thus linearity is not maintained. Since it requires least frequency and bending vibration is easy to carry out than stretching vibration.

Q. 2. For a molecule A_2B, the following spectroscopic data was obtained. Assign and explain the various transitions involved and predict the geometry of the molecule.

Wavenumber (cm^{-1})	IR	Raman
589	Active (PQR)	Inactive
1285	Active (PR)	Active (Polarised)
2224	Active (PR)	Active (Depolarised)

(OR)

A molecule AB_2 has the following IR and Raman spectra

Wavenumber (cm^{-1})	IR	Raman
589	Strong (PQR)	—
1285	Very strong (PR)	Very strong (polarised)
2224	Very strong (PR)	Strong (depolarised)

(i) Assign these bands to various vibrations.

(ii) Predict the geometry of A_2B molecule with reasons. Make suitable comments about each observation.

(iii) Calculate the zero point energy of AB_2 molecule.

(OR)

The spectroscopic data for N_2O molecule is given below. Predict the geometry of the N_2O molecule.

Wavenumber (cm^{-1})	IR	Raman
1285	Active (PR)	Active (polarized)
2224	Active (PR)	Active (depolarized)
589	Active (PQR)	Inactive

Ans. The IR and Raman spectra include some common lines and the molecule does not contain the centre of symmetry. Thus, the structure of N_2O is N≡N=O, i.e. O is not in the centre.

Wavenumber (cm^{-1})	Mode of Vibration	Explanation (N_2O is a Linear Molecule)
1285	Symmetric stretching N—N—O ← →	Since the transition includes PR contour, thus molecule is linear. As Raman line is polarised, thus it is assigned symmetric vibration.
2224	Anti-Symmetric stretching N—N—O → ← →	Since the transition includes PR contour, thus molecule is linear. As Raman line is depolarised, thus it is assigned asymmetric stretch.
589	Bending ↑ N — N — O ↓ ↓	This transition includes PQR contour, thus linearity of the molecule is lost. This transition is assigned bending vibration as minimum energy is required for bending.

Q. 3. A triatomic molecule AB_2 gave the following IR and Raman vibration spectral data. Deduce the structure of the molecule giving proper explanations and assign the wavenumbers to the appropriate modes of vibration.

Wavenumber (cm^{-1})	IR	Raman
519	Active (PQR)	Active (Polarised)
1151	Active (PQR)	Active (Polarised)
1361	Active (PQR)	Active (Depolarised)

(OR)

A molecule AB_2 has the following infrared and Raman spectra

$\bar{v}$ (cm^{-1})	IR	Raman
519	Active	Active (Polarised)
1151	Active	Active (Polarised)
1361	Active	Active (Depolarised)

Giving proper explanation to arrive at the geometry of the molecule. Assign the wavenumber to specific vibrations.

(OR)

The spectroscopic data for SO_2 molecule is given below. Predict the geometry of the SO_2 molecule.

Wavenumber (cm^{-1})	IR	Raman
1151	Active (‖) (PQR)	Active (Polarised)
1361	Active (⊥) (PQR)	Active (Depolarised)
519	Active (‖) (PQR)	Active (Polarised)

Ans. Since all the transitions include PQR contour, thus the molecule is non-linear.

- The IR and Raman spectra involve common lines and hence there is absence of centre of symmetry in the molecule.
- But two of the Raman lines are polarized which indicates that the symmetry of the molecule is preserved during the vibration. Thus, the structure assigned to SO_2 is 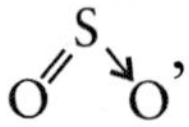, with two oxygen atoms placed symmetrically with respect to sulphur atom.

Wavenumber (cm^{-1})	Modes of Vibration	Explanation
1151	Symmetric stretching 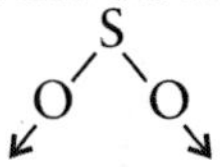	Raman line is polarised thus symmetric vibration
1361	Anti-symmetric stretching 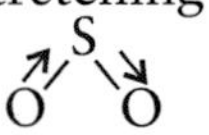	Raman line is depolarised, thus anti-symmetric vibration
519	Bending 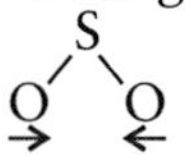	Raman line is polarised. Minimum energy is required for bending of a bond.

Q. 4. A triatomic molecule DB_2 shows the following lines in Raman and IR spectra

Wavenumber (cm^{-1})	IR	Raman
2671	Active	Active
1178	Active	Active
2788	Active	Active

Deduce the structure of this molecule and assign these bands to various vibrational modes explaining why those modes are IR or Raman active.

Ans. [Same as Ans. 3]

- The IR and Raman spectra involve common lines and hence there is absence of symmetry in the molecule. Thus, the molecule is a linear triatomic molecule with no centre of symmetry.
- Since

$\nu_{bending} < \nu_{symm.stretching} < \nu_{antisymm.stretching}$

Wavenumber (cm^{-1})	IR	Raman
2671	Symmetric stretching 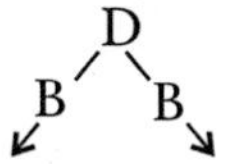	Active
2788	Anti-symmetric stretching 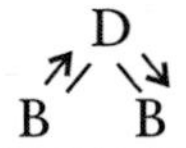	Active
1178	Bending 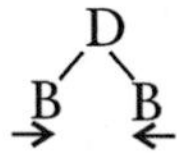	Active

Q. 5. A molecule AB_2 shows IR and Raman spectra as given below:

Wavenumber (cm^{-1})	IR	Raman
653	—	Active (Polarised, very strong)
1533	Active (very strong)	—
397	Active (very strong)	—

Predict the shape of the molecule and assign various observed lines to the appropriate normal modes of vibration.

Ans. [Same as Ans. 1]

The IR and Raman spectra do not include any common line and hence the molecule contains the centre of symmetry. Hence, the structure of AB_2 is B — A — B.

Wavenumber (cm^{-1})	Mode of Vibration	Explanation
653	Symmetric stretching ← B — A — B →	This vibration is IR inactive, i.e. dipole moment does not change. Also this vibration is Raman active as polarizability changes during vibration. Thus, this must be symmetric stretching.
1533	Anti-symmetric stretching B — A — B → ← →	This vibration is IR active, since dipole moment changes. Also it is Raman inactive since its polarizability remains unchanged. Also anti-symmetric stretching has higher frequency than symmetric stretching.
397	Bending ↑ B — A — B ↓ ↓	Since bending requires least frequency and bending vibration is easy to carry out than stretching vibration.

Q. 6. A molecule XY_2, has the following IR and vibration Raman spectral data.

Wavenumber (cm^{-1})	IR	Raman
1243	Inactive	Active
2920	Active (PR)	Inactive
786	Active (PQR)	Inactive

Give proper explanations, arrive at the geometry of the molecule. Assign the wavenumbers to specific vibrations.

Ans. (Same as Ans. 1)

Wavenumber (cm^{-1})	Modes of Vibration
1243	Symmetric stretching ← Y — X — Y →
2920	Anti-symmetric stretching Y — X — Y → ← →
786	Bending ↑ Y — X — Y ↓ ↓

Q. 7. Water molecule shows following IR and Raman spectrum.

Wavenumber (cm^{-1})	IR	Raman
3755	Active	Active
3650	Active	Active
1595	Active	Active

(i) Assign and show these lines to various vibrations.

(ii) We know the structure of water but deduce this structure from above spectrum.

Ans. [Same as Ans. 3]

Q. 8. For acetylene, the following spectroscopic data are available. Deduce its structure.

Wavenumber (cm^{-1})	IR	Raman
3374	—	Active (polarized)
1974	—	Active (polarized)
3287	Active (PR)	—
612	—	Active (weak)
729	Active (PQR)	—

Ans. Since IR and Raman spectra do not contain any common line and hence the molecule contains the centre of symmetry.

Wavenumber (cm^{-1})	Mode of Vibration	Explanation
3374	H—C≡C—H ← → ← → Symmetric stretching	The vibration is IR inactive, i.e. dipole moment does not change. Also, this vibration is Raman active as polarizability changes during vibration. Thus, this must be symmetric stretching.
1974	H—C≡C—H ← ← → → Symmetric stretching	The vibration is IR inactive, i.e. dipole moment does not change. Also, this vibration is Raman active as polarizability changes during vibration. Thus, this must be symmetric stretching.
3287	H—C≡C—H ← → → ← Anti-symmetric stretching	This vibration is IR active, since dipole moment changes. Also it is Raman inactive since its polarizability remains unchanged. Since the transition includes PR contour, thus molecule is linear.
612	↑ ↑ H—C≡C—H ↓ ↓ Bending (in plane)	This vibration is IR inactive, i.e. dipole moment does not change. Since it requires least frequency and bending vibration is easy to carry out than stretching vibration.
729	↑ ↑ H—C≡C—H ↓ ↓ Bending (out of plane)	This transition includes PQR contour, thus linearity is destroyed. Also it is a perpendicular vibration, thus linearity is not maintained. Also bending vibration needs less frequency.

Q. 9. A molecule A_2B_2 shows IR and Raman spectra as tabulated below:

Wavenumber (cm^{-1})	IR	Raman
3374	—	Strong, polarized
1974	—	Strong, polarized
3287	Active (PR)	—
612	—	Weak, polarized
729	Active (PQR)	—

Predict the shape of the molecule and assign the various observed lines to the appropriate normal modes of vibrations.

Ans. (Same as Ans. 8)

6

Electronic Spectroscopy

6.1 Introduction

An electron in a molecule can be excited from an occupied molecular orbital to an empty or partially filled higher molecular orbital. These transitions are called **electronic transition.** These transitions involve high energy, so there is absorption of radiations falling in the visible and ultraviolet regions. Thus, electronic spectroscopy is also called as **Ultraviolet Spectroscopy**. While electronic spectra in the visible region span 12,500–25,000 cm^{-1}, those in the ultraviolet region span 25,000–70,000 cm^{-1}. As electronic transition in a molecule is accompanied by vibrational and rotational transitions, resulting each band in the spectra to have number of lines and thus making the electronic spectra of molecules highly complex. The change in electronic arrangement in a molecule is always accompanied by change in dipole moment, so the molecules like H_2 or N_2, which show no vibrational-rotational spectra in the infrared region show vibrational and rotational spectrum in their electronic spectra. Thus, major advantage of electronic spectroscopy is that every molecule gives electronic spectra irrespective of whether it has a permanent dipole moment or not as well whether it shows a change in dipole moment or not.

If the electronic excitation in the molecule had occurred without any accompanying changes in vibrational and rotational sub-levels, the spectrum would have consisted of sharp lines with each line representing a change in electronic state. But due to mixing of vibrational and rotational changes with electronic changes in the molecules, there will be a large number of possible transitions requiring only slightly different energies. This will require the absorption of a large number of wavelengths which are quite close together resulting in the formation of broad bands (shown in Fig. 6.1) in the spectrum.

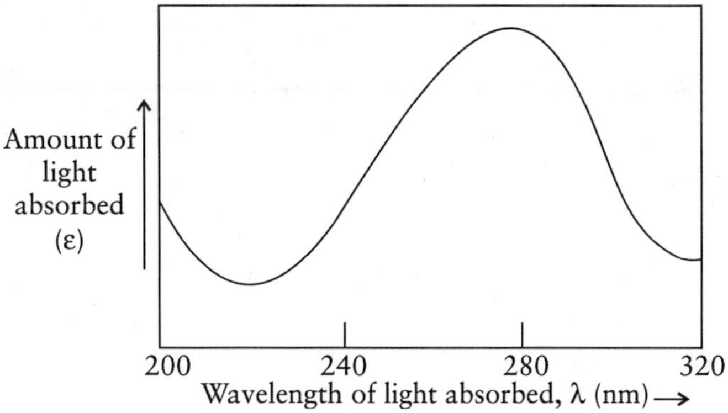

Fig. 6.1: The Ultraviolet spectrum of acetone in hexane

PROBLEMS BASED ON ELECTRONIC SPECTRA

Q. 1. Why is it that all the molecules show electronic spectra but not necessarily a vibration spectrum?

Ans. This is because a change in a electron distribution in a molecule are always accompanied by a change in dipole. Thus, the homonuclear molecules like H_2 or N_2 which show no rotation or vibrational-rotational spectra, do give the electronic spectrum and show vibrational and rotational structure in their spectra from which rotational constant and bond vibrational frequencies can be calculated.

Q. 2. What change does the absorption of UV-visible radiation causes in the molecule?

Ans. On absorption of UV-visible radiation, an electron in a molecule can be excited from an unoccupied molecular orbital to an empty or partially filled higher molecular orbital. The change in electronic arrangement in a molecule is always accompanied by change in dipole moment. So the homonuclear diatonic molecules (H_2 or N_2) which show no vibrational-rotational spectra in the infrared region show vibrational and rotational spectrum in their electronic spectra.

6.2 The Born-Oppenheimer Approximation for Electronic Spectra

The Born-Oppenheimer approximation in the present context may be written as:

$$E_{Total} = E_{es} + E_{vib} + E_{rot}$$

This implies that the electronic, vibrational and rotational energies of a molecule are completely independent of each other.

A change in the total energy of a molecule may be written as

$$\Delta E_{Total} = \Delta E_{es} + \Delta E_{vib} + \Delta E_{rot} \text{ (Joules)}$$

$$\Delta \in_{Total} = \Delta \in_{es} + \Delta \in_{vib} + \Delta \in_{rot} \text{ (cm}^{-1}\text{)} \qquad ...(6.1)$$

and $$\psi_{Total} = \psi_{es} \cdot \psi_{vib} \cdot \psi_{rot}$$

The approximate orders of magnitude of these changes are: $\in_{es} >> \in_{vib} >> \in_{rot}$

$$\Delta \in_{vib} \approx 10^3 \, \Delta \in_{rot}$$

$$\Delta \in_{es} \approx 10^3 \, \Delta \in_{vib}$$

Thus, $$\Delta \in_{es} \approx 10^6 \, \Delta \in_{rot}$$

The electronic transitions are accompanied by change in vibrational and rotational energy levels. We see that vibrational changes will produce a 'coarse structure' and rotational changes will produce a 'fine structure' on the spectra of electronic transitions.

6.3 Coarse Structure of Electronic-Vibration Transition

Since $\Delta \in_{es} \approx 10^6 \in_{rot}$, i.e. compared to rotation, electronic energy levels is 10^6 times, thus one can neglect ψ_{rot} and $\in_{rot}$ for convenience. Thus for a transition of the vibrational coarse structure of spectra, we may write

$$\Delta \in_{Total} = \Delta \in_{es} + \Delta \in_{vib} \text{ (in cm}^{-1}\text{)} \qquad ...(6.2)$$

$$\Delta \in_{Total} = \Delta \in_{es} + \bar{\nu}_e \left(v + \frac{1}{2} \right) - x_e \, \bar{\nu}_e \left(v + \frac{1}{2} \right)^2 \qquad ...(6.3)$$

and $$\psi_{Total} = \psi_{es} \cdot \psi_{vib}$$

The energy levels for these electronic-vibrational transitions is shown in Fig. 6.2. The ground state electronic and vibrational levels are represented as $(\in''_{es}, v'')$, while the upper state electronic and vibrational level are represented as $(\in'_{es}, v')$.

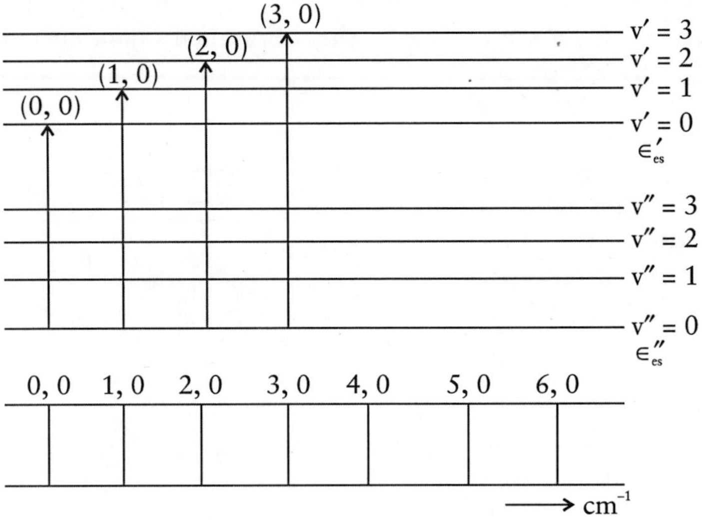

Fig. 6.2: The Vibrational coarse structure during electronic absorption from ground state to upper state

From the Fig. 6.2, we can see that:

(1) The spacing in upper state vibrational levels is small as compared to ground state vibrational levels because the molecule in an electronic excited state have a weaker bond and a smaller value of $\bar{\nu}_e$.

(2) There is no specific selection rule for electronic transition and every transition from $v'' \rightarrow v'$ have some probability and thus a number of lines are expected in the electronic spectrum.

(3) The set of electronic transitions are represented as (v', v''). Therefore, (0, 0), (1, 0), (2, 0), (3, 0) appears.

(4) Also equation (6.3) can be written as

$$\Delta\in_{Total} = (\in'_{es} - \in''_{es}) + \bar{\nu}_e\left(v+\frac{1}{2}\right) - x_e\,\bar{\nu}_e\left(v+\frac{1}{2}\right)^2 \text{ (in cm}^{-1}\text{)}$$

(5) The transitions are considered from electronic ground state only because at room temperature maximum population is in ground state. Also we have,

$\Delta v = 0, \pm 1, \pm 2$ .....

$\Delta v = 0$ is also allowed because there is simultaneous change in electronic energy.

(6) The vibrational coarse structure during an electronic transition can be shown in Fig. 6.3, as in terms of Morse curve.

The Morse curve in excited electronic state is a little shallower in comparison to ground state electronic curve, as the molecule is less stable in excited state. Also the spacing between different vibrational level is small in excited state in comparison to ground electronic state.

(7) For a given electronic transition there is no quantum mechanical restrictions (i.e. the selection rules) on the change in the vibrational quantum number. Since in the ground electronic-state most molecules occupy the lowest vibrational level (v″ = 0), all types of transitions from this lowest vibrational level to any of vibrational levels of the higher electronic state are possible. Experimentally, it is known that all such transitions are not equally intense. In some spectra, (0, 0) transition is the strongest, while in some other spectra, intensity is maximum at some other v′ value and in some other spectra, few vibrational lines are seen followed by a continuum. All these behaviour are explained with the help of **Frank Codon Principle.**

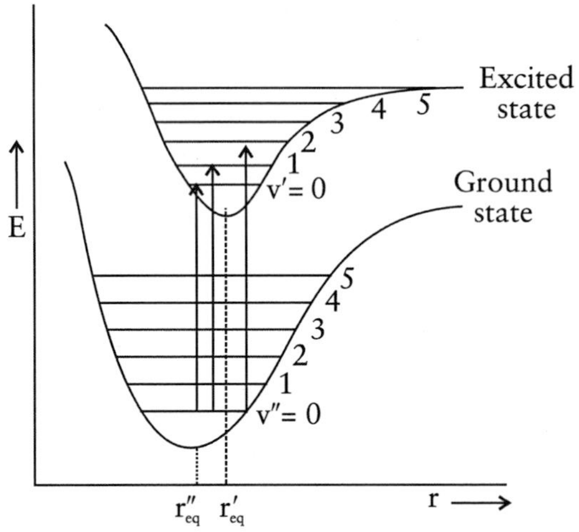

Fig. 6.3: Morse P.E. Curve $r'_{eq} > r''_{eq}$ (Since the Molecule in Excited Electronic State is less Stable in Comparison to ground Electronic state)

6.4 Frank Codon Principle

The relative intensity of molecular spectra, both in absorption and emission, may be interpreted in terms of the Franck Codon Principle.

According to this principle, the time required for an electronic transition is much shorter than the time required for a vibrational transition and hence during the transition, the internuclear distance of the vibrating molecule does not undergo any appreciable change. It states that **"The electronic transition occurs so quickly that a vibrating molecule does not change its internuclear distance during transition."**

Thus, the instantaneous internuclear separation in the excited electronic state is the same as it was in the initial ground state at the time of electronic transition.

An electronic transition occurs in about 10^{-16} s, that is very short as compared to the vibration of nuclei (10^{-13} s). This is because the electrons move at a faster rate as compared to heavier nuclei so that during transition atoms/nuclei do not change their position.

During absorption or emission of radiation, a molecule is being transformed into a different energy state represented by a different potential energy curve. Since the internuclear distance remains same, so the transition can be represented by a vertical line in the potential energy curve.

This principle also explains the occurrence of lines of variable intensities. The intensity of transition at a particular internuclear separation depends on the probability of finding the molecule in the two vibrational levels at the given internuclear separation. A transition which involves the maximum probability of finding the molecule in the ground vibrational level and excited vibrational level will be the most intense. Other transitions will have lesser probability and lesser intensities. Transition labelled as 'b' in Fig. 6.4 will be more intense than the other two shown transitions, i.e. a and c.

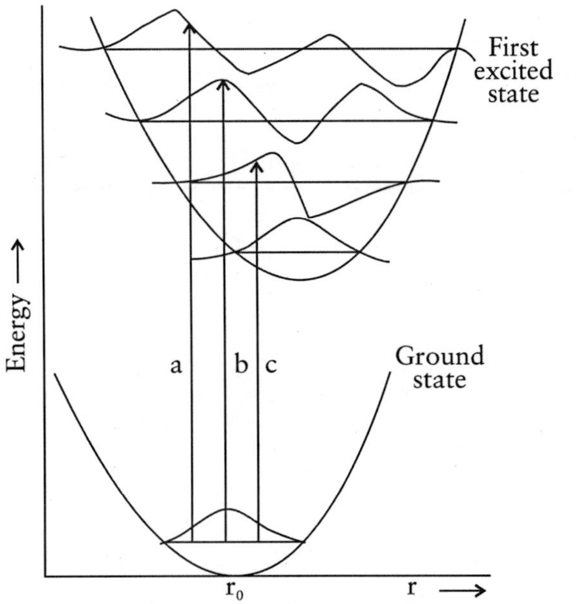

Fig. 6.4: Electronic transition occurring along vertical line, illustrating Frank Codon Principle

In anharmonic oscillator,

$\Delta v = \pm 1, \pm 2, \pm 3, \pm 4$ ... i.e. all the transitions are possible or no restrictions on the change in the vibrational quantum number during an electronic transition. The Frank Codon Principle, governs that which Δv is going to be most intense or probable. The most probable transition is the most probable initial state. At room temperature, all molecules are likely to be present in the vibrational ground state for which $\Delta v = 0$. The final vibrational level in the most probable transition depends on the relative position of two potential energy curves. Thus, the level which corresponds to the point of intersection of the vertical line and the potential energy curve will be the final level is the most probable transition. So, there are no selection rules for Δv and the transition will occur between the positions at which the wavefunction is maximum.

The exact nature of the electronic spectrum depends on the internuclear separation of the molecule in the excited state relative to ground state. In general, the internuclear separation in the excited state is slightly greater in comparison to ground state and the force constant is slightly smaller for excited state. Thus, in a molecule, bond in the excited state is slightly weaker than in the ground state. Therefore, the minimum in the potential energy curve is displaced towards larger internuclear distance and spacing between different vibrational level is smaller and the curve is little shallower.

The different cases showing excited electronic states of a diatomic molecule are:

Case I: When excited electronic state have the same equilibrium internuclear distance (r_e) as the ground state

i.e. $r_e(\text{E.S.}) = r_e(\text{G.S.})$

For this case, the Morse potential energy curve is similar for ground state and is shown in Fig. 6.5.

For this the strongest spectral line is indicated as (0, 0) i.e. vertical transition from $v'' = 0$ to $v' = 0$.

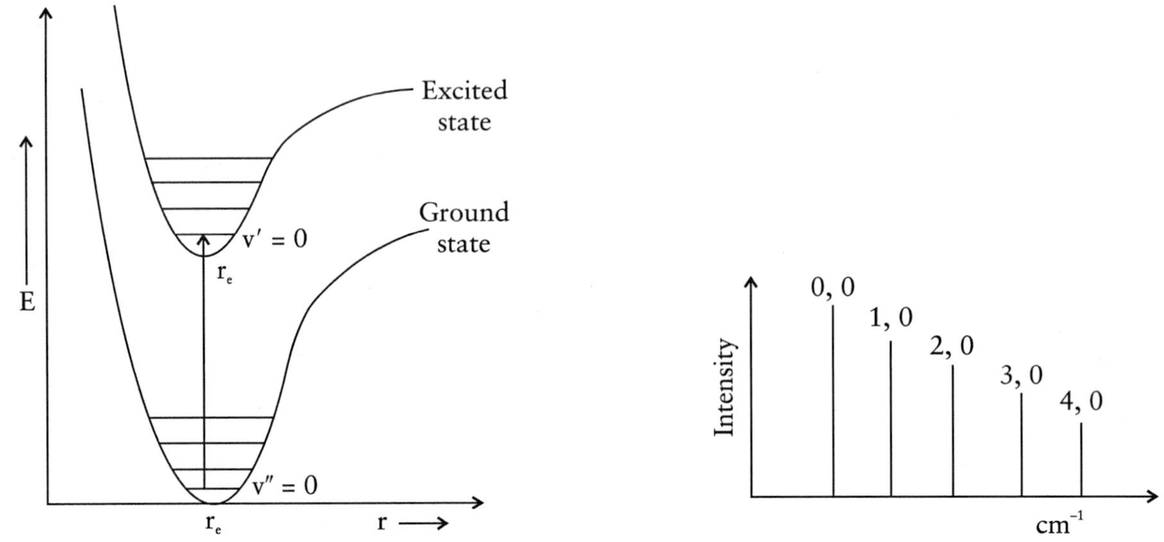

Fig. 6.5: r_e(E.S.) = r_e(G.S.)

Case II: When the excited electronic state has a slightly smaller internuclear distance than the ground state

i.e. $r_e(E.S.) < r_e(G.S.)$

For this the vertical transition from v″ = 0 → v′ = 2 would be most probable transition having maximum intensity as shown in Fig. 6.6.

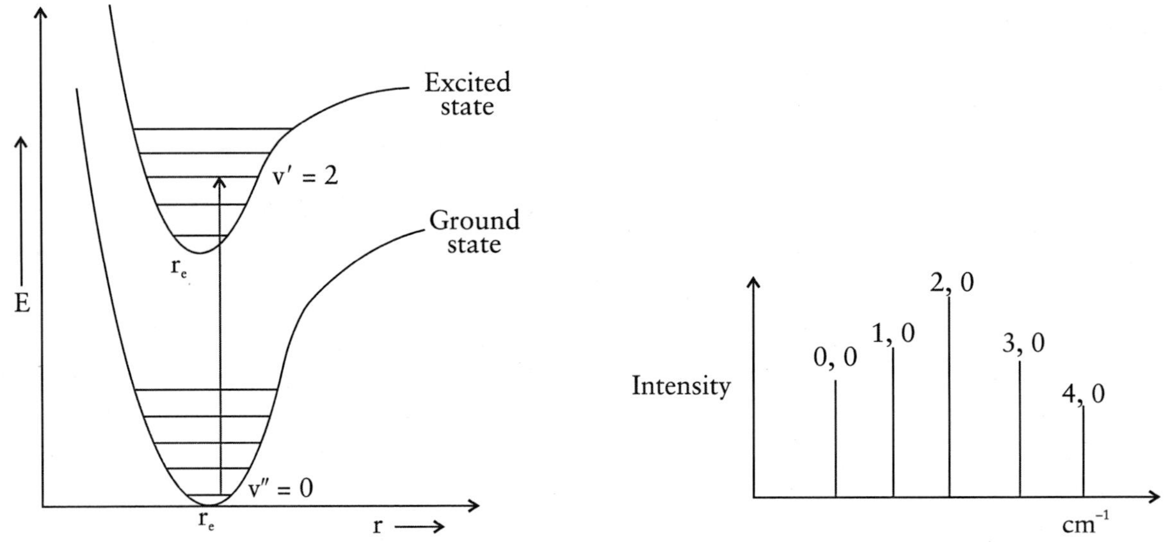

Fig. 6.6: r_e(E.S.) < r_e(G.S.)

Case III: When the excited electronic state has a slightly larger internuclear distance than the ground state

i.e. $r_e(E.S.) > r_e(G.S.)$

For this also the vertical transition from v″ = 0 → v′ = 2 would be most probable transition as shown in Fig. 6.7.

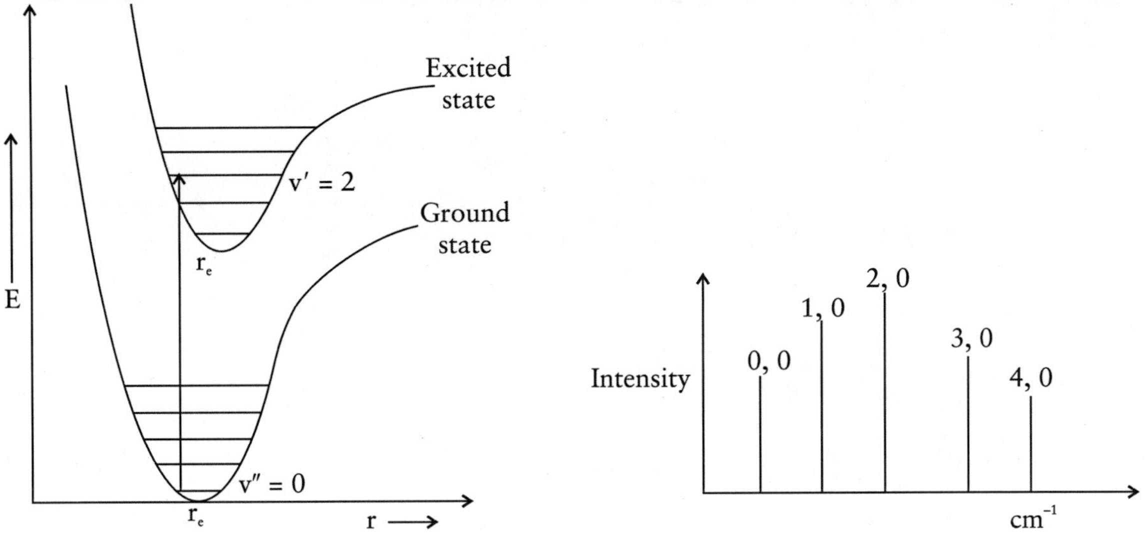

Fig. 6.7: $r_e(E.S.) > r_e(G.S.)$

Case IV: When the excited electronic state has a considerably larger equilibrium internuclear distance than the ground state

i.e. $r_e(E.S.) >> r_e(G.S.)$

In this case the vibrational level to which the transition take place have a high value of 'v'' shown in Fig. 6.8 and at this 'v'' value, the excited molecule has energy in excess of its own dissociation energy, i.e. all radiations are absorbed causing the molecule to dissociate. For such states, the molecule will dissociate without any vibrations and since the atoms which are formed may take up any value of kinetic energy, the transitions are not quantized and continuum results. In other words, energy excess of dissociation limit is utilized in imparting kinetic energies to the dissociated atoms.

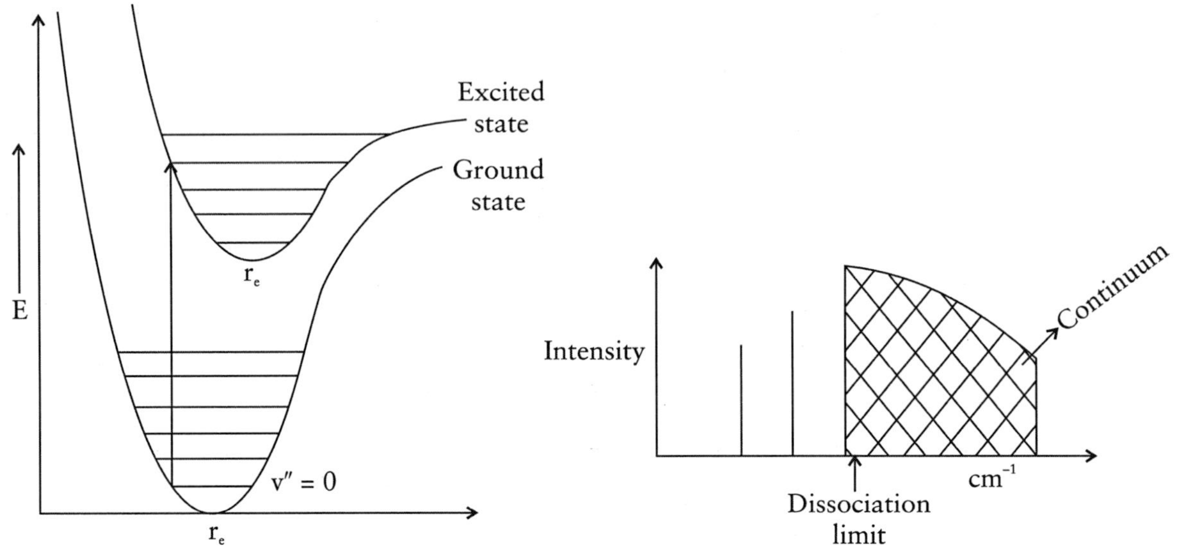

Fig. 6.8: r_e(E.S.) >> r_e(G.S.)

PROBLEMS BASED ON FRANK CODON PRINCIPLE

Q. 1. Electronic transitions are called vertical transitions. Explain.

Q. 2. What is Frank Codon Principle?

(or)

State and explain Frank Codon Principle?

(or)

Discuss how Frank Codon Principle helps to explain the intensity of spectral lines.

(or)

Explain the significance of Frank Codon Principle.

Q. 3. Applying Frank Codon Principle, predict the intensities of lines in the vibration-electronic spectrum if the bond length in a diatomic molecule is slightly greater in the upper electronic state than in the lower electronic state.

Q. 4. Discuss the electronic spectrum obtained in the following cases using appropriate diagrams.

(i) When the equilibrium internuclear distance in the ground state and excited state are same.

(ii) When the equilibrium internuclear distance in the excited state is much larger than in the ground state.

6.5 Dissociation of Molecule (Continuum at High Wavenumber)

Sometimes the electronic excitation passes an amount of energy that lifts the molecule in a state above the highest vibrational levels and dissociation occurs.

In such case, excitation transfers a molecule in a state containing more energy than the separated components, and dissociation takes place. During this, the internuclear separation in the excited state is much larger than in the lower level.

The vertical transition ends at a point where the electronic energy of the molecule in the excited state is greater than the dissociation energy. The electronic spectrum here consists of a coarse structure which ends as a continuum, i.e. respective spectrum displays a blurred zone beyond the so-called **dissociation limit**. Above the dissociation limit, the kinetic energy is not quantized and energy continuum appears instead of discrete energy levels.

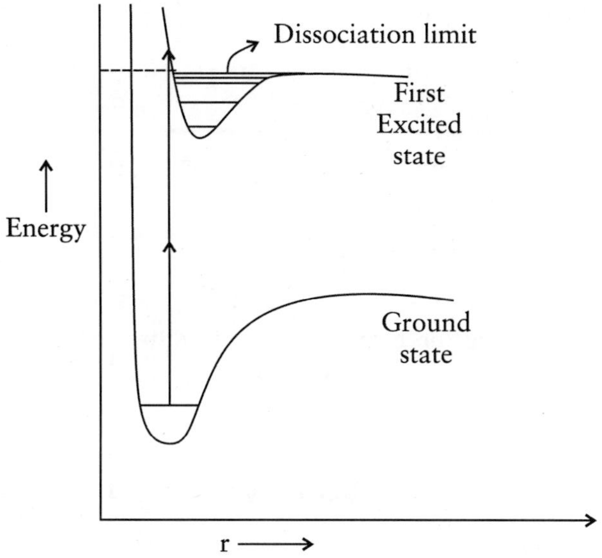

Fig 6.9: Representation of dissociation limit

Case I: Usually, in most of the cases, on dissociation only one of the dissociated atoms produced as a result of absorption of energy (E_D) goes to excited electronic state and the other atom remains in ground electronic state. From Fig. 6.10, we can write this as:

$$E_D = D''_0 + E_{ex} \quad ...(6.4)$$

where, E_D = energy of dissociation limit or total dissociation energy of molecule

D''_0 = dissociation energy of the molecule in the ground electronic state

E_{ex} = excitation of one of the atoms produced on dissociation

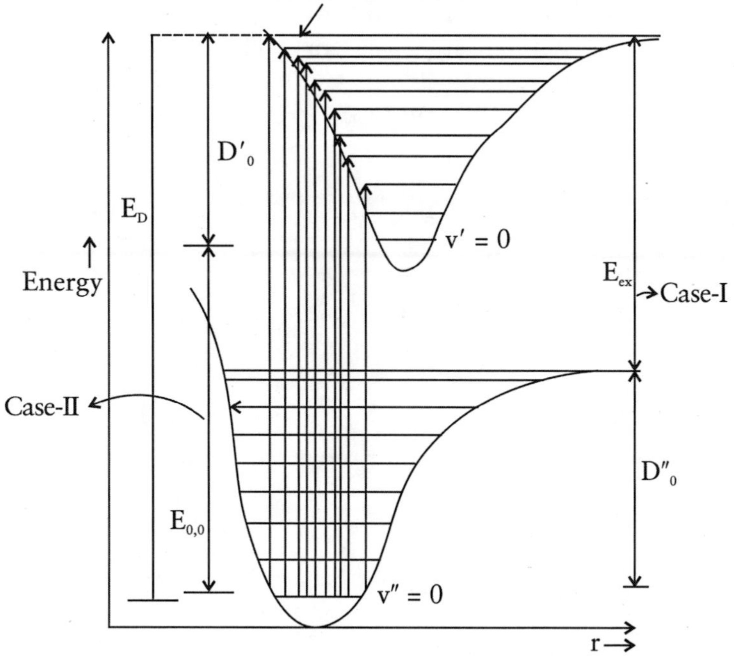

Fig. 6.10: Representation of dissociation of molecule on excitation

Case II: It may happen that molecule in ground state absorbs energy ($E_{0,0}$) and goes to excited state, where it dissociates (D'_0). Thus, referring Fig. 6.10, the E_D can also be written as:

$$E_D = E_{0,0} + D'_0 \qquad \text{...(6.5)}$$

Thus equating equations (6.4) and (6.5):

$$E_{0,0} + D'_0 = E_{ex} + D''_0$$

Hence, we can calculate D'_0 and D''_0:

Dissociation energy of molecule in excited electronic state

$$D'_0 = (E_{ex} + D''_0) - E_{0,0}$$

$$D'_0 = E_D - E_{0,0} \qquad \text{...(6.6a)}$$

Dissociation energy of molecule in ground electronic state

$$D''_0 = (E_{0,0} + D'_0) - E_{ex}$$

$$D''_0 = E_D - E_{ex} \qquad \text{...(6.6b)}$$

PROBLEMS BASED ON DISSOCIATION ENERGY

Q. 1. The vibrational coarse structure of the UV-spectrum of oxygen molecule is a continuum beyond 56876 cm^{-1}. The upper electronic state dissociate into one ground state atom and one excited state atom with excitation energy 15875 cm^{-1}. Estimate the dissociation energy of the ground state of oxygen molecule in cm^{-1} and kJ/mole.

(or)

The absorption spectrum of O_2 shows vibrational coarse structure which ends with a continuum at 56876 cm^{-1}. The excitation energy is 15875 cm^{-1} and the molecule dissociates into one ground state atom and the other at excited state. Calculate the dissociation energy of the ground state in kJ/mole.

Ans. Convergence limit = $\bar{v}$ = 56876 cm^{-1}

Total dissociation energy of molecule or energy of photon absorbed = E_D

$$E_D = h\nu = \frac{hc}{\lambda} = hc\bar{v}$$

$$E_D = (6.626 \times 10^{-34}\ \text{Js}) \times (3 \times 10^{10}\ \text{cms}^{-1}) \times (56876\ \text{cm}^{-1})$$

$$E_D = 11.395 \times 10^{-19}\ \text{J}$$

Also, $\bar{v}_{ex}$ = 15875 cm^{-1}

Excitation energy of one of the atoms produced on dissociation = E_{ex}

$$E_{ex} = h\nu = \frac{hc}{\lambda} = hc\bar{v}_{ex}$$

$$E_{ex} = (6.626 \times 10^{-34}\ \text{Js}) \times (3 \times 10^{10}\ \text{cms}^{-1}) \times (15875\ \text{cm}^{-1})$$

$$E_{ex} = 3.15 \times 10^{-19}\ \text{J}$$

Using equation 6.6 (b); we can calculate,

Dissociation energy of the molecule in the ground electronic state = D''_0

$$D''_0 = E_D - E_{ex}$$

$$D''_0 = (11.395 \times 10^{-19} - 3.15 \times 10^{-19})J$$

$$D''_0 = 8.245 \times 10^{-19}\ J$$

Dissociation energy of one mole of the molecule in the ground electronic state

$$D''_0 = (8.245 \times 10^{-19}\ J) \times (6.023 \times 10^{23}\ mol^{-1})$$

$$D''_0 = 49.651 \times 10^4\ J/mol$$

$$D''_0 = 496.5\ kJ/mol$$

Q. 2. The spectrum arising out of transitions from the ground state of a molecule to its excited state shows that $\bar{v}_{0,0}$ line at 19378 cm^{-1} and a convergence limit at 39231 cm^{-1}. The dissociation is into one ground state atom and one excited atom and the excitation energy of the latter is 10,308 cm^{-1}. Calculate the dissociation energies of the molecule in both the states.

Ans. (a) Convergence limit = $\bar{v}$ = 39231 cm^{-1}

Total dissociation energy of molecule = E_D

$$E_D = h\nu = \frac{hc}{\lambda} = hc\bar{v}$$

$$= (6.626 \times 10^{-34}\ Js) \times (3 \times 10^{10}\ cms^{-1}) \times (39231\ cm^{-1}) = 7.80 \times 10^{-19}\ J$$

Also, $\bar{v}_{0,0} = 19378\ cm^{-1}$

Energy absorb for transition of molecule from ground state to excited state = $E_{0,0}$

$$E_{0,0} = h\nu = \frac{hc}{\lambda} = hc\bar{v}_{0,0}$$

$$= (6.626 \times 10^{-34}\ Js) \times (3 \times 10^{10}\ cms^{-1}) \times (19378\ cm^{-1}) = 3.85 \times 10^{-19}\ J$$

$\therefore$ Dissociation of molecule in excited electronic state = D'_0

$$D'_0 = E_D - E_{0,0} = (7.80 \times 10^{-19} - 3.85 \times 10^{-19})J$$

$$D'_0 = 3.95 \times 10^{-19}\ J$$

(b) Convergence limit = $\bar{v}$ = 39231 cm^{-1}

Total dissociation energy of molecule = E_D

$$E_D = h\nu = \frac{hc}{\lambda} = hc\bar{v}$$

$$E_D = (6.626 \times 10^{-34}\ Js) \times (3 \times 10^{10}\ cms^{-1}) \times (39231\ cm^{-1})$$

$$E_D = 7.80 \times 10^{-19}\ J$$

Also, $\bar{v}_{ex} = 10308\ cm^{-1}$

Excitation energy of one of the atoms produced on dissociation = E_{ex}

$$E_{ex} = h\nu = \frac{hc}{\lambda} = hc\,\bar{v}_{ex}$$

$$E_{ex} = (6.626 \times 10^{-34}\ Js) \times (3 \times 10^{10}\ cms^{-1}) \times (10308\ cm^{-1}) = 2.04 \times 10^{-19}\ J$$

Dissociation energy of molecule in ground electronic state = D''_0

$$D''_0 = E_D - E_{ex}$$

$$D''_0 = (7.80 \times 10^{-19} - 2.04 \times 10^{-19})J$$

$$D''_0 = 5.76 \times 10^{-19}\ J$$

6.6 Predissociation of Molecule

In some spectra, the line structure disappears for energy below the dissociation limit but reappears again at higher energies. This finding is explained by a phenomenon called **predissociation.**

Predissociation is found for a molecule where potential curves for a bound and an unbound state cross each other. A small amount of energy lifts the molecule in a state where the electrons can rearrange to an unbound state (internal conversion). In the spectral range that corresponds to these energies, a blurred zone appears. Beyond this zone, there are again discrete lines. Thus, there is a continuum in the central region. The continuum in the middle of spectrum shown in Fig. 6.12 is due to predissociation. Whereas the continuum which appear at high wavenumber corresponds to complete dissociation.

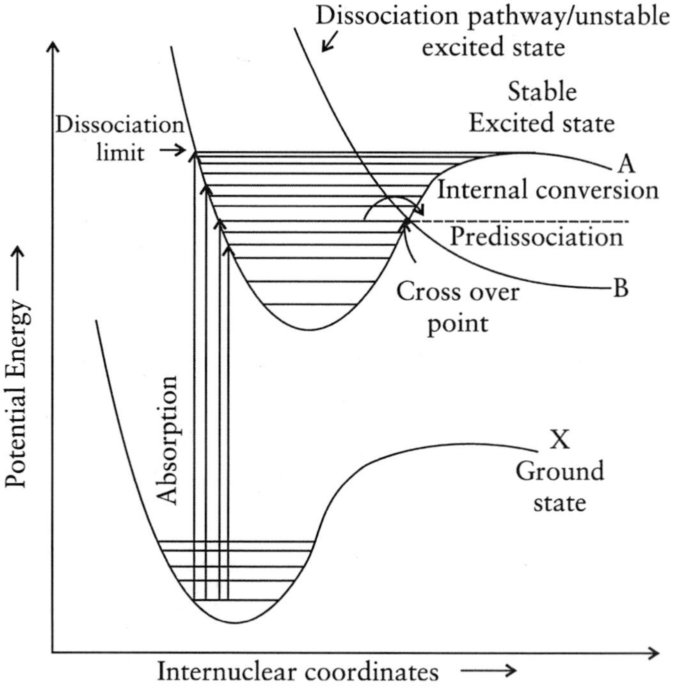

Fir. 6.11: Representation of predissociation of molecule on excitation

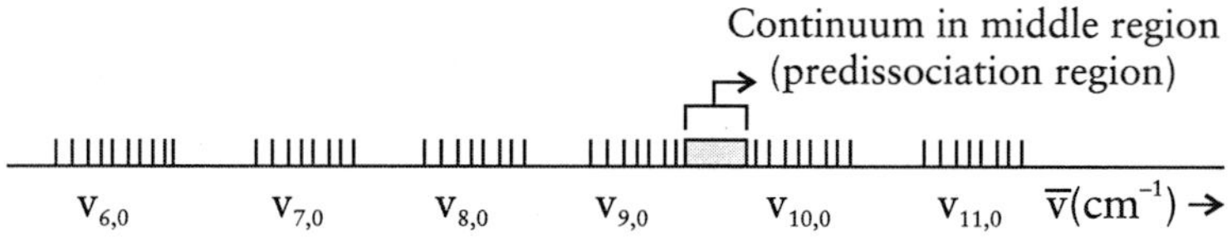

Fig. 6.12: Continuum in the middle region due to predissociation

For a particular geometry near to the end of the molecular vibration, the energy of the molecule in the stable electronic excited state is identical with that in the unstable electronic state, thus the two energy curves (A and B curves in Fig. 6.11) intersect each other at this geometry (marked as cross over point in Fig. 6.11). At this geometry, the molecule during vibration may shift over from the stable to unstable configuration. There is high probability of predissociation, since the molecule spends large time near the extreme ends of vibrational motion for larger value of vibrational quantum number. But for distance smaller and larger than this geometry, the probability of crossing over decreases rapidly, as the probability of finding the molecule at other distance is small.

Thus, if the electronic excitation ends with the low and high vibrational quantum levels in the excited state (above and below cross over point), it is expected to show normal coarse and fine structures, but if electronic excitation occurs at the vibration level where inter-crossing occurs, the molecule can cross over from stable to unstable electronic state and molecule will cross over to continuous curve B and thus dissociates and causes predissociation.

This process is radiation less transfer and is faster than the time taken by molecule to rotate but usually slower than the vibrational time. If the time is faster than vibrational time, a complete continuum will appear in the spectrum.

In most of the usual cases, transitions to higher vibrational level give rise to normal vibrational electronic spectrum with rotational fine lines because as we move to higher vibrational levels, probability of molecule to spend maximum time at the extreme increases and it spends a very little time in the centre. So usually during transition from ground to higher levels, molecule spends less time near the cross over point and there is not predissociation and a normal spectrum appears.

PROBLEMS BASED ON PREDISSOCIATION

Q. 1. What is the difference between dissociation and predissociation?

Q. 2. Explain with the help of potential energy diagram, the phenomena of predissociation.

Q. 3. Write short note on predissociation spectra.

Q. 4. What is predissociation? Under what conditions it is observed?

6.7 Rotational Fine Structure of Electronic-Vibration Transition

The vibrational coarse structure observed has broad and diffuse lines. If the resolution is sufficiently good, each line appear as a cluster of very close lines. This spectrum is rotational fine structure.

Therefore, $\in_{Total} = \in_{es} + \in_{vib} + \in_{rot}\ cm^{-1}$

$$\in_{Total} = \underline{(\in_{es} + \in_{vib}} + \bar{B}J(J+1)cm^{-1} \qquad ...(6.7)$$

$$\Delta \in_{Total} = \underline{\Delta(\in_{es} + \in_{vib})} + \Delta[\bar{B}J(J+1)]cm^{-1}$$

$$\Delta \in_{Total} = \bar{\nu}_{(v',v'')} + \Delta[\bar{B}J(J+1)]cm^{-1} \qquad ...(6.8)$$

where, $\bar{\nu}_{(v',v'')}$ is the wavenumber of electronic-vibration transition

6.7.1 Selection Rule

The selection rule for J depends on the type of electronic transition that a molecule undergo.

(a) If both lower and upper excited states have no electronic angular momentum around the internuclear axis. The selection rule is

$\Delta J = \pm 1$ (i.e. only P and R branches appear)

(b) For all other transitions where either lower or upper state or both state have angular momentum around the internuclear axis, the selection rule is

$\Delta J = 0, \pm 1$ (i.e. only P, Q, R branches appear).

$\Delta J = 0$, but $J = 0 \not\leftrightarrow J = 0$

(i.e. J = 0 to J = 0 level transition is not allowed)

Therefore from equation (6.8).

$$\Delta \in_{Total} = \bar{v}_{(v',v'')} + \bar{B}'J'(J'+1) - B''J''(J''+1) \qquad ...(6.9)$$

where B′, J′ refer to upper electronic state and

B″, J″ refer to ground electronic state

The $\bar{B}$ value in the ground electronic state and upper electronic state is different. Quiet often the bond in the excited state is weaker in comparison to bond in ground state. Hence, the moment of inertia will be larger and $\bar{B}'$ is smaller than $\bar{B}''$.

(1) **For P Branch,** $\Delta J = -1$, i.e. $J' - J'' = -1$ or $J'' = J' + 1$

Substitute the value in equation (6.9):

$$\Delta \in_{Total} = \bar{v}_p = \bar{v}_{(v',v'')} + \bar{B}'J'(J'+1) - \bar{B}''(J'+1)(J'+1+1)$$

$$\bar{v}_p = \bar{v}_{(v',v'')} + \bar{B}'(J'^2 + J') - \bar{B}''(J'^2 + 3J' + 2)$$

Add and subtract $(J' + 1)$ in the middle terms and simplify.

$$\bar{v}_p = \bar{v}_{(v',v'')} + \bar{B}'[J'^2 + J' + (J'+1) - (J'+1)] - \bar{B}''[J'^2 + 3J' + 2 + (J'+1) - (J'+1)]$$

$$\bar{v}_p = \bar{v}_{(v',v'')} + \bar{B}'[(J'^2 + 2J' + 1) - (J'+1)] - \bar{B}''[(J'^2 + 2J' + 1) + (J'+1)]$$

$$\bar{v}_p = \bar{v}_{(v',v'')} + \bar{B}'[(J'+1)^2 - (J'+1)] - \bar{B}''[(J'+1)^2 + (J'+1)]$$

$$\bar{v}_p = \bar{v}_{(v',v'')} + (\bar{B}' - \bar{B}'')(J'+1)^2 - (\bar{B}' + \bar{B}'')(J'+1)\ cm^{-1} \qquad ...(6.10)$$

where, J′ = 0, 1, 2, 3

(2) **For R Branch,** $\Delta J = +1$, i.e. $J' - J'' = 1$ or $J' = J'' + 1$

Substitute the value of J′ in equation (6.9) and on simplifying.

$$\Delta \in_{Total} = \bar{v}_R = \bar{v}_{(v',v'')} + (\bar{B}' - \bar{B}'')(J''+1)^2 + (\bar{B}' + \bar{B}'')(J''+1)cm^{-1} \qquad ...(6.11)$$

where, J″ = 0, 1, 2, 3....

On combining equations (6.10) and (6.11), we get

$$\bar{\nu}_{PR} = \bar{\nu}_{(v',v'')} + (\bar{B}' + \bar{B}'')m \pm (\bar{B}' - \bar{B}'')m^2 cm^{-1} \quad ...(6.12)$$

where m = – ve for P Branch

m = + ve for P Branch

m ≠ 0 (if m = 0, J′ + 1 = 0 or J′ = – 1, which is not possible)

From equation (6.12) we conclude that R-branch appear on higher wavenumber from ($\bar{\nu}_{v',v''}$) [Band origin]

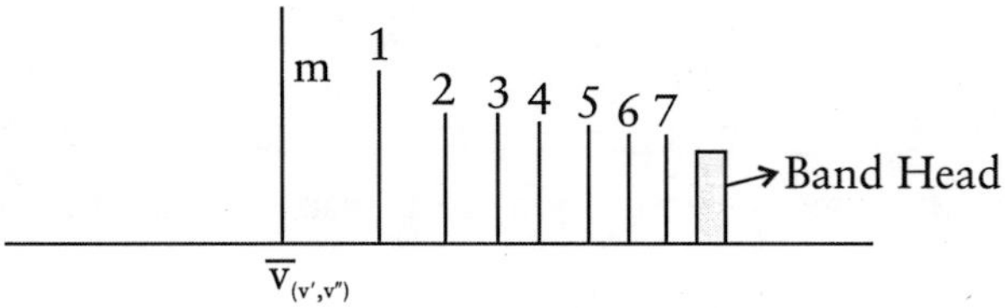

Fig. 6.13: R-Branch in rotational fine structure

Similarly from equation (6.10), we conclude that P-branch appears on lower wavenumber from band centre and spacing lines goes on increasing.

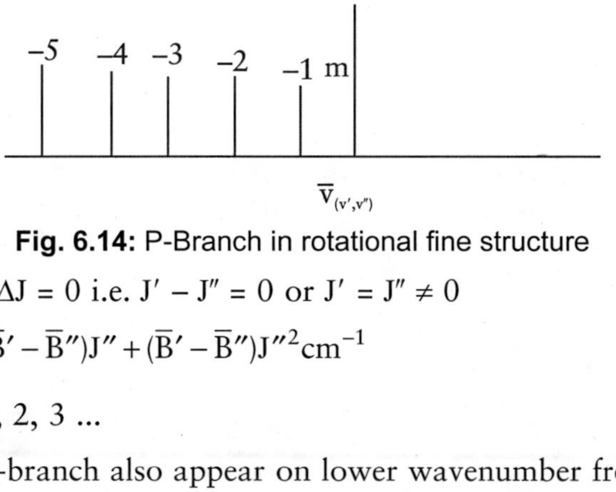

Fig. 6.14: P-Branch in rotational fine structure

(3) For **Q-branch**, ΔJ = 0 i.e. J′ – J″ = 0 or J′ = J″ ≠ 0

$$\bar{\nu}_Q = \bar{\nu}_{(v',v'')} + (\bar{B}' - \bar{B}'')J'' + (\bar{B}' - \bar{B}'')J''^2 cm^{-1}$$

where J″ = 0, 1, 2, 3 ...

Since $\bar{B}' < \bar{B}''$, the Q-branch also appear on lower wavenumber from band origin.

PROBLEM BASED ON ROTATIONAL FINE STRUCTURE

Q. 1. What is the position of P, Q, R lines if B′ > B″.

Ans. (a) P-Branch will appear on lower wavenumber side and band head appears.
(b) Q-Branch will appear on lower wavenumber from band centre.
(c) R-Branch will appear on higher wavenumber from band centre.

6.8 Electronic Spectra of Conjugated Polyene Molecule using Free Electron Molecular Orbital (FEMO) Theory

"Particle in a box" model provides satisfactory interpretation of electronic spectra (UV or visible) of conjugated organic molecules like linear conjugated polyene molecule. The experimental spectral data available for these molecules agree fairly well with the values calculated using "particle in a box" model.

A linear conjugated polyene molecule may be regarded as one-dimensional box and π electrons of molecule as a particle. The π electrons move from one end of the conjugated chain to the other end just as particle in a box continue to move in a to and fro motion within the box. Although there are more than one π electron moving in a box, but the coulombic repulsion among the π electrons is neglected and each electron is supposed to move freely (i.e. $V = 0$). The π electrons occupies the various energy levels in accordance with the rules of electronic distribution in atoms and molecules. The energies of π electrons in various energy levels is calculated by the expression of energy,

$$E_n = \frac{n^2h^2}{8ma^2}$$

Also, the length (a) of the box is taken equal to the sum of all the C—C bond distances of the molecule.

An electron in an atom or molecule can jump from a lower energy level to a higher energy by absorption of appropriate wavelength of light, or it can jump from a higher to a lower level by emission of light. The wavelength (λ) of the light absorbed or emitted is related to the energy gap between the two levels.

$$\Delta E = \frac{hc}{\lambda}$$

This energy gap further depends on the mass and the length of the molecules, i.e. length of the conjugated chain. If the chain length of the molecule is small, hence the mass is less (like butadiene), then the energy gap would be large and the absorption takes place in ultraviolet region (lower λ). While, if the chain length of the molecule is large, hence the mass is large, then the energy gap would be small and the absorption takes place in visible region (higher λ). These above observations can be explained by using the expressions of "particle in a box" model.

Let the energy gap (ΔE) between highest occupied molecular orbital (n) and the lowest unoccupied molecular orbital (n + 1) is:

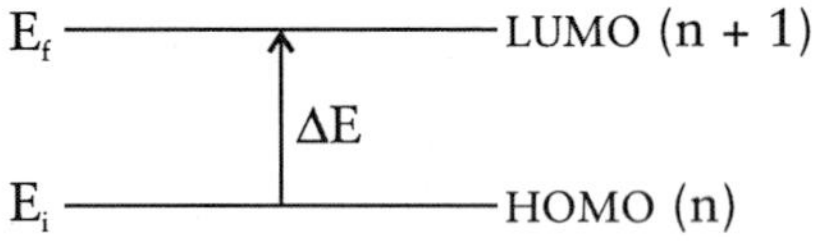

$$\Delta E = E_f - E_i$$

$$\Delta E = E_{n+1} - E_n$$

$$\Delta E = \frac{(n+1)^2h^2}{8ma^2} - \frac{n^2h^2}{8ma^2}$$

$$\Delta E = \frac{(\cancel{n^2} + 1 + 2n - \cancel{n^2})h^2}{8ma^2}$$

$$\Delta E = \frac{(2n+1)h^2}{8ma^2} \quad \text{...(6.13)}$$

$$\cancel{h}\nu = \frac{(2n+1)h^{\cancel{2}}}{8ma^2} \quad (\Delta E = h\nu)$$

$$\nu = \frac{(2n+1)h}{8ma^2}$$

$$\lambda = \frac{c}{\nu} = \frac{8ma^2c}{(2n+1)h} \quad ...(6.14)$$

$$\underset{\text{(wave number)}}{\bar{v}} = \frac{1}{\lambda} = \frac{(2n+1)h}{8ma^2c} \quad ...(6.15)$$

Example: If we take a specific case of butadiene molecule and treat the π-electrons of this conjugated system as particles moving in a one-dimensional box. Then the ground state energy, hence, the lowest absorption frequency and the wavelength of absorbed light for the butadiene molecule can be calculated as follows:

There are 4π electrons in butadiene molecule: $\overset{1}{C}H_2{=}\overset{2}{C}H{-}\overset{3}{C}H{=}\overset{4}{C}H_2$ and these π-electrons are distributed in various energy levels by Aufbau rule, as shown in Fig. 6.15.

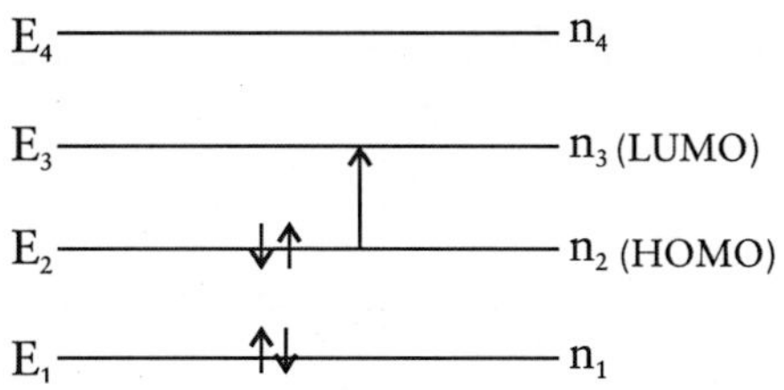

Fig. 6.15: Electron distribution in butadiene molecule

The minimum energy or minimum frequency or maximum wavelength is required for the transition from the highest occupied energy level (HOMO) to the lowest occupied energy level (LUMO).

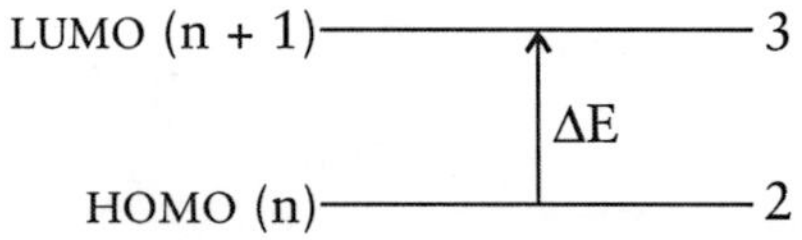

Also, the length of the butadiene molecule is given by

$$a = 1 \text{ (single bond length)} + 2 \text{ (double bond length)} + 2\frac{\text{(single bond length)}}{2}$$

$$a = R_{C-C} + R_{C=C} + 2\left(\frac{R_{C-C}}{2}\right)$$

Whenever, length of molecule is to be determined, we add extra 1/2 R_{C-C} to each end of the molecule. This is called **End effect length.**

Here, in case of butadiene.

$a = 1(154) + 2(135) + 2(154/2)$

$= 578 \text{ pm} = 578 \times 10^{-12} \text{ m} = 5.78 \times 10^{-10} \text{ m}$

m = mass of electron = 9.11×10^{-31} kg

h = Planck's constant = 6.6×10^{-34} Js

$$\Delta E = \frac{(2n+1)h^2}{8ma^2}$$

$$\Delta E = \frac{[2(2)+1]\times(6.6\times10^{-34})^2}{8\times(9.11\times10^{-31})\times(5.78\times10^{-10})^2}$$

$$\Delta E = 8.945 \times 10^{-19}\ \text{J}$$

The wavenumber of spectral line in butadiene molecule is:

$$\bar{v} = \frac{(2n+1)h}{8ma^2c}$$

$$\bar{v} = \frac{[(2\times2)+1]\times(6.6\times10^{-34})}{8\times(9.11\times10^{-31})\times(5.78\times10^{-10})^2\times(3\times10^8)}$$

$$\bar{v} = 4.52 \times 10^6\ \text{m}^{-1}$$

$$\bar{v} = 4.52 \times 10^4\ \text{cm}^{-1}$$

The experimental value of wavenumber of spectral line in case of butadiene molecule ($n_2 \rightarrow n_3$) is found to be 4.61×10^4 cm^{-1}, which is agreed by calculated value, indicating that treating a linear polyene molecule like a particle in a box is justified.

PROBLEMS BASED ON FEMO THEORY

Q. 1. Calculate the wavelength of the light absorbed when an electron in a linear molecule of length 10 Å, makes a transition from energy level, n = 1 to n = 2 using FEMO Theory.

Ans.

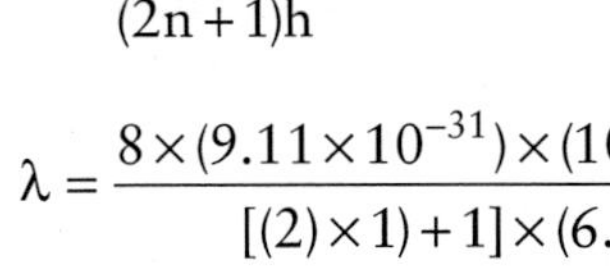

Here,

$h = 6.6 \times 10^{-34}$ Js

$m = 9.11 \times 10^{-31}$ kg

$a = 10$ Å $= 10 \times 10^{-10}$ m $= 10^{-9}$ m

$$\lambda = \frac{8ma^2c}{(2n+1)h}$$

$$\lambda = \frac{8\times(9.11\times10^{-31})\times(10^{-9})^2\times(3\times10^8)}{[(2)\times1)+1]\times(6.6\times10^{-34})} = 1.1\times10^{-6}\,\text{m}$$

Q. 2. Calculate the wavelength in nm, corresponding to the transition from HOMO to LUMO in 1, 3, 5-Hexatriene molecule. Using FEMO theory, the C—C and C=C bond lengths are 154 pm and 135 pm respectively.

Ans. $\overset{1}{H_2C}=\overset{2}{HC}-\overset{3}{HC}=\overset{4}{HC}-\overset{5}{HC}=\overset{6}{CH_2}$

The 6π electrons in 1, 3, 5-Hexatriene molecule are distributed in various energy levels by Aufbau rule as:

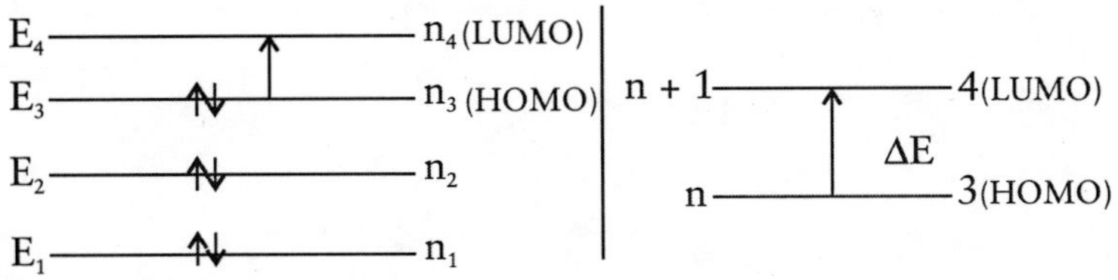

The length of hexatriene molecule is given by

$$a = 2\text{ (single bond length)} + 3\text{ (double bond length)} + 2\ \frac{\text{(single bond length)}}{2}$$

$$a = 2(154) + 3(135) + 2\frac{(154)}{2}$$

$$a = 867\text{ pm} = 867 \times 10^{-12}\text{ m}$$

$$m = 9.11 \times 10^{-31}\text{ kg}$$

$$h = 6.6 \times 10^{-34}\text{ Js}$$

$$c = 3 \times 10^{8}\text{ ms}^{-1}$$

$$\lambda = \frac{8ma^2c}{(2n+1)h}$$

$$\lambda = \frac{8 \times (9.11 \times 10^{-31}) \times (867 \times 10^{-12})^2 \times (3 \times 10^8)}{[2(3)+1] \times (6.6 \times 10^{\ 34})}$$

$$\lambda = 3.55 \times 10^{-7}\text{ m} = 3.55 \times 10^{-2}\text{ nm}$$

Q. 3. Treat the π-electrons in butadiene as particle in a box problem of length $4R_{CC}$ (as in this case an extra $1/2R_{CC}$ is added to each end of molecule, take $R_{CC} = 140$ pm). Calculate the absorption frequency for the transition of an electron from the ground state (i.e. valence state) to the lowest unoccupied orbital using FEMO description.

Ans. $\overset{1}{C}H_2{=}\overset{2}{C}H{-}\overset{3}{C}H{=}\overset{4}{C}H_2$

The 4π electrons in butadiene molecule are distributed in various energy levels by Aufbau rule as:

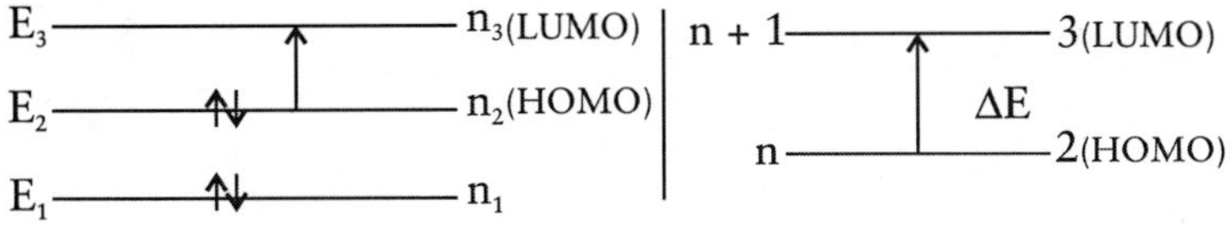

The length of butadiene molecule is given by

$$a = 4R_{CC} + 2\left(\frac{1}{2}R_{CC}\right)$$

$$a = 4(140) + \cancel{2}\left(\frac{1}{\cancel{2}} \times 140\right)$$

$a = 700 \text{ pm} = 700 \times 10^{-12} \text{ m}$

$m = 9.11 \times 10^{-31} \text{ kg}$

$h = 6.6 \times 10^{-34} \text{ Js}$

$c = 3 \times 10^{8} \text{ ms}^{-1}$

$$\nu = \frac{(2n+1)h}{8ma^2}$$

$$\nu = \frac{[2(2)\times 1]\times(6.6\times 10^{-34})}{8\times(9.11\times 10^{-31})\times(700\times 10^{-12})^2}$$

$$\nu = \frac{5\times(6.6\times 10^{-34})}{8\times(9.11\times 10^{-31})\times(700\times 10^{-12})^2}$$

$\nu = 9.24 \times 10^{14} \text{ ms}^{-1}$

Q. 4. Show that the model of particle in a one-dimensional box can be applied to calculate π electron transition energy, when such an electron is excited from HOMO $\rightarrow$ LUMO of a butadiene molecule using FEMO theory.

Ans. $\overset{1}{C}H_2{=}\overset{2}{C}H{-}\overset{3}{C}H{=}\overset{4}{C}H_2$

Thus, 4π electrons in butadiene molecule are distributed in various energy levels by Aufbau rule as:

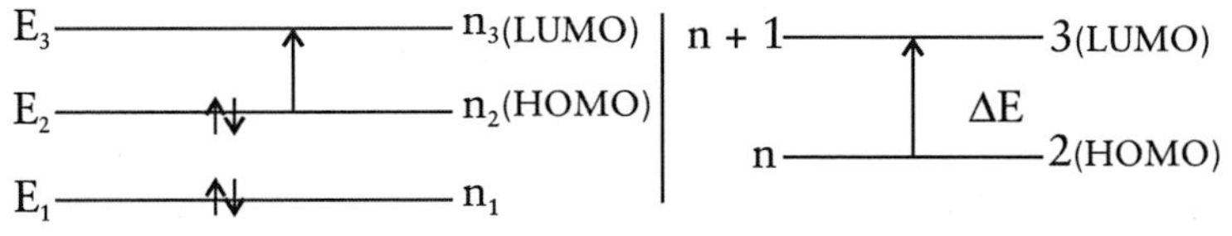

The length of butadiene molecule is given by

$$a = 1 \text{ (single bond length)} + 2\text{(double bond length)} + 2\,\frac{\text{(single bond length)}}{2}$$

$$a = 1(154) + 2(135) + 2\frac{(154)}{2}$$

$a = 578 \text{ pm} = 578 \times 10^{-12} \text{ m}$

$$\Delta E = \frac{(2n+1)h^2}{8ma^2}$$

$$\Delta E = \frac{[2(2)+1]\times(6.6\times 10^{-34})^2}{8\times(9.11\times 10^{-31})\times(5.78\times 10^{-12})^2}$$

$\Delta E = 8.945 \times 10^{-19} \text{ J}$

The wavenumber of spectral line is:

$$\bar{v} = \frac{(2n+1)h}{8ma^2c}$$

$$\bar{v} = \frac{[2(2)+1]\times(6.6\times10^{-34})}{8\times(9.11\times10^{-31})\times(5.78\times10^{-12})^2\times(3\times10^8)}$$

$\bar{v}$ = 4.52 × 10^6 m^{-1} = 4.52 × 10^4 cm^{-1}

The experimental value of wavenumber of spectral line in case of butadiene molecule ($n_2 \rightarrow n_3$) is found to be 4.61 × 10^4 cm^{-1}, which is agreed that treating a linear polyene molecule (butadiene) like a particle in a box is justified.

Q. 5. The π-electrons of a conjugated molecule can be regarded as moving in a particle in a box, where the box length is somewhat more than the length of the conjugated chain. For butadiene, take this length as 7.0 Å. Estimate the wavelength of the light absorbed when a π-electron is excited from a highest occupied to lowest vacant level. The experimental value is 217 nm.

Ans. $\overset{1}{C}H_2=\overset{2}{C}H-\overset{3}{C}H=\overset{4}{C}H_2$

The 4π electrons in butadiene molecule are distributed in various energy levels by Aufbau rule as:

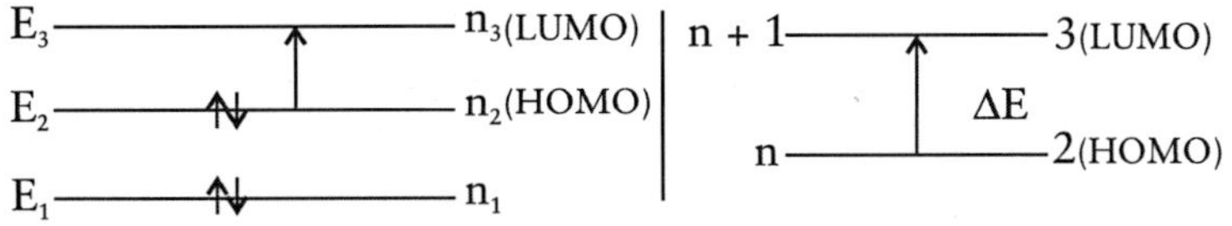

The length of the butadiene molecule is given as

a = 7Å = 7 × 10^{-10} m = 7 × 10^{-10} × 10^{12} pm = 700 pm

$$\lambda = \frac{8ma^2c}{(2n+1)h}$$

$$\lambda = \frac{8\times(9.11\times10^{-31})\times(700\times10^{-12})^2\times(3\times10^8)}{[2(2)+1]\times(6.6\times10^{-34})}$$

λ = 3,246,472.7 × 10^{-13} m

λ = 324.6 × 10^{-9} m ⇒ λ = 324.6 nm

Q. 6. Predict the wavenumber of the lowest energy absorption band in the conjugated Octatetraene molecule. Given that the average C—C bond distance is 0.141 nm.

Ans. $H_2\overset{1}{C}=H\overset{2}{C}-H\overset{3}{C}=\overset{4}{C}H-H\overset{5}{C}=\overset{6}{C}H-H\overset{7}{C}=\overset{8}{C}H_2$

The 8π electrons in Octatetraene molecule are distributed in various energy levels by Aufbau rule as:

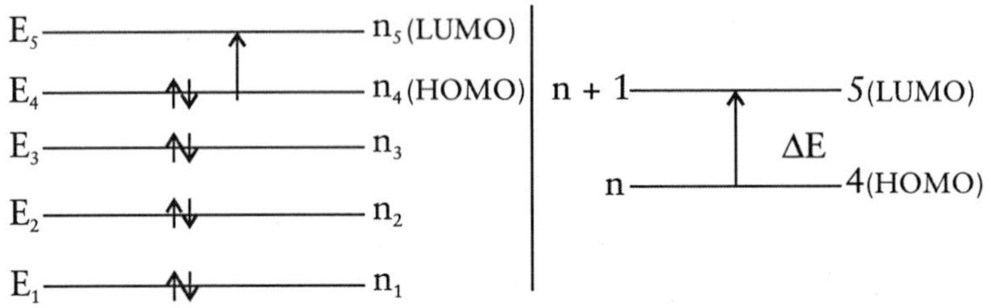

Here, $h = 6.6 \times 10^{-34}$ Js

$m = 9.11 \times 10^{-31}$ kg

$a = 8 \times (0.141)$ nm $= 1.128$ nm

$= 1.128 \times 10^{-9}$ m

$= 1128 \times 10^{-12}$ m

$n = 4$

$$\lambda = \frac{8ma^2c}{(2n+1)h}$$

$$\lambda = \frac{8 \times (9.11 \times 10^{-31}) \times (1128 \times 10^{-12}) \times (3 \times 10^{8})}{[2(4)+1] \times (6.6 \times 10^{-34})}$$

$\lambda = 466.5 \times 10^{-9}$ m

Q. 7. Consider the free electron molecular orbital description of the linear conjugated molecule hexatriene

$CH_2{=}CH{-}CH{=}CH{-}CH{=}CH_2$

What is the minimum energy given that

C—C bond length = 154 pm

C=C bond length = 135 pm

Ans.

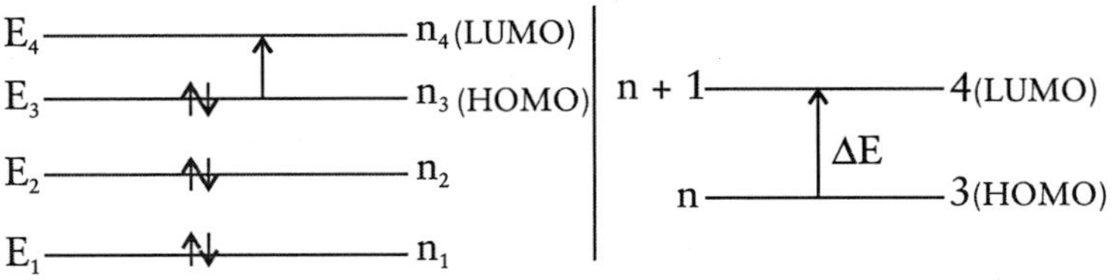

$$a = 2(154) + 3(135) + 2\frac{(154)}{2}$$

$a = 867$ pm $= 867 \times 10^{-12}$ m

$m = 9.11 \times 10^{-31}$ kg

$h = 6.6 \times 10^{-34}$ Js

$c = 3 \times 10^{8}$ ms^{-1}

$n = 3$

$$\Delta E = \frac{(2n+1)h^2}{8ma^2}$$

$$\Delta E = \frac{[2(3)+1](6.6\times10^{-34})^2}{8\times(9.11\times10^{-31})\times(867\times10^{-12})^2}$$

$\Delta E = 5.6 \times 10^{-19}$ J

$$\Delta \in = \frac{\Delta E}{hc} = \frac{(5.6\times10^{-19})}{(6.6\times10^{-34})\times(3\times10^{8})}$$

$\Delta \in = 2.81 \times 10^{6}\ m^{-1}$

$\Delta \in = 2.81 \times 10^{4}\ cm^{-1}$

Q. 8. Consider the free electron molecular orbital (FEMO) description of the linear molecule, hexatriene. What is the minimum excitation energy, given that the average C—C bond distance is 140 pm?

Ans.

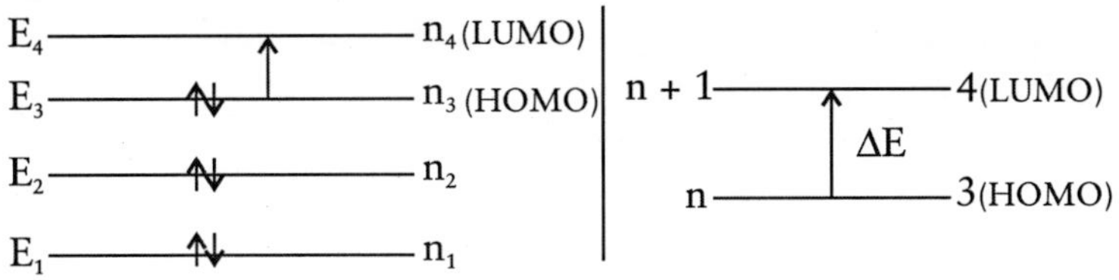

$\overset{1}{H_2C}=\overset{2}{HC}-\overset{3}{HC}=\overset{4}{HC}-\overset{5}{HC}=\overset{6}{CH_2}$

$a = 6 \times (140)$ pm = 840 pm

$a = 840 \times 10^{-12}$ m

$n = 3$

$m = 9.11 \times 10^{-31}$ kg

$h = 6.6 \times 10^{-34}$ Js

$c = 3 \times 10^{8}\ ms^{-1}$

$$\Delta E = \frac{(2n+1)h^2}{8ma^2}$$

$$\Delta E = \frac{[2(3)+1](6.6\times10^{-34})^2}{8\times(9.11\times10^{-31})\times(840\times10^{-12})^2}$$

$\Delta E = 5.97 \times 10^{-19}$ J

$$\Delta \in = \frac{\Delta E}{hc} = \frac{(5.97\times10^{-19})}{(6.6\times10^{-34})\times(3\times10^{8})}$$

$\Delta \in = 3 \times 10^{6}\ m^{-1}$

$\Delta \in = 3 \times 10^{4}\ cm^{-1}$

6.9 Application of Electronic Spectroscopy to Organic Molecules

Electronic spectroscopy involves the measurement of absorption of light in the visible and ultraviolet regions 200–800 nm by the substances under investigation. It involves the promotion of electrons (σ, π, n) from the ground state to the higher energy state.

For recording the electronic or ultraviolet spectrum, the given compound is generally dissolved in some suitable solvent which does not itself absorb light in the region under examination. The commonly employed solvents are 95% ethanol, hexane and water. However, it should be noted that the positions of the absorption peaks of a compound may be shifted slightly if different solvents are used.

The solution of the compound is placed in a suitable transparent container which does not absorb light in the region being studied; generally a quartz cell of 1 cm path length is used for this purpose (glass cannot be used, since it absorbs strongly in ultraviolet region). The solution is then exposed to the ultraviolet and visible radiations in a spectrophotometer in which a hydrogen discharge lamp is usually employed as a source of light in the ultraviolet region (200–400 nm) while a tungsten filament lamp is used for the visible region (400–800 nm) of the radiation.

The spectrophotometer measures the wavelength of maximum absorption of light and the intensity of absorption. The spectrum obtained consists of one or more broad absorption peaks or humps. The wavelength corresponding to the top of the hump is referred to as λ_{max} (absorption maxima) and the intensity of an absorption in that position is called **molar extinction coefficient** (ε_{max}).

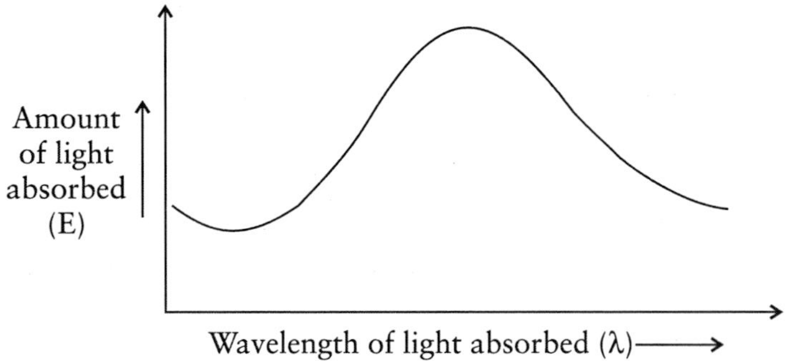

Fig. 6.16: Absorption spectrum

6.9.1 Types of Molecular Orbitals

The various types of **molecular orbitals present in a molecule are:** σ and σ^*; π and π^*; n(non-bonding) orbitals. Since non-bonding electrons do not form bonds, there are no antibonding orbitals associated with them.

The energy levels of all these molecular orbitals in increasing order of energy are shown in Fig. 6.17.

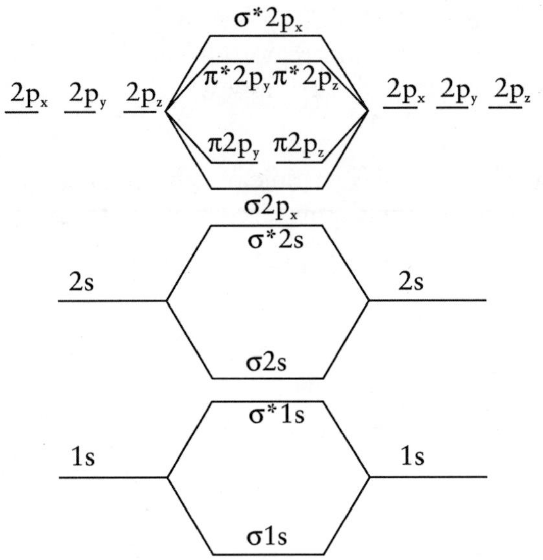

Fig. 6.17: Schematic molecular orbital energy level diagram

6.9.2 Electronic Transitions in the Molecular Orbitals

According to molecular orbital theory, when a molecule is excited by the absorption of energy (UV or visible light), its electrons are promoted from a bonding molecular orbital (BMO) to an antibonding molecular orbital (ABMO). The various possible transitions of the electron between the molecular orbitals are shown in Fig. 6.18.

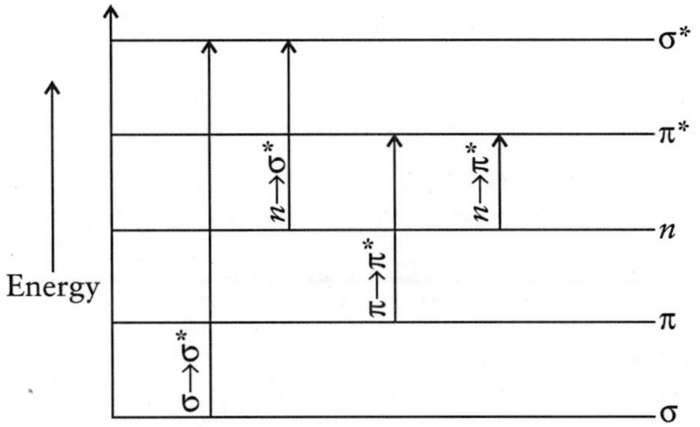

Fig. 6.18: Relative energies of molecular orbitals and the type of transitions between them

The electronic spectra of polyatomic molecules depends on the nature of electrons that is under excitation. Electronic excitation occurs in the range of 200–800 nm. The usual spectroscopic technique cannot be used below 200 nm since oxygen (present in air) begins to absorb strongly. To study such high energy electronic transition, the entire path must be evacuated. Thus, the region below 200 nm is called **vacuum ultraviolet region.**

(a) **$\sigma \rightarrow \sigma^*$ transition:** The energy for this electronic transition is very high and absorption band occurs in far UV or vacuum region. It is a high energy process since σ bond is very strong and requires light of very low wavelength.

Example: Alkanes.

(b) ***n*** **$\rightarrow \sigma^*$ transition:** This type of electronic transition takes place in saturated compounds containing one hetero atom with unshared pair of electrons or non-bonding electrons

$$\left(R\ddot{X},\ R-\ddot{O}H,\ R-\ddot{O}-R,\ R-CH\ddot{O},\ R-\underset{\underset{:O:}{\|}}{C}-R,\ R\dot{N}H_2\right)$$

This transition requires comparatively less energy than required for $\sigma \rightarrow \sigma^*$ transition. It occurs in near UV region.

Example:

(i) In saturated $R\ddot{X}$, the energy required for such a transition decreases with the increase in the size of halogen atom (or decrease in electronegativity of atom).

Like: Absorption maxima for CH_3Cl = 175 nm

and Absorption maxima for CH_3I = 258 nm.

These values of λ_{max} shows that 'n' electrons on iodine atom are loosely bound and hence n $\rightarrow \sigma^*$ transition is more easy in CH_3I, thus light of higher wavelength is sufficient for excitation.

Also greater the probability of a particular transition, greater the value of its molar extinction coefficient (ε_{max}). Thus as λ_{max} increases, the value of ε_{max} also increases.

(ii) Amines ($R\dot{N}H_2$) absorbs at higher absorption maxima (λ_{max}) as compared to alcohol ($R\ddot{O}H$) and hence ε_{max} for amines will be greater. This is because nitrogen is less electronegative than oxygen.

Now ***n*** **$\rightarrow \sigma^*$ transitions are very sensitive to hydrogen bonding.** Alcohols as well as amines form hydrogen bonding with the solvent molecules. Such association occurs due to the presence of non-bonding electrons on the hetero atom and thus electronic transition requires greater energy. Hydrogen bonding shifts the absorption band to shorter wavelength, i.e. more energy would be required for n $\rightarrow \sigma^*$ transition after 'n' electrons gets involved in hydrogen bonding.

(c) **$\pi \rightarrow \pi^*$ transition:** This type of electronic transition occurs in the compounds containing double or triple bonds and also in aromatics.

Example: Alkenes, Alkynes, Carbonyls, Cyanide, Azo, Nitro compounds. The excitation of π electrons requires smaller energy and hence this electronic transition occurs at longer wavelength (λ_{max} is high). This transition occurs in ordinary UV region.

(d) ***n*** **$\rightarrow \pi^*$ transition:** This electronic transition requires least amount of energy out of all the electronic transitions and hence longer wavelength is required.

Thus, the energy required for various transitions obey the following order:

$$\sigma \rightarrow \sigma^* > n \rightarrow \sigma^* > \pi \rightarrow \pi^* > n \rightarrow \pi^*$$

6.9.3 Allowed and Forbidden Electronic Transitions

Depending upon the symmetry and the value of extinction coefficient (ε_{max}), transitions can be classified as allowed transition and forbidden transition.

(A) Depending on the value of ε_{max}:

(a) **Allowed Transition:** The transition with values of $\varepsilon_{max} > 10^4$ are called allowed transition. These transitions have high intensity and they arise due to $\pi \rightarrow \pi^*$ transition.

(b) **Forbidden Transition:** The transition with value of $\varepsilon_{max} < 10^3$ are called forbidden transition. These transitions have very low intensity and they arise due to $n \rightarrow \pi^*$ transition.

Example: In Benzophenone, the two transitions occurs at

$\pi \rightarrow \pi^*$, 252 nm, $\varepsilon_{max} = 20{,}000$; Allowed

and $n \rightarrow \pi^*$, 325 nm, $\varepsilon_{max} = 180$; Forbidden

(B) Symmetry restrictions in electronic transitions

The transitions (allowed or forbidden) is related with the geometries of the lower and the higher energy molecular orbital and also on the symmetry of the molecule as a whole. Symmetrical molecules have more restrictions on their electronic transition.

Example: Benzene (highly symmetrical molecule) has many restrictions on the electronic transitions and its electronic absorption spectra is simple. For a totally unsymmetrical molecule, no symmetry restrictions apply to the electronic transition and thus transitions may be observed among all its molecular orbitals. Therefore, a complex electronic absorption spectrum is observed for unsymmetrical molecule.

6.9.4 Chromophores and Auxochromes

Chromophores: The term chromophore is defined as any isolated covalently bonded group that shows a characteristic absorption maxima in the ultraviolet or visible region, i.e. they have characteristic λ_{max} value. The chromophore is responsible for imparting colour to the compound.

Like:

(i) Nitro compounds are yellow in colour because nitro group (—NO_2) is a chromophore which imparts yellow colour.

(ii) Aryl conjugated azo group is a chromophore for providing colour to the azo dyes.

It may be seen that chromophores are of two types:

(a) Chromophores such as $>C=C<$ and —C≡C— which contains π electrons and undergo $\pi \rightarrow \pi^*$ transitions.

Example: Ethylene, Acetylene, Benzene, etc.

(b) Chromophores such as $>C=O$, —N=N— and —NO_2 which contains both π and n electrons. They can undergo $\pi \rightarrow \pi^*$ as well as $n \rightarrow \pi^*$ transitions.

Example: Carbonyls, Nitriles, Azo compounds, Nitro compounds

Auxochromes: An auxochrome is a group which itself does not act as a chromophore but when attached to a chromophore shifts the absorption maxima towards longer wavelength along with an increase in the intensity of absorption. Auxochrome does not have a characteristic λ_{max} value.

Like: $\ddot{O}$H, —NH_2, —OR, —NHR, —NR_2, —SH.

An auxochrome group is called **colour enhancing group**. The effect of the auxochrome is due to its ability to extend the conjugation of the chromophore by the sharing of non-bonding electrons. Thus, a new chromophore results which has a different value of the absorption maxima as well as the extinction coefficient.

Example: Benzene shows an absorption maxima at 255 mμ (ε_{max} = 203) whereas Aniline absorbs at 280 mμ (ε_{max} = 1430). Hence, amino group is an auxochrome.

6.9.5 Changes in Position and Intensity of Absorption

For isolated chromophore groups such as >C=C< and —C≡C—, absorption takes place at far ultraviolet region which cannot be easily studied. But the position of absorption maximum and the intensity of absorption can be modified in different ways by some structural changes or change of solvent as given below:

(i) **Bathochromic shift or Red shift: It involves the shift of absorption maximum towards longer wavelength because of the presence of certain groups such as OH and NH_2 called auxochromes or by change of solvent.**

For example, decreasing the polarity of solvent causes a red shift in the n → π* absorption of carbonyl compounds. Bathochromic shift is also produced when two or more chromophores are present in conjugation in a molecule. Like, ethylene shows a π → π* transition at 170 nm whereas 1, 3-butadiene (where two double bonds are in conjugation) shows λ_{max} at 217 nm.

(ii) **Hypsochromic shift or Blue shift: It involves the shift of absorption maximum towards shorter wavelength** and may be caused by removal of conjugation in a system or by change of solvent. The absorption shift towards shorter wavelength is also called **blue shift.** It may be caused by the removal of conjugation and also by changing the polarity of the solvent.

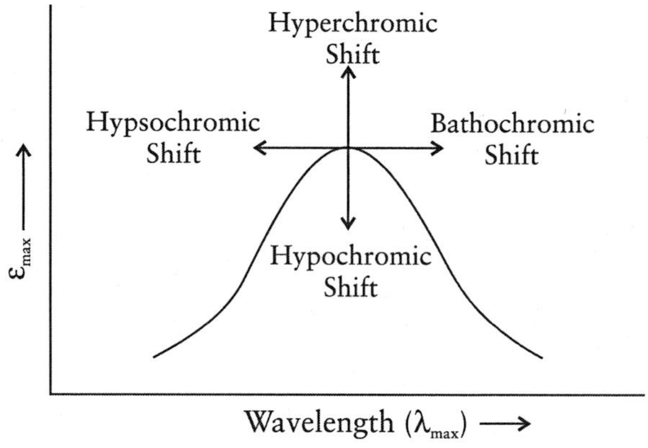

Fig. 6.19: Absorption and intensity shifts

For example, in the case of aniline, absorption maximum takes place at 280 mμ because the pair of electrons on nitrogen atom is in conjugation with the π bond system of the benzene ring. In acidic solutions, a blue shift is caused and absorption takes place at short wavelength (λ_{max} = 200 mμ). In 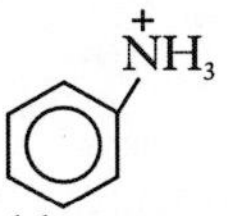 ion which is formed in acidic solution, the electron pair is no longer present and hence conjugation is removed.

(iii) **Hyperchromic effect: The effect involves an increase in the intensity of absorption** and is usually brought about by introduction of an auxochrome.

For example, introduction of methyl group in position 2 of pyridine increases ε_{max} (λ_{max} = 262 nm) from 2750 to 3560 for π-π* transition.

(iv) **Hypochromic effect: It involves a decrease in the intensity of absorption** and is brought by groups which are able to distort the geometry of the molecule.

For example, when a methyl group is introduced in position 2 of biphenyl group hypochromic effect occurs because of distortion caused by methyl group.

PROBLEMS BASED ON ELECTRONIC TRANSITIONS

Q. 1. What are bathochromic and hypsochromic shift?

Q. 2. Write short note on Chromophore and Auxochrome.

Q. 3. How would you identify n → π* and π → π* transition in a given molecule?

Q. 4. A bathochromic shift is observed for the π → π* transition in 1, 3-Butadiene as compared to ethene, $CH_2=CH_2$.

6.10 Electronic Excited State: Singlet and Triplet

In the ground state, most molecules have an even number of electrons and thus in the ground state, all the electrons are spin paired. When the spins are paired (↑↓), the upward orientation of the electron spin is cancelled by the downward spin, so that $S = s_1 + s_2 = +1/2 - 1/2 = 0$

Spin multiplicity = 2S + 1 = 2(0) + 1 = 1

Thus, the spin multiplicity of the molecule is 1. We express it by saying that the molecule is in the **singlet ground state.**

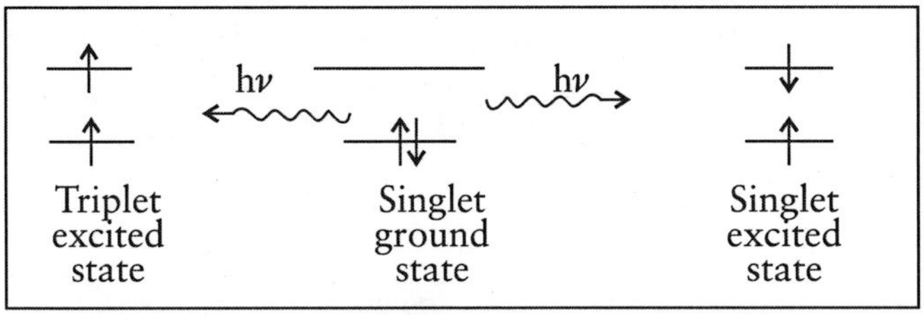

Fig. 6.20: Electronic excited state on absorption of light

On absorption of a photon of a suitable energy, hν, one of the paired electron (present in singlet ground state) goes to a higher energy level (excited state), the spin orientations of the two single electrons may be either parallel (↑↑) or anti-parallel (↑↓) as shown in Fig. 6.20.

(a) If in the excited state, the spins are parallel
then, $S = s_1 + s_2 = +1/2 + 1/2 = 1$
So that, $2S + 1 = 2(1) + 1 = 3$

Thus, the spin multiplicity of the molecule is 3. This is expressed by saying that the molecule is in the **triplet excited state.**

(b) If in the excited state, the spins are anti-parallel
then, $S = s_1 + s_2 = +1/2 - 1/2 = 0$
So that, $2S + 1 = 2(0) + 1 = 1$

Thus, the spin multiplicity of the molecule is 1. This is expressed by saying that the molecule is in the **singlet excited state.**

Since the electron can jump to any of the higher electronic states, so we get a series of singlet excited states (S_1, S_2, S_3 ... etc.) known as first singlet excited state, second singlet excited state, etc. and series of triplet excited states (T_1 T_2, T_3 ... etc.) known as first triplet excited state, second triplet excited state, etc.

PROBLEMS BASED ON SINGLET AND TRIPLET STATE

Q. 1. If there are two electrons in the same orbital, it is not possible to have a situation where only one of these changes its spin i.e. $\boxed{\uparrow\downarrow}$ to $\boxed{\uparrow\uparrow}$ does not occur.

Ans. This is according to Pauli Exclusion Principle which states that, in an atom or molecule, no two electrons can have the same set of four quantum numbers.

Q. 2. Why is ground state of H_2 molecule singlet state, while its excited state is triplet state?

Ans. In ground state of H_2 electronic configuration = $(\sigma)^2$.
Total spin = $S = s_1 + s_2 = 1/2 - 1/2 = 0$
Spin multiplicity = $2S + 1 = 1$ (singlet state)

In **excited state,** electronic configuration = $(\sigma)^1 (\sigma^*)^1$
Total spin = $S = s_1 + s_2 = 1/2 + 1/2 = 1$
Spin multiplicity = $2S + 1 = 3$ (triplet state)

6.11 Transition Dipole Moment

In electronic spectroscopy, we get a broad spectra because electronic spectra consists of vibrational coarse structure and rotational fine structure.

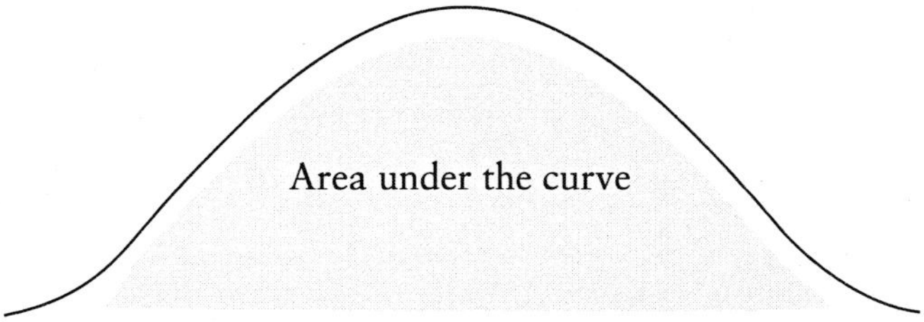

Fig. 6.21: Transition dipole moment, $|\mu_{\ell m}|^2$

Area under the curve (A) α Transition dipole moment $|\mu_{lm}|^2$

Transition dipole moment is a theoretical quantity

Now,

$$\psi_{Total} = \psi_{es} \cdot \psi_{vib}$$

$$|\mu_{lm}| = \int \psi_{es}^{ex} \cdot \psi_{vib}^{ex} (\hat{\mu}_n + \hat{\mu}_e) \psi_{es}^{gr} \cdot \psi_{vib}^{gr} \, d\tau_n d\tau_{es}$$

where, $\hat{\mu}_n$ (nuclear part) operates only on ψ_{vib}

and, $\hat{\mu}_e$ (electronic part) operates only on ψ_{es}.

$$|\mu_{lm}| = \underbrace{\int \psi_{es}^{ex} \psi_{es}^{gr} d\tau_{es}}_{A} \underbrace{\int \psi_{vib}^{ex} \cdot \hat{\mu}_n \cdot \psi_{vib}^{gr} d\tau_n}_{B} + \underbrace{\int \psi_{vib}^{ex} \cdot \psi_{vib}^{gr} d\tau_n}_{C} \underbrace{\int \psi_{es}^{ex} \cdot \hat{\mu}_e \cdot \psi_{es}^{gr} d\tau_{es}}_{D} \quad ...(6.16)$$

In the above equation (6.16),

Term (A) is zero, since ψ_{es}^{ex} and ψ_{es}^{gr} are orthogonal to each other.

Term (C) is not equal to zero, since ψ_{vib}^{ex} and ψ_{vib}^{gr} are not orthogonal to each other. This term is called **Frank Codon Factor**.

Term (D) determines the selection rules.

6.12 Selection Rules

Using the term (D) from above equation (6.16)

$$|\mu_{lm}| = \int \psi_{es}^{ex} \cdot \hat{\mu}_e \cdot \psi_{es}^{gr} d\tau_{es}$$

To obtain the selection rules, we follow certain assumptions:

(i) Orbital part/electronic part is independent of spin part, i.e. ψ_e and ψ_s are independent of each other

$$\psi_{es} = \psi_e \cdot \psi_s$$

$$d\tau_{es} = d\tau_e \, d\tau_s$$

(ii) Electronic dipole moment part ($\hat{\mu}_e$) is independent of spin part, i.e. μ does not depend upon spin but only on the position of electron.

$$|\mu_{lm}| = \int \psi_{es}^{ex} \cdot \hat{\mu}_e \cdot \psi_{es}^{gr} d\tau_{es}$$

$$|\mu_{lm}| = \int \psi_e^{ex} \cdot \psi_s^{ex} \cdot \hat{\mu}_e \cdot \psi_e^{gr} \cdot \psi_s^{gr} d\tau_e d\tau_s \quad ...(6.17)$$

$$|\mu_{lm}| = \int \psi_s^{ex} \cdot \psi_s^{gr} d\tau_s \int \psi_e^{ex} \cdot \hat{\mu}_e \psi_e^{gr} d\tau_e \quad ...(6.18)$$

Using this expression, we will obtain the selection rule:

1. First Selection Rule: (Spin Selection Rule)

If, $\psi_s^{ex} \cdot \psi_s^{gr} d\tau_s = 0$ in the above equation (6.18), then the transition dipole moment becomes zero, which is not possible. This integral is zero when the spin multiplicity changes during the transition.

Therefore, $\left.\begin{matrix} S \not\rightarrow T \\ T \not\rightarrow S \end{matrix}\right]$ Forbidden transition

Thus always transition takes place between the levels having same spin multiplicity.

Therefore, $\left.\begin{matrix} S \rightarrow S \\ T \rightarrow T \end{matrix}\right]$ Allowed transition

So, $\boxed{\Delta S = 0}$ called **spin selection rule,** i.e. total spin cannot change

The above selection rule, i.e. $\Delta S = 0$ is frequently violated. This happens when orbital part is not separated from spin part, i.e. there is large spin-orbit coupling, which is seen in the case of heavier atoms like S and Se.

2. Second Selection Rule: (Laporte Selection Rule)

$\int \psi_e^{ex} \cdot \hat{\mu}_e \cdot \psi_e^{gr} d\tau_e$ in the above equation (6.18) is taken to derive the Laporte selection rule.

Molecule that have centre of symmetry, will belong to 'g' or 'u' wavefunction.

$u \times g = u$
$g \times u = u$
$u \times u = g$
$g \times g = g$

$$\underbrace{|\mu_{lm}|}_{\text{make it 'g'}} = \int \underbrace{\psi_e^{ex}}_{\text{'u'}} \cdot \underbrace{\hat{\mu}_e}_{\text{always 'u'}} \cdot \underbrace{\psi_e^{gr}}_{\text{always 'g'}} d\tau_e$$

$$\hat{\mu}_e \cdot \psi_e^{gr}: u \times g = u \qquad \psi_e^{ex} \cdot (\hat{\mu}_e \cdot \psi_e^{gr}): u \times u = g$$

Thus to make $|\mu_{lm}|$ as 'g', the ψ_e^{ex} should be 'u'.

Therefore, $\left.\begin{matrix} g \not\rightarrow g \\ u \not\rightarrow u \end{matrix}\right]$ Forbidden transition

$\left.\begin{matrix} g \rightarrow u \\ u \rightarrow g \end{matrix}\right]$ Allowed transition

So, $\boxed{\Delta \ell = \pm 1}$ called **Laporte selection rule,** i.e. change in total orbital angular momentum can be $\Delta \ell = 0, \pm 1$, but $\ell = 0 \rightarrow \ell = 0$ transition is not allowed.

- Laporte selection rule ($g \rightarrow u$) is violated when $d \rightarrow d$ transition are observed. The reason for violation is that orbital part is not independent of spin part.
- Although there is violation of laporte selection rule during $d \rightarrow d$ transition, but even then $d \rightarrow d$ transition is seen because

1. There is mixing of d – p or d – f orbitals, so d → d is not pure.
2. Due to vibration, there if symmetry breakdown, hence d → d transition is not pure.

3. Third Selection Rule:

Instead of quantum number 'ℓ' in quantum mechanics, we have ∧ in molecular spectroscopy.

ℓ	0	1	2	3
∧	Σ	π	Δ	ϕ

We have, $\boxed{\Delta\wedge = \pm 1}$, i.e. change in total angular momentum can be $\Delta\wedge = 0, \pm 1$, but $\wedge = 0 \rightarrow \wedge = 0$ transition is not allowed.

$$\left.\begin{array}{l}\Delta \rightarrow \Delta \\ \pi \rightarrow \pi \\ \Sigma \rightarrow \Sigma \\ \phi \rightarrow \phi\end{array}\right] \text{Forbidden transition} \quad \text{and} \quad \left.\begin{array}{l}\Sigma \rightarrow \pi \\ \pi \rightarrow \Delta \\ \Delta \rightarrow \phi\end{array}\right] \text{Allowed transition}$$

To represent the molecular state, we discuss certain examples.

E.g. (a) **N_2 molecule ($14e^-$)**

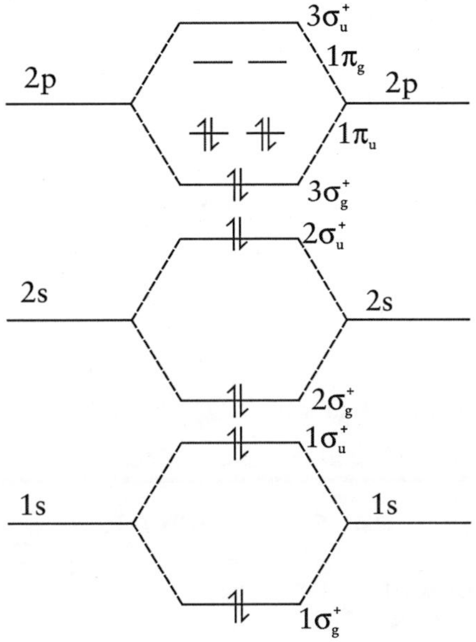

Electronic configuration would

$$(1\sigma_g^+)^2(1\sigma_u^+)^2(2\sigma_g^+)^2(2\sigma_u^+)^2(3\sigma_g^+)^2(1\sigma_u)^4$$

To represent the molecular state, let us take $(1\pi_u)^4$.

(i) $\pi_u \times \pi_u$ means $u \times u = g$

(ii) Spin of $(1\pi_u)^4$ is zero, since all the electrons are paired. | ↑↓ | ↑↓ |

∴ $2S + 1 = 2(0) + 1 = 1$

(iii) $\wedge = \lambda_1 + \lambda_2$

$\wedge = (2 \times 1) + (-2 \times 1) = 0$

$\wedge = (2$	$\times$	$1)$	$+$	$(-2$	$\times$	$1) = 0$
↓		↓		↓		↓
No. of electrons with spin up		Present in π and value of π is 1		No. of electrons with spin down		Present in π and value of π is 1

$\wedge = 0$ which means 'Σ'.

Thus, the molecular state is $^1\Sigma_g$.

We can state that when all the orbitals are completely filled, the configuration would be completely symmetrical and molecular state is $^1\Sigma_g$.

(b) **O_2 molecule ($16e^-$)**

Electronic configuration would be

$$(1\sigma_g^+)^2 (1\sigma_u^+)^2 (2\sigma_g^+)^2 (2\sigma_u^+)^2 (3\sigma_g^+)^2 (1\pi_u)^4 (1\pi_g)^2$$

To present the molecular state, let us take $(1\pi_g)^2$

(i) $\pi_g \times \pi_g$ means $g \times g = g$

(ii) Spin of $(1\pi_g)^2$ is one, since both the electrons are unpaired. | ↑ | ↑ |

$S = s_1 + s_2 = +1/2 + 1/2 = 1$

$\therefore \; 2S + 1 = 2(1) + 1 = 3$

(iii) $\wedge = \lambda_1 + \lambda_2$

$\wedge = (2 \times 1) + (0 \times 1) = 2$

$\wedge = 2$, means Δ

$\therefore$ Molecular state is $^3\Delta_g$.

(c) **C_2 molecule ($12e^-$)**

Electronic configuration would be

$$(1\sigma_g^+)^2 (1\sigma_u^+)^2 (2\sigma_g^+)^2 (2\sigma_u^+)^2 (1\pi_u)^4 (3\sigma_g^+)^0$$

Since all the orbitals are completely filled, the molecular state is $^1\Sigma_g$.

PROBLEMS BASED ON SELECTION RULES

Q. 1. Are the given transitions allowed or not?

(i) $^3\Delta_g \rightarrow {}^3\pi_u$

Ans.

(a) $g \rightarrow u$, allowed transition.

(b) $\Delta, \wedge = 2 \rightarrow \pi, \wedge = 1$

$\Delta\wedge = -1$, allowed transition.

(c) $3 \rightarrow 3$, i.e. $T \rightarrow T$, allowed transition

As all the three selection rules are followed, thus above transition is allowed.

(ii) ${}^1\Delta_u \rightarrow {}^1\Sigma_g$

Ans.

(a) $u \rightarrow g$, allowed transition.

(b) $\Delta, \wedge = 2 \rightarrow \Sigma, \wedge = 0$

$\Delta\wedge = -2$, forbidden transition.

(c) $1 \rightarrow 1$, i.e. $S \rightarrow S$, allowed transition

As one of the above selection rules is not followed, thus above transition is not allowed.

Q. 2. Give reason why singlet to triplet transitions are forbidden?

Ans. $|\mu_{lm}| = \int \psi_s^{ex*} \psi_s^{gr} d\tau_s \int \psi_e^{ex*} \hat{\mu}_e \psi_e^{gr} d\tau_e$

If, $\int \psi_s^{ex*} \psi_s^{gr} d\tau_s = 0$ in the above equation, then the transition dipole moment becomes zero, which is not possible. This integral is zero only when the spin multiplicities in the ground and excited state are different. Therefore,

$$\left.\begin{array}{l} S \nrightarrow T \\ T \nrightarrow S \end{array}\right] \text{Forbidden transition}$$

Thus, always transition takes place between the levels having same spin multiplicity.

Therefore, $$\left.\begin{array}{l} S \rightarrow S \\ T \rightarrow T \end{array}\right] \text{Allowed transition}$$

6.13 Vibronic Coupling

Vibronic coupling in a molecule involves the interaction between electronic and nuclear vibrational motion.

$$|\mu_{lm}| = \int \psi_{vib}^{ex} \psi_{es}^{ex} \cdot \underset{\substack{\downarrow \\ \text{always} \\ 'u'}}{\hat{\mu}} \cdot \underset{\substack{\downarrow \\ \text{always} \\ 'g'}}{\psi_{vib}^{gr}} \cdot \underset{\substack{\downarrow \\ \text{always} \\ 'g'}}{\psi_{es}^{gr}} \, d\tau_{es} d\tau_n$$

If $\psi_{vib}^{ex} \equiv u$ and $\psi_{es}^{ex} \equiv g$, then coupling between electronic and vibrational part occurs and this coupling between these two states is called **vibronic coupling.**

$$|\mu_{lm}| = \int \underset{\substack{\downarrow \\ u}}{\psi_{vib}^{ex}} \, \underset{\substack{\downarrow \\ g}}{\psi_{es}^{ex}} \cdot \underset{\substack{\downarrow \\ u}}{\hat{\mu}} \cdot \underset{\substack{\downarrow \\ g}}{\psi_{vib}^{gr}} \cdot \underset{\substack{\downarrow \\ g}}{\psi_{es}^{gr}} \, d\tau_{es} d\tau_n$$

$g \times g = g$

$u \times g = u$

$g \times u = u$

$u \times u = g$

Thus $|\mu_{lm}| \equiv g$, which is allowed.

Example: Benzene shows vibronic coupling.

6.14 Consequences of Light Absorption: The Jablonski Diagram

The important difference between the lowest excited singlet state (S_1) and the lowest excited triplet state (T_1) is the difference in energy of the two states. The triplet state is lower in energy, as a result of the greater electronic repulsion in the excited singlet state (Hund's rule of maximum multiplicity).

Accordingly, the energy sequence is as follows:

$E_{S_1} > E_{T_1}$; $E_{S_2} > E_{T_2}$; $E_{S_3} > E_{T_3}$ and so on.

On absorption of light photon, the electron of the absorbing molecule may jump from S_0 to S_1, S_2 or higher singlet excited state depending upon the energy of the light photon absorbed. (Transition always takes place between the levels having same spin multiplicity, i.e. $S \rightarrow S$ and $T \rightarrow T$ allowed.)

For each singlet state (S_1, S_2, S_3, etc.) there is a corresponding triplet excited state, (T_1, T_2, T_3 etc.).

The molecule, whether in singlet or triplet excited state, is said to be activated. The activated molecule returns to the ground state by dissipating its energy through the following types of processes:

6.14.1 Non-Radiative Transition: Internal Conversion and Intersystem Crossing

These transitions do not involve the emission of any radiations and thus referred to as non-radiative or radiationless transitions. It can be shown by the two phenomena:

(a) **Internal Conversion (IC):** It is the predominant non-radiative process, i.e. the exchange of energy without change of spin state via a downward cascading process through vibrationally excited states, typically occurring in less than a few nanoseconds.

The molecule after electronic transition is in high vibrational state and lost its vibrational energy by intermolecular collisions. The vibrational energy is converted to kinetic energy and appears as heat. Such transfer between energy levels is radiationless and is called internal conversion (IC). These transitions involve the return of the activated molecule from the higher excited states (S_3, S_2, or T_3, T_2) to the first excited state (S_1 or T_1).

(b) **Intersystem Crossing (ISC):** The molecule may also lose energy by another process called inter-system crossing. This process involves transition between states of different spins, i.e. different multiplicities. These transitions involve movement from S_2 to T_2 or S_1 to T_1. These transitions are also non-radiative. Although such transitions with different multiplicity are spectroscopically forbidden but there is no restriction if the transfer between excited state takes place kinetically, i.e. through radiationless transfer induced by collisions. However, they do occur, though at relatively slow rates.

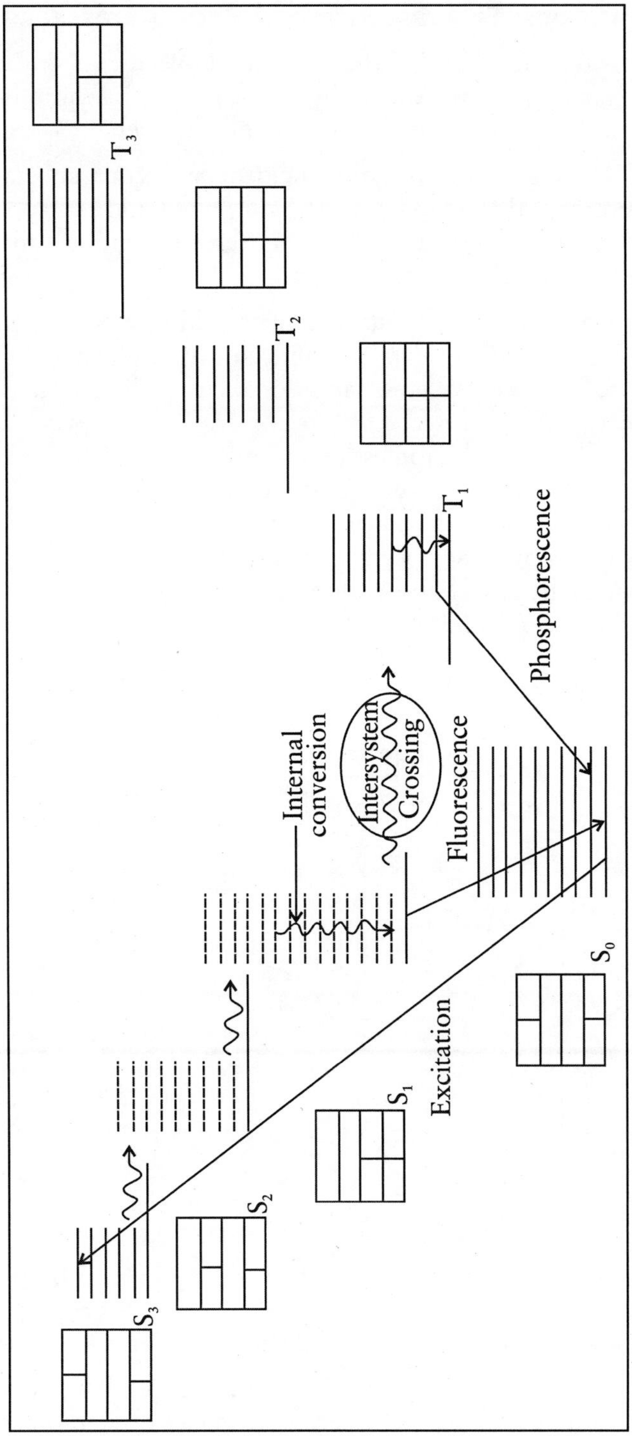

Fig. 6.22: Jablonski diagram

6.14.2 Radiative Transitions: Fluorescence and Phosphorescence

These transitions involve the return of the activated molecule from the singlet excited state (S_1) and triplet excited state (T_1) to the ground state (S_0). Such transitions are accompanied by the emission of radiation. It can be shown by the two phenomena.

(a) **Fluorescence:** When a beam of light is incident on certain substance, they emit visible light of radiation and they stop emitting light or radiation as soon as the incident light is cut off. This phenomena is known as **fluorescence** and the substances showing this fluorescence is called **fluorescent substances.**

Fluorescence is a radiative transition that is taking place from singlet excited state to singlet ground state ($S_1 \rightarrow S_0$). The molecule after electronic transition is in higher vibrational state and lose its excess vibrational energy by intermolecular collisions. The vibrational energy is converted to kinetic energy and appears as heat. Such transfer between energy levels is radiationless and is called **internal conversion.** When the excited state molecule reached to a lower vibrational level ($v' = 0$), it may then emit radiation and return to the ground state. The emitted radiations are of lower frequency than the absorbed radiations and the spectrum thus obtained is called **fluorescence spectrum.** The time between the initial absorption and returning back to the ground state is very small, i.e. of the order of 10^{-8} sec.

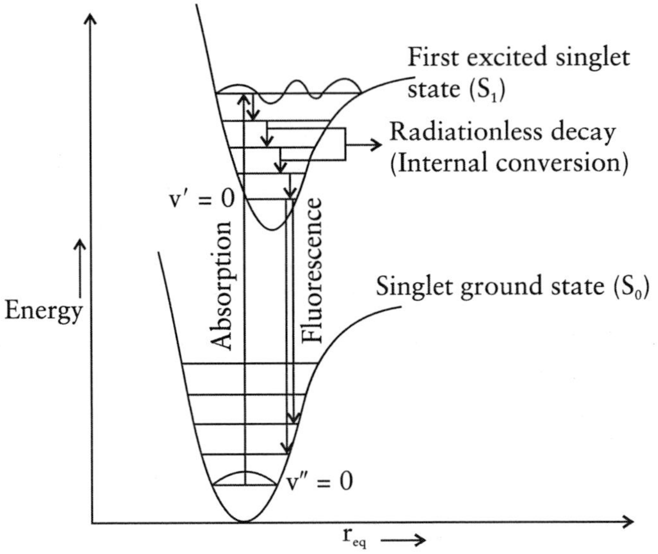

Fig. 6.23: Fluorescence spectrum

(b) **Phosphorescence:** When light is incident on certain substances, they emit light continuously even after the incident light is cut off. This type of **delayed fluorescence** is called **phosphorescence** and the substance are called **phosphorescent substances.**

Phosphorescence is a forbidden transition that takes place from excited triplet state (T_1) to singlet ground state (S_0). This takes place when two excited states of different total spin ($S_n \rightarrow T_n$) have comparable energies. The energy in the two excited states (singlet and triplet) is identical and the two curves intersect each other. Although the selection rule, $\Delta S = 0$, forbids spectroscopic transition from excited singlet state to excited triplet state, there is no restriction if the transfer between excited state takes place kinetically, i.e. through radiationless transfer induced by collisions. Such transfer is known as **intersystem crossing** and can occur close to the cross over point of the two potential energy curves.

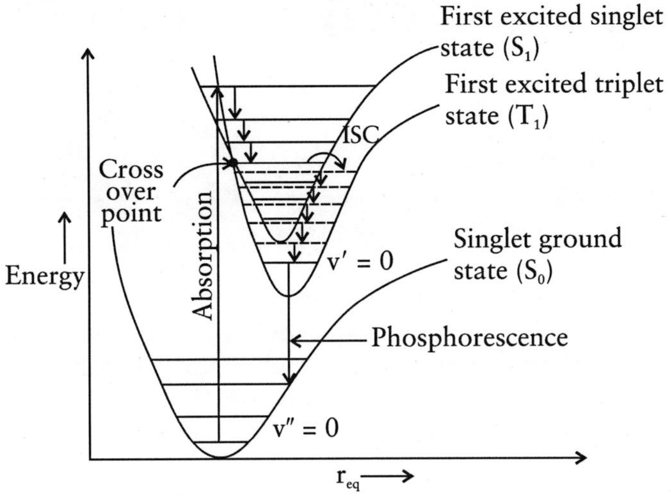

Fig. 6.24: Phosphorescence spectrum

Once the molecule arrived in the triplet state, it undergoes some loss of vibrational energy through internal conversion and return to v′ = 0 level of the excited triplet state. The transition from excited triplet state (T_1) to singlet ground state (S_0) is spectroscopically forbidden, it may takes place slowly in comparison to allowed electronic transition. That is why phosphorescent material may emit radiations for seconds, minutes or even hours after initial absorption. Since there is $T_1 \rightarrow S_0$ transition, which is forbidden and hence phosphorescence is weak.

PROBLEM BASED ON FLUORESCENCE AND PHOSPHORESCENCE

Q. 1. Discuss the process by which a molecule loses its excess energy after undergoing an electronic transition.

7

Nuclear Magnetic Resonance (NMR) Spectroscopy

7.1 Introduction

The NMR spectroscopy lies in the radiofrequency region of electromagnetic radiation. It is very useful in structure elucidation of organic molecules. NMR is concerned with the magnetic properties of certain atomic nuclei, mainly the nucleus of hydrogen atom (proton) and that of the carbon-13 isotope of carbon. The NMR spectroscopy helps us to record the magnetic properties of the various magnetic nuclei present and to deduce the positions of these nuclei within the molecule. Thus, we can determine how many different kinds of environments there are in the molecule, and also measure how many atoms are present in each of these environments.

The most single example of NMR spectrum is toluene. The toluene has two groups of hydrogen atoms-the methyl hydrogens and the ring hydrogens. Two signals appear on the spectrum. The areas under each signal are in the ratio of the number of protons in each part of the molecule, i.e. the ratio of these areas is 5:3.

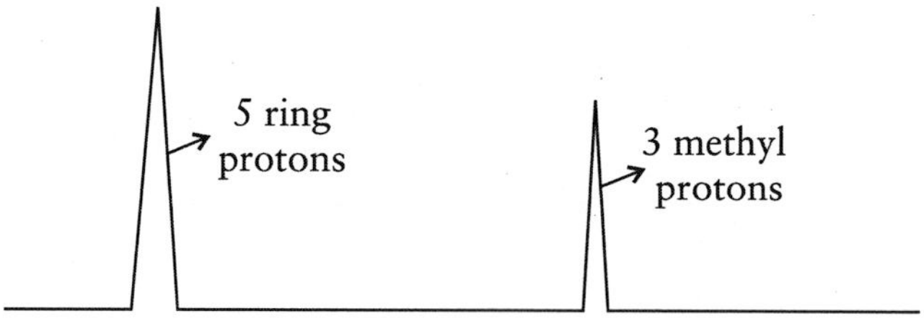

Fig. 7.1: NMR spectra of toluene

7.2 The Spinning Nucleus

All moving charges are associated with electric as well as magnetic fields and behave as tiny bar magnets. The spinning of charged particle generates a magnetic field. The nucleus of a hydrogen atom (proton) behaves as a spinning bar magnet because it possesses both electric and magnetic spin. Like any other spinning charged body, the nucleus of hydrogen atom generates a magnetic field with magnetic dipole moment (Fig. 7.2).

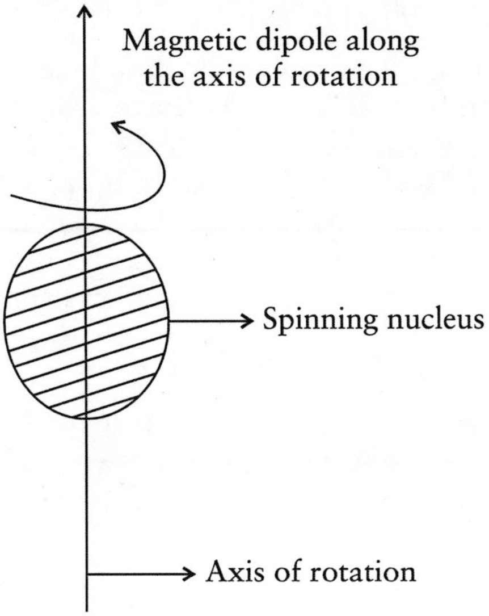

Fig. 7.2: Magnetic dipole generated by spinning nucleus

7.3 Angular Momentum

Just as linear momentum represents the tendency of an object move in a straight line, the angular momentum represents the tendency of an object to move in angular motion or rotate. Since momentum has both a magnitude and a direction, it is a vector quantity. Linear momentum is represented by a vector in the direction of motion ($\vec{p}$). Whereas the angular momentum is represented through the use of the Right Hand Rule (RHR): when the fingers of the right hand are curled in the direction of circular motion, the thumb points in the direction of the angular momentum vector. Mathematically, angular momentum (L) is the cross product of the position (r) and linear momentum (p) vectors

$$\vec{L} = \vec{r} \times \vec{p}.$$

There are two kinds of classical angular momentum: **orbital angular momentum** and **spin angular momentum**, as shown in Fig. 7.3.

(a) **Orbital angular momentum** is the circular motion of an object about a point, like the earth orbiting about the sun once a year.

(b) **Spin angular momentum** is the spinning motion of an object about its own axis, like the earth spinning about its axis once a day.

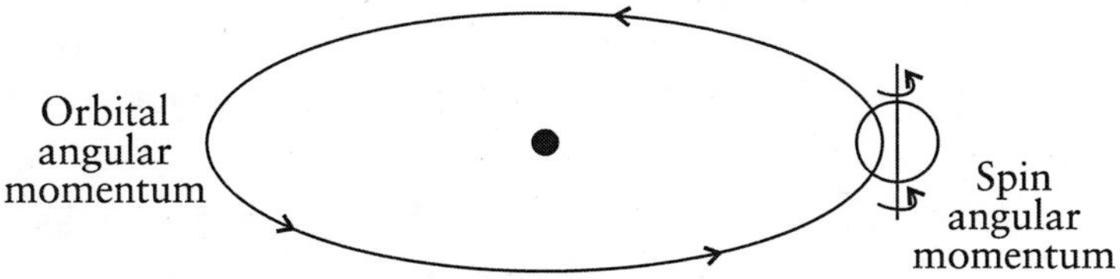

Fig. 7.3: Spin and orbital angular momentum

7.3.1 Quantisation of Angular Momentum

Just as energy is known to be quantized at the atomic level, angular momentum is also quantized. For example, an electron in an atom may have orbital angular momentum quantum numbers, $\ell = 0, 1, 2, ...$ (which are more commonly denoted as s, p, d, ...) and orbital magnetic quantum numbers, $m_\ell = -\ell, -\ell + 1, ..., \ell$ (which lead to p_x, p_y and p_z, for $\ell = 1$ and to the five d-orbitals for $\ell = 2$). Also, the electron has spin angular momentum quantum number s = 1/2 and spin magnetic quantum numbers, $m_s = -1/2, +1/2$ (which are also known as spin down and spin up). All types of angular momentum obey the same quantum mechanical rules.

7.3.2 Spin Angular Momentum of the Nucleus

Spin angular momentum is a specific type of angular momentum possessed by some nuclei. There are two quantum numbers (I and m_I) associated with the spin angular momentum of nuclei that determine its properties.

(a) **Spin quantum number (I):** All nuclei have a spin quantum number, I, which may be integral (including zero) or half-integral, but never negative. The value of I is characteristic of a given nucleus and may vary between isotopes. Like, all ^1H, hydrogen nuclei have I = 1/2, but all ^2H, deuterium nuclei have I = 1. The magnitude of spin angular momentum of a nuclei is determined by the quantum number, I, and is given by

$$\text{Magnitude of spin angular momentum} = L = \hbar\sqrt{I(I+1)} \quad ...(7.1)$$

Thus, all nuclei with I > O have spin angular momentum.

(b) **Spin magnetic quantum number (m_I):** All nuclei with a spin angular momentum (i.e. all nuclei with non-zero value of I) have a magnetic moment with constant magnitude and an orientation determined by spin magnetic quantum number (m_I). (To say that a nucleus has a magnetic moment means that in some ways it behaves like a small bar magnet. This is because like any other spinning charged body, the nucleus of atom generates a magnetic field with magnetic moment.)

The quantum number m_I determines the component of the spin angular momentum on an arbitrary axis normally taken as the z-axis. (The arbitrary axis may be defined by the field direction of an external magnetic field.)

$$\text{z-component of spin angular momentum} = m_I\hbar$$

where, m_I can take values from + I to –I (m_I = I, I – 1, I – 2, ... – I)

This indicates that for a given I, the magnetic moment of a nucleus can adopt (2I + 1) different orientations relative to a defined axis.

Like, a proton (^1H, hydrogen nucleus), has I = 1/2, and thus its spin may adopt two different orientations ($m_I = +1/2$ or $m_I = -1/2$). The state with $m_I = +1/2$ is denoted as 'α', while the state $m_I = -1/2$ is denoted as 'β'.

Note: Very common isotopes, ^{12}C and ^{16}O, have I = 0, so have no spin angular momentum, no magnetic moment and hence are invisible in NMR studies.

7.4 Magnetically Active and Inactive Nuclei

Absorption of radiation of a suitable wavelength when a nucleus is placed in a magnetic field occurs only for those nuclei which have non-zero spin angular momentum. Such type of nuclei are known **as magnetic active nuclei.** Whereas the nuclei which have zero spin angular momentum are known **as magnetic inactive nuclei.** Such type of nuclei cannot interact with the external magnetic field and thus do not show any absorption of radiation.

Magnetically Inactive nuclei includes

(a) Nuclei having both charge number and mass number even (i.e. having both protons and neutrons even in number) have zero spin quantum number

Example: $_2^4He$, $_6^{12}C$ and $_8^{16}O$

Magnetically Active nuclei includes

(a) Nuclei having odd charge number and even mass number (i.e. both protons and neutrons are odd in nature) have integral spin quantum number

Example: ^2H (I = 1), ^{14}N (I = 1) and ^{10}B (I = 3)

(b) Nuclei with odd mass number have half-integral spin

Example: ^1H, ^{15}N, ^{19}F, ^{13}C, ^{31}P (I = 1/2); ^{17}O (I = 5/2); ^{35}Cl (I = 3/2)

PROBLEM BASED ON MAGNETICALLY ACTIVE NUCLEI

Q. 1. The $_6^{12}C$ nucleus does not give rise to an NMR peak, whereas the $_6^{13}C$ nucleus does.

Ans. For a nuclei to be magnetically active, it must have a non-zero spin angular momentum ($I \neq 0$). Now, nuclei having both atomic number and atomic mass even, have zero spin quantum number (I = 0) and hence magnetically inactive. This is why $_6^{12}C$ does not give NMR peak.

Whereas the nuclei with odd mass number have half-integral spin (I = 1/2) and are magnetically active. That is why $_6^{13}C$ gives NMR peak.

7.5 Effect of External Magnetic Field

NMR is an absorption spectroscopy in which sample absorb electromagnetic radiation in radio-frequency region (3 MHz to 30,000 MHz). Like all bar magnets, the proton will also respond to the influence of an external magnetic field, and will tend to align itself with that field.

There is a magnetic moment associated with the angular momenta. Thus, the motion of a charged particle produces a magnetic moment. So the spinning of a charged particle generates a magnetic field. (The magnitude and direction of magnetic field can be described by a vector called a **magnetic moment.**) Thus, the spinning nuclei behave as a tiny bar magnet with a north pole and a south pole.

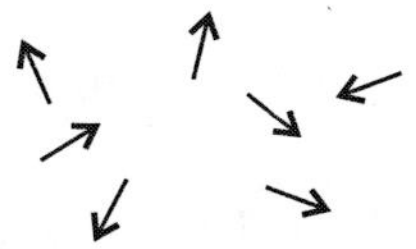

Fig. 7.4: Nuclear magnetic dipole with no magnetic field

In the absence of any external magnetic field, these nuclear magnetic dipoles or magnetic moments are randomly oriented (shown in Fig. 7.4). When a uniform magnetic field 'B' is applied (nuclei with spin quantum number, $I \neq 0$), the nuclei can adopt $2I + 1$ orientations with respect to the direction of the applied magnetic field 'B'.

Like, for a nucleus with spin $I = 1/2$ (^1H, ^{13}C, ^{19}F, ^{31}P), number of orientations with be $2I + 1 = (2 \times 1/2) + 1 = 2$. Hence, there will be two possible arrangements for the nuclear magnetic dipoles, these will either align with the applied magnetic field or against the applied magnetic field (Fig. 7.5).

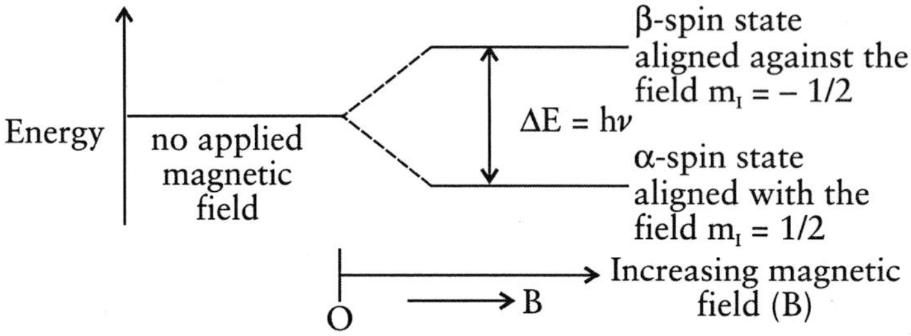

Fig. 7.5: Orientation of nuclear magnetic dipoles in an external magnetic field B

Thus in the presence of external magnetic field, the proton will respond in the same manner of a compass needle in the earth's magnetic field.

7.6 Magnetic Moment of the Nucleus

The nuclear magnetic moment is the magnetic moment of an atomic nucleus and arises from spin of the protons and neutrons.

On a microscopic level, the magnetism of a particle or nucleus is better represented as an electromagnet which results from an electrical current travelling in a circular orbit.

The spinning of a nucleus is equivalent to the circulation of positive charge around the axis of spinning. This, in turn, produces a tiny magnet placed along the spin axis. The magnetic moment (μ_m) of the generated magnet can be calculated with the help of Ampere's Law

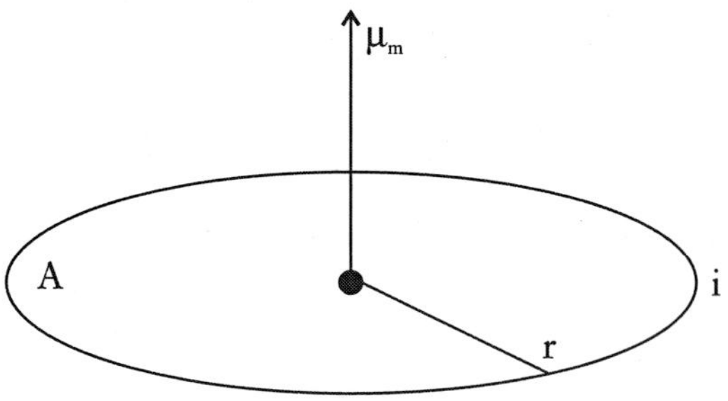

Fig. 7.6: Electromagnet resulting from electrical current

$$\mu_m = iA \quad ...(7.2)$$

where, i is the current circulating in the coil per unit time and A is the area of cross-section of the coil.

If 'r' is the radius of the coil and 'v' is the velocity with which the charge is spinning, then i = (Number of revolutions per unit time) × (Positive charge on the nucleus)

$$i = \left(\frac{v}{2\pi r}\right)e$$

and $A = \pi r^2$

Substituting the expression of 'i' and 'A' in equation 7.2; we get

$$\mu_m = iA = \left(\frac{ve}{2\pi r}\right) \times (\pi r^2) \quad \Rightarrow \quad \mu_m = (vr)\frac{e}{2}$$

Multiplying and dividing by mass of proton (m_p):

$$\mu_m = (m_p vr)\left(\frac{e}{2m_p}\right)$$

$$\mu_m = L\left(\frac{e}{2m_p}\right) \qquad ...(7.3)$$

where, L is the angular momentum.

Since the charge on the nucleus is positive, it follows from equation (7.3) that both angular momentum (L) and magnetic moment (μ_m) vectors point towards in the same direction. Experimentally, it is found that, equation (7.3) is not applicable as such but requires the addition of one more factor 'g'. This factor is known as the **nuclear splitting factor** or **nuclear g value.** The numerical value of 'g' cannot be calculated theoretically and is determined from the experimental data. **For proton, the value of g is 5.585.**

Thus equation (7.3) becomes

$$\mu_m = g\left(\frac{e}{2m_p}\right)L$$

Substituting the expression of 'L' from equation (7.1) in equation (7.3):

$$\mu_m = g\left(\frac{e}{2m_p}\right)\hbar\sqrt{I(I+1)}$$

$$\mu_m = g\left(\frac{eh}{4\pi m_p}\right)\sqrt{I(I+1)} \quad \left(\because \hbar = \frac{h}{2\pi}\right)$$

$$\mu_m = g\mu_N\sqrt{I(I+1)} \qquad ...(7.4)$$

where, μ_N is the basic unit of magnetic moment called **Nuclear Magneton.**

The value of nuclear magneton for the hydrogen nucleus can be calculated as:

$$\mu_N = \left(\frac{eh}{4\pi m_p}\right)$$

$$\mu_N = \frac{(1.602\times10^{-19}\,C)\times(6.626\times10^{-34}\,Js)}{4\times(3.14)\times(1.673\times10^{-27}\,kg)}$$

$$\mu_N = 5.047 \times 10^{-27}\ J/T$$

PROBLEMS BASED ON ANGULAR MOMENTUM AND MAGNETIC MOMENT OF THE NUCLEUS

Q. 1. What is Bohr Magneton?

Ans. Magnetic moment of an electron is caused by its orbital or spin angular momentum. The physical quantity which represent this magnetic moment which is caused by either its orbital or spin angular momentum is called **Bohr Magneton** (μ_B)

$$\mu_B = \frac{eh}{4\pi m_e} = \frac{(1.60\times10^{-19}\,C)\times(6.6\times10^{-34}\,Js)}{4\times(3.14)\times(9.11\times10^{-31}\,kg)}$$

$$= 9.273 \times 10^{-24}\ JT^{-1}$$

where, e is the electronic charge and m_e is the mass of the electron.

Q. 2. What is Nuclear Magneton?

Ans. The Nuclear Magneton is the natural unit for expressing magnetic moments of heavy particles such as nucleons and atomic nuclei

$$\mu_N = \frac{e\hbar}{2m_p} = \frac{eh}{4\pi m_p}$$

where, m_p is the mass of hydrogen nucleus and e is charge on nucleus.

Q. 3. Is Bohr Magneton greater than Nuclear Magneton? Give reasons.

Ans. Bohr Magneton is for electrons and Nuclear Magneton is for protons. Both these numbers are inversely proportional to their respective masses. Since a proton is 2000 times heavier than an electron, therefore the nuclear magneton is 2000 times smaller than the Bohr magneton.

Q. 4. Given that the nuclear spin quantum number of ${}^{12}_{6}C$ is zero and that of ${}^{1}_{1}H$ is half.

(i) Calculate the nuclear angular momentum of these nuclei.

Ans. Using equation (7.1), magnitude of spin angular momentum $= L = \hbar\sqrt{I(I+1)}$

For ${}^{12}_{6}C$, $I = 0$, $L = 0$

For ${}_1^1H$, $I = 1/2$, $L = \hbar\sqrt{1/2(1/2+1)} = h/2\pi\sqrt{1/2(1/2+1)}$

$$L = \frac{(6.626 \times 10^{-34}\ Js)}{2 \times 3.14}\sqrt{3/4}$$

$$L = 0.914 \times 10^{-34}\ Js$$

(ii) How many different energy states do these nuclei have in a magnetic field?

Ans. Total number of states = (2I + 1)

For ${}_6^{12}C$, I = 0, 2I + 1 = 1

For ${}_1^1H$, I = 1/2, 2I + 1 = 2 (1/2) + 1 = 2

(iii) Calculate the magnetic moment of ${}_1^1H$, given g = 5.585

Ans. Using equation (7.4):

Nuclear magnetic moment $= \mu_m = g\mu_N\sqrt{I(I+1)}$

Here, g = 5.558

$\mu_N = 5.047 \times 10^{-27}\ JT^{-1}$

I = 1/2 (for ${}_1^1H$)

$$\mu_m = (5.585) \times (5.047 \times 10^{-27}) \times \sqrt{1/2(1/2+1)}$$

$$\mu_m = 2.44 \times 10^{-26}\ JT^{-1}$$

7.7 Precessional Motion

Consider the behaviour of a spinning top, not only it shows the spinning motion but the top will also perform a slower waltz-like motion, in which the spinning axis of the top moves slowly around the vertical. This is **precessional motion** and the top is said to be precessing around the vertical axis of the earth's gravitational field. The precession arises from the interaction of spin with earth's gravity acting vertically downwards. Only a spinning top will precess, a static top will merely fall over.

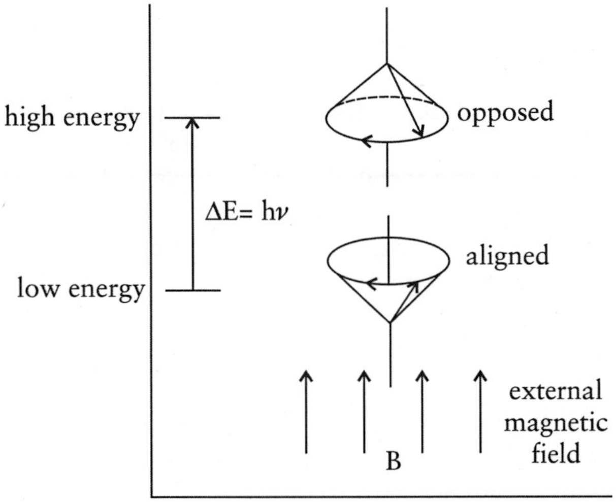

Fig. 7.7: Precessional motion of the spinning nucleus

As the proton is a spinning magnet, so just like the top, it will precess around the axis of an applied external magnetic field. The proton can do so in two principal orientations, either aligned with the field or opposed to the field (shown in Fig. 7.7).

7.8 Precessional Frequency/Larmor Frequency

The spinning frequency of the nucleus does not change. But at the same time, the speed of the precession changes.

The **Larmor** or **precessional frequency** refers to the rate of precession of the magnetic moment of the proton around the external magnetic field. In other words, when placed in a magnetic field, charged particles will precess about the magnetic field. In NMR, the charged nucleus, will then exhibit precessional motion at a characteristic frequency known as the **Larmor frequency**. The Larmor frequency is specific to each nucleus. It is measured during the NMR experiment, as it is dependent on the magnetic field that the nucleus experiences. The precessional frequency, ν, is directly proportional to the strength of the external magnetic field, B, that is

$$\nu \propto B$$

Like, a proton exposed to an external magnetic field of 1.4 T will precesses at a frequency, $\nu \approx 60$ MHz. Whereas a proton exposed to an external magnetic field of 2.3 T, will precess at a frequency, $\nu \approx 600$ MHz.

7.9 Flipping of the Proton

If a proton is placed in a magnetic field, then it starts precessing at a certain frequency and capable of taking up one of the two orientations with respect to the axis of the applied external field.

If we irradiate the precessing nuclei with a beam of electromagnetic radiation of desired frequency, then the proton which is precessing in the aligned orientation will pass into the opposed orientation by absorbing energy. From the high energy opposed orientation, it comes back to the low energy aligned orientation by loosing energy. This transition from one energy state to the other is called **flipping** of the proton. The transition can be brought about between the two energy states by absorption of a quantum of electromagnetic radiation in the radiowave region with energy hν.

The energy required to bring about the transition ($\Delta E = h\nu$) or to flip the proton depends upon the strength of the external field. Stronger is the external magnetic field applied, greater will be the tendency of the nucleus to remain aligned up with it and higher will be the frequency of the electromagnetic radiation needed to flip the proton to the higher energy state. Thus, the energy difference between the two states ΔE or the energy required for the transition is a function of applied magnetic field 'B'. As the size of atomic nucleus is very small, therefore, its interaction even with the strongest applied magnetic field would be small. As a result, NMR transitions are induced by very low frequency radiations, i.e. radio frequency region (3 MHz–30,000 MHz) of the electromagnetic spectra. This implies that frequency of NMR transition or the energy difference between the $\alpha(m_I = +1/2)$ and $\beta(m_I = -1/2)$ spin state is directly proportional to the applied field, i.e.

$$\nu \propto B$$

Stronger the magnetic field, greater is the frequency of radiation required. In the absence of external magnetic field, no NMR effect is observed as there is no energy difference between the two states (Fig. 7.8).

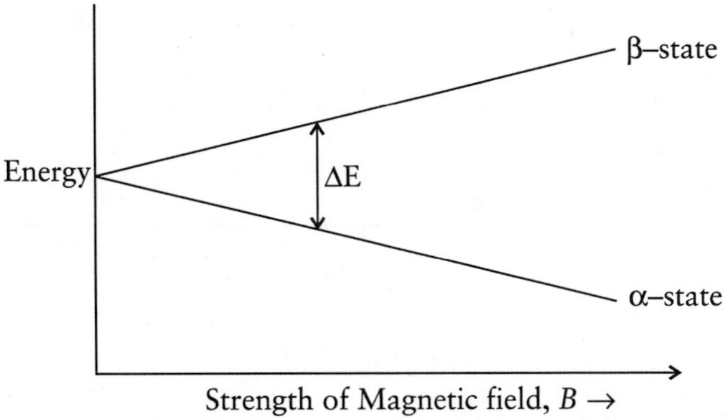

Fig. 7.8: Relationship of the applied magnetic field and energy

7.10 Condition for Energy Transition

The precessing proton will absorb energy from the radiofrequency source only if the precessing frequency is the same as the frequency of the radiofrequency beam, i.e. when the quantum energy ($h\nu$) of electromagnetic radiation matches up the energy difference between the two energy states at the external applied field strength, B.

When this occurs, the nucleus and the radiofrequency beam are said to be in resonance. Thus, the term **nuclear magnetic resonance** (NMR).

Thus,

Precessing frequency of nucleus = Frequency of the radiofrequency beam

7.11 Potential Energy of a Nucleus in a Magnetic Field

Potential energy 'E' of interaction of a nucleus with external magnetic field 'B' is given by:

$$E = -B\mu_z \quad ...(7.5)$$

where, μ_z is the z-component of the magnetic moment vector.

Let θ be the angle between the magnetic moment vector and z-axis, then

$$\mu_z = \mu_m \cos\theta \quad ...(7.6)$$

Substituting equation (7.4) in equation (7.6), we get

$$\mu_z = \left(g\mu_N\sqrt{I(I+1)}\right)\cos\theta \quad ...(7.7)$$

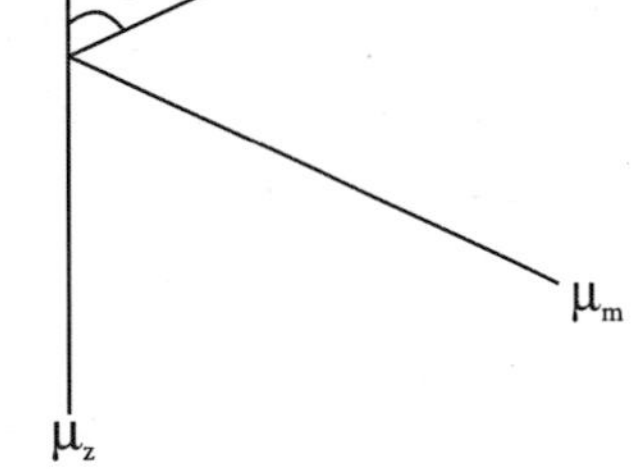

Fig. 7.9: Angle between magnetic moment and its z-component

The angle θ cannot have any arbitrary value, but only a few discrete values are allowed. The allowed values must satisfy the condition of quantisation of the component of angular momentum vector along the axis of magnetic field.

i.e. $$\left(\sqrt{I(I+1)}\frac{h}{2\pi}\right)\cos\theta = m_I\frac{h}{2\pi}$$

$$\sqrt{I(I+1)}\cos\theta = m_I \qquad ...(7.8)$$

Substituting equation (7.8) in equation (7.7):

$$\mu_z = g\mu_N\left(\sqrt{I(I+1)}\cos\theta\right)$$

$$\mu_z = g\mu_N m_I \qquad ...(7.9)$$

Substituting the expression of μ_z from, equation (7.9) in equation (7.5).

$$E = -B\mu_z$$

$$E = -B\mu_N g m_I \qquad ...(7.10)$$

Multiply and divide by $2\pi/h$:

$$E = -\left(\mu_N g \times \frac{2\pi}{h}\right) \times \frac{h}{2\pi} \times Bm_I$$

$$E = -\gamma\frac{h}{2\pi}Bm_I \qquad ...(7.11)$$

where, $\gamma = \dfrac{2\pi\mu_N g}{h}$ is a nuclear constant and is know as **magnetogyric** or **gyromagnetic ratio.**

The equation (7.11) gives the potential energy of a nucleus in a magnetic field.

7.11.1 Representation of Directions of Magnetic Moment and its z-component

The expressions of magnetic moment and its z-component are:

$$\mu_m = g\mu_N\sqrt{I(I+1)} \text{ (equation 7.4)}$$

and $\mu_z = g\mu_N.m_I$ (equation 7.9)

where I is spin quantum number and m_I is spin magnetic quantum number.

The permitted value of m_I are + I, I – 1, ... –I.

Also, the magnitude of magnetic moment (μ_m) and its z-component (μ_z) can never be same because m_I can never be equal to $\sqrt{I(I+1)}$. This means that magnetic moment vector can never be directed along the magnetic axis and will always be oriented with respect to z-axis. The angle of orientation between the magnetic moment vector and z-axis is given by:

$$\mu_z = \mu_m \cos\theta \text{ (equation 7.6)}$$

$$\cos\theta = \frac{\mu_z}{\mu_m}$$

$$\cos\theta = \frac{g\mu_N m_I}{g\mu_N\sqrt{I(I+1)}}$$

$$\cos\theta = \frac{m_I}{\sqrt{I(I+1)}}$$

For example, for a proton (I = 1/2)

$$\cos\theta = \frac{1/2 g\mu_N}{\sqrt{3}/2 g\mu_N} = \frac{1/2}{\sqrt{3}/2} = 0.5447$$

$$\theta = 54.72°$$

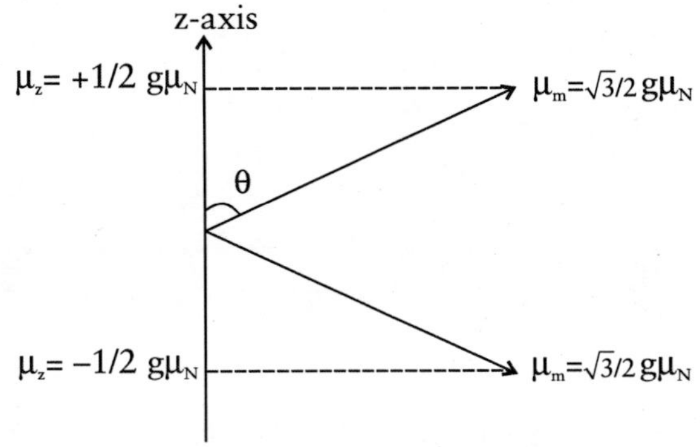

Fig. 7.10: Orientation of magnetic moment and its z-component of a proton

7.11.2 Potential Energy of a Proton in a Magnetic Field

For a proton, the value of spin quantum number, I = 1/2. Thus, m_I can have two values of +1/2 and −1/2.

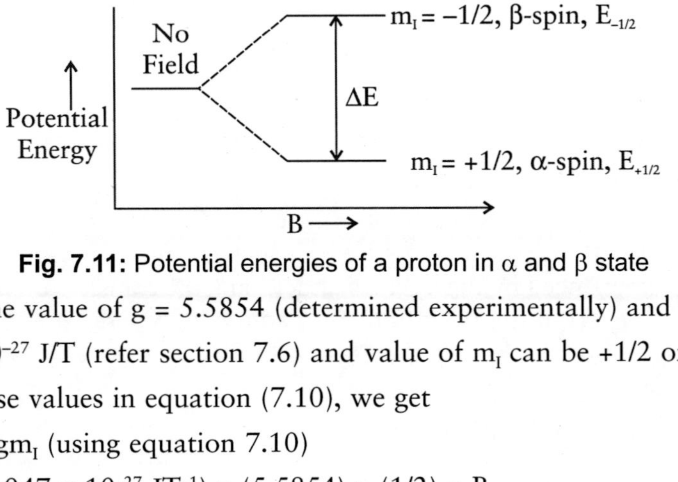

Fig. 7.11: Potential energies of a proton in α and β state

For a proton, the value of g = 5.5854 (determined experimentally) and value of $\mu_N = 5.047 \times 10^{-27}$ J/T (refer section 7.6) and value of m_I can be +1/2 or −1/2.

Substituting these values in equation (7.10), we get

$E = -B\mu_N g m_I$ (using equation 7.10)

$$E_{+1/2} = -(5.047 \times 10^{-27} \text{ JT}^{-1}) \times (5.5854) \times (1/2) \times B$$
$$= -(1.410 \times 10^{-26} \text{ JT}^{-1}) B$$
$$= -(1.410 \times 10^{-26}) \text{ B Joules}$$

$$E_{-1/2} = -(5.047 \times 10^{-27} \text{ JT}^{-1}) \times (5.5854) \times (-1/2) \times B$$
$$= +(1.410 \times 10^{-26} \text{ JT}^{-1}) B$$
$$= +(1.410 \times 10^{-26}) \text{ B Joules}$$

The energy difference between the two levels is given by

$$\Delta E = B\mu_N g \qquad ...(7.12)$$

$$\Delta E = E_{-1/2} - E_{+1/2}$$

$$= (1.410 \times 10^{-26})B - (-1.410 \times 10^{-26})B$$

$$= (2.820 \times 10^{-26} JT^{-1})B$$

$$= (2.820 \times 10^{-26})B \text{ Joules}$$

From the energy difference between the two nuclear levels, the frequency required for flipping the proton can also be calculated.

$$\Delta E = h\nu$$

$$\nu = \frac{\Delta E}{h} = \frac{(2.820 \times 10^{-26} JT^{-1}) \times B(T)}{6.626 \times 10^{-34} Js}$$

$$\nu = (4.26 \times 10^{-7} T^{-1} s^{-1})B$$

Thus, a proton in the lower energy level (m_I = + 1/2) can be promoted to the higher energy level (m_I = –1/2) in the presence of a magnetic field, B, by supplying the necessary energy in the form of electromagnetic radiation of frequency, ν, which is $(4.26 \times 10^7)B\ s^{-1}$ is case of proton.

The magnitude of the magnetic field, B, normally used is 1 Tesla. Thus for the field strength of 1 Tesla, the frequency of radiation required for the flipping of spin of proton is

$$\nu = (4.26 \times 10^7\ T^{-1}s^{-1}) \times (1 \text{ Tesla})$$

$$\nu = 4.26 \times 10^7 s^{-1}$$

$$\nu = 42.6 \times 10^6 \text{ Hz}$$

$$\nu = 42.6 \text{ MHz}$$

The above frequency lies in the radiofrequency region of the electromagnetic spectrum ($3 \times 10^6 - 3 \times 10^{10}$ Hz).

In NMR spectroscopy, if protons are involved, it may be called as **proton magnetic resonance** (PMR) spectroscopy. Also, the frequency required for such a transition depends on the strength of the magnetic field, B, experienced by the nucleus undergoing the transition. Further, the precessing proton will absorb energy from the radiofrequency source only when the quantum energy ($h\nu$) of electromagnetic radiation matches up the energy difference between the two energy states at the external applied field strength, B.

Thus, we can say that for NMR to occur:

Precessing frequency of nucleus = Frequency of the radiofrequency beam

7.11.3 Precession Frequency using Gyromagnetic Ratio

$$\gamma = \frac{2\pi\mu_N g}{h} \text{ (gyromagnetic ratio)}$$

$$\mu_N g = \left(\frac{\gamma}{2\pi}\right)h$$

Also, $\Delta E = B\mu_N g$ (using equation 7.12)

$$\Delta E = \left(\frac{\gamma}{2\pi}\right) hB$$

Now, $\Delta E = h\nu$

$$\nu = \frac{\Delta E}{h}$$

$$\nu = \left(\frac{\gamma}{2\pi}\right) \frac{hB}{h}$$

$$\nu = \left(\frac{\gamma}{2\pi}\right) B \qquad \text{...(7.13)}$$

7.11.4 Larmor Angular Frequency

(i) From the energy difference (ΔE), the **Larmor angular frequency** (in radians per second) required for flipping the proton can also be calculated

$$\omega = 2\pi\nu \qquad \text{...(7.14)}$$

$$\omega = 2\pi\left(\frac{\Delta E}{h}\right)$$

$$\omega = 2\pi\left(\frac{B\mu_N g}{h}\right) \text{ (using equation 7.12, } \Delta E = B\mu_N g)$$

$$\omega = \left(\frac{2\pi\mu_N g}{h}\right) B \qquad \text{...(7.15)}$$

$$\omega = \gamma B \qquad \text{...(7.16)}$$

where, γ is the gyromagnetic ratio

(ii) The Larmor angular frequency is also given by

$$\omega = \gamma B$$

$$\omega = \left(\frac{2\pi\mu_N g}{h}\right) B$$

$$\omega = \left(\frac{g\mu_N}{h/2\pi}\right) B$$

Multiply and divide by $\sqrt{I(I+1}$

$$\omega = \left[\frac{g\mu_N\sqrt{I(I+1)}}{h/2\pi\sqrt{I(I+1)}}\right] B$$

Using equation (7.4); $\mu_m = g\mu_N\sqrt{I(I+1)}$ and

equation (7.1); $L = h/2\pi\sqrt{I(I+1)}$, we have

$$\omega = \left(\frac{\mu_m}{L}\right) B \qquad \text{...(7.17)}$$

PROBLEMS BASED ON PRECESSIONAL FREQUENCY, MAGNETIC FIELD STRENGTH AND GYROMAGNETIC RATIO

Q. 1. Write a short note on Larmor precessional frequency.

Q. 2. What do you understand by Larmor precession. How does the precession frequency vary with the applied field strength?

Q. 3. What do you understand by Larmor precession and Larmor frequency?

Q. 4. Calculate the precessional frequency of a proton in a field of 1.5 T. The 'g' factor for the proton is 5.585 ($e = 1.602 \times 10^{-19}$ C and $m_p = 1.673 \times 10^{-27}$ kg).

Ans. $\mu_N = \frac{eh}{4\pi} m_p$

$$\mu_N = \frac{(1.602 \times 10^{-19}\,C) \times (6.626 \times 10^{-34}\,Js)}{4 \times (3.14) \times (1.673 \times 10^{-27}\,kg)}$$

$$\mu_N = 5.047 \times 10^{-27}\,J/T$$

Precessional frequency $= \nu = \frac{\Delta E}{h}$

$$\nu = \frac{B\mu_N g}{h}$$

$$\nu = \frac{(1.5T) \times (5.047 \times 10^{-27}\,JT^{-1}) \times (5.585)}{(6.626 \times 10^{-34}\,Js)}$$

$\nu = 6.38 \times 10^{7}\,s^{-1}$

Q. 5. A proton precess at a frequency of $6.38 \times 10^{-7}\,s^{-1}$. Calculate the magnetic field strength in Tesla?

Ans. $\nu = \frac{B\mu_N g}{h}$

For a proton, $\mu_N = 5.047 \times 10^{-27}\,JT^{-1}$

and $g = 5.585$

$\nu = 6.38 \times 10^{7}\,s^{-1}$ (given)

$$B = \frac{\nu h}{\mu_N g} = \frac{(6.38 \times 10^{7}\,s^{-1}) \times (6.626 \times 10^{-34}\,Js)}{(5.047 \times 10^{-27}\,JT^{-1}) \times (5.585)}$$

$B = 1.5$ T

Q. 6. The ^{14}N nucleus has a spin, I = 1.

(i) Draw an energy level diagram to show interaction of this spin with an external magnetic field B_z.

Ans.

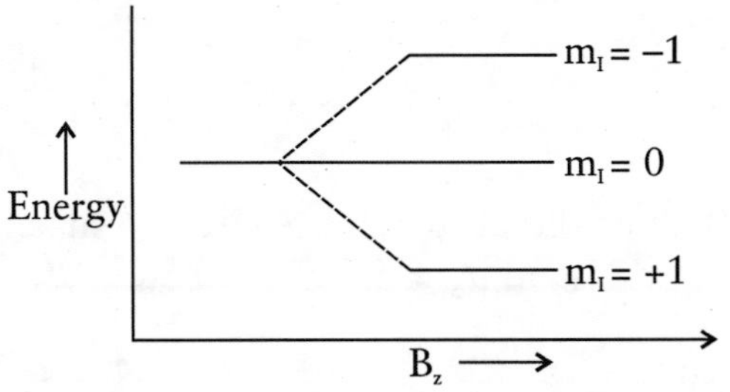

(ii) Calculate the frequency of the incident radiation required to observe the ^{14}N NMR spectrum. Given that B_z is 1.4 Tesla and the nuclear g-factor, $g_N = 0.403$.

Ans. $B_z = 1.4$ Tesla

$g_N = 0.403$

$\mu_N = 5.047 \times 10^{-27}\ JT^{-1}$

$$\nu = \frac{B\mu_N g}{h} = \frac{(1.4T)\times(5.047\times10^{-27}\ JT^{-1})\times(0.403)}{(6.626\times10^{-34}\ Js)}$$

$\nu = 4.29 \times 10^6\ s^{-1}$.

Q. 7. Calculate the magnetic field strength if a proton magnetic resonance at 220 MHz. Given, the factor 'g' for proton is 5.585.

Ans. $\nu = 220$ MHz $= 220 \times 10^6$ Hz $= 220 \times 10^6\ s^{-1}$

$g = 5.585$

$\mu_N = 5.047 \times 10^{-27}\ JT^{-1}$

$$\nu = \frac{B\mu_N g}{h}$$

$$B = \frac{\nu h}{\mu_N g} = \frac{(220\times10^6\ s^{-1})\times(6.626\times10^{-34}\ Js)}{(5.047\times10^{-27}\ JT^{-1})\times(5.585)}$$

$B = 5.17$ T

Q. 8. Calculate the NMR frequency (in MHz) of the proton (^1_1H) in a magnetic field of intensity 1.5 T. Given that $g_N = 5.585$ and $\mu_N = 5.047 \times 10^{-27}\ JT^{-1}$

Ans. $$\nu = \frac{B\mu_N g}{h} = \frac{(1.5\ T)\times(5.047\times10^{-27}\ JT^{-1})\times(5.585)}{(6.626\times10^{-34}\ Js)}$$

$\nu = 60 \times 10^6\ s^{-1} = 60 \times 10^6$ Hz $= 60$ MHz

Q. 9. Calculate the magnetic field strength required to observe fluorine (^{19}F) to precess at NMR frequency of 60 MHz. Given g_N for $^{19}F = 5.585$ and $\mu_N = 5.047 \times 10^{-27}\ JT^{-1}$.

Ans. $\nu = 60$ MHz $= 60 \times 10^6$ Hz $= 60 \times 10^6\ s^{-1}$

$$\nu = \frac{B\mu_N g}{h}$$

$$B = \frac{\nu h}{\mu_N g} = \frac{(60 \times 10^6\ s^{-1}) \times (6.626 \times 10^{-34}\ Js)}{(5.047 \times 10^{-27}\ JT^{-1}) \times (5.585)}$$

B = 1.49 T

Q. 10. Calculate the NMR frequency (in MHz) of the proton ^{1}H NMR in a magnetic field of intensity 1.4092 T. Given that $g_N = 5.585$ and $\mu_N = 5.047 \times 10^{-27}\ JT^{-1}$.

Ans. $$\nu = \frac{B\mu_N g}{h} = \frac{(1.4092T) \times (5.047 \times 10^{-27}\ JT^{-1}) \times (5.585)}{(6.634 \times 10^{-34}\ Js)}$$

$\nu = 59.85 \times 10^6\ s^{-1}$

$\nu \approx 60$ MHz

Q. 11. Calculate precessional frequency of a proton in 5 T magnetic field. Given $(\gamma_H/2\pi) = 42.57$ MHz T^{-1}.

Ans. $$\nu = \left(\frac{\gamma}{2\pi}\right)B$$

$= (42.57\ MHz\ T^{-1}) \times 5T$

$= 212.85$ MHz

Thus to excite this proton we require frequency of 212.85 MHz.

Q. 12. Calculate the required magnetic field to resonate ^{19}F nucleus at 500 MHz. Given ($\gamma_F/2\pi = 40.052$ MHz T^{-1})

Ans. $$\nu = \left(\frac{\gamma}{2\pi}\right)B$$

$500 = 40.052 \times B$

$$B = \frac{500}{40.052} = 12.5 \text{ Tesla}$$

Q. 13. The precessional frequency of a proton is 300 MHz. Calculate precessional frequency of ^{13}C of same instrument.

Given, $(\gamma_H/2\pi) = 42.57$ MHz T^{-1}

$(\gamma_c/2\pi) = 10.78$ MHz T^{-1}

Ans. Since the instrument is same, thus value of B is same

$$\frac{\nu_H}{\nu_C} = \frac{\left(\frac{\gamma_H}{2\pi}\right)\cancel{B}}{\left(\frac{\gamma_C}{2\pi}\right)\cancel{B}}$$

$$\frac{300\,\text{MHz}}{\nu_c} = \frac{42.57}{10.78}$$

$\nu_c = 75.4$ MHz

Thus, $\nu_c = 1/4\ \nu_H$.

Q. 14. For a magnetic field strength, free proton spin transition occurs at 700 MHz. Keeping the magnetic field strength same, the ^{14}N will resonate at what frequency?

Given $(\gamma_H/2\pi) = 42.57$ MHz T^{-1}

$(\gamma_N/2\pi) = 3.077$ MHz T^{-1}

Ans. $$\left(\frac{\nu_N}{\nu_H}\right) = \frac{\left(\frac{\gamma_N}{2\pi}\right)\cancel{B}}{\left(\frac{\gamma_H}{2\pi}\right)\cancel{B}}$$

$$\frac{\nu_N}{700\,\text{MHz}} = \frac{3.077}{42.57}$$

$\nu_N = 50.5$ MHz.

Q. 15. The 'g' factor of ^1H and ^{13}C are 5.6 and 1.4, respectively. For the same value of magnetic field strength, if the ^1H resonates at 600 MHz, at what frequency would ^{13}C resonate.

Ans. $\nu = \left(\frac{\gamma}{2\pi}\right)B$ (using equation 7.13)

Also, $\gamma = \frac{2\pi\mu_N g}{h}$

$$g = \left(\frac{\gamma}{2\pi}\right)\frac{h}{\mu_N}$$

where, h and γ_N, are constants

Thus, $g \, \alpha \, \frac{\gamma}{2\pi}$

Thus, the above equation (7.13) can be written as

$\nu = (g)B$

Since the same value of magnetic field strength is used.

$$\frac{\nu_C}{\nu_H} = \frac{\left(\frac{\gamma_C}{2\pi}\right)\cancel{B}}{\left(\frac{\gamma_H}{2\pi}\right)\cancel{B}} = \frac{g_C}{g_H}$$

$$\frac{\nu_c}{600} = \frac{1.4}{5.6}$$

$\nu_c = 150$ MHz

Q. 16. The nuclear g factor for ^1H and ^{14}N are 5.6 and 0.40, respectively. If the magnetic field is set as such that proton resonate at 700 MHz, the ^{14}N nucleus will resonate at what frequency?

Ans. $\frac{\nu_N}{\nu_H} = \frac{g_N}{g_H}$

$$\frac{x}{700} = \frac{0.14}{5.6}$$

$$x = 50 \text{ MHz}$$

Q. 17. Given γ (^{1}H) = 2.7 × 10^8 T^{-1} s^{-1}. Calculate the resonance frequency of a proton in magnetic field of 12.6 T.

Ans. $\nu = \left(\frac{\gamma}{2\pi}\right)B$ (using equation 7.13)

$$\nu = \frac{2.7\times10^8 T^{-1}s^{-1}\times12.6T}{2\times3.14} = 5.4 \times 10^8 \text{ Hz}$$

$$\nu = 540 \text{ MHz}$$

Q. 18. In a spectrometer operating at 1 Tesla, the NMR frequency of ^{19}F is 40.06 MHz. Calculate the gyromagnetic ratio.

Ans. ν = 40.06 MHz = 40.06 × 10^6 Hz = 40.06 × 10^6 s^{-1}

B = 1 Tesla

$\omega = \gamma B$

$2\pi\nu = \gamma B$

$$\text{Gyromagnetic ratio } = \gamma = \frac{2\pi\nu}{B} = \frac{(2)\times(3.14)\times(40.06)\times(10^6 s^{-1})}{1T}$$

$\gamma = 2.517 \times 10^8 T^{-1} s^{-1}$

7.12 Relative Population of Energy Levels

The relative population of nuclei in energy levels is given by the Boltzmann equation.

$$\frac{n_2}{n_1} = \exp\left(\frac{-\Delta E}{kT}\right)$$

where, n_1 and n_2 are the number of nuclei in the lower and higher energy levels respectively. ΔE is the energy difference between the two levels and k is the Boltzmann constant ($k = R/N_A$). The relative population, i.e. the term n_2/n_1, depends on the ratio $\Delta E/kT$.

The excitation of a nucleus from the lower energy level to the higher energy level can occur only if the lower level is more populated than the higher one. The most stable arrangement of nuclei (i.e. the arrangement with minimum potential energy) in the magnetic field is the one in which all nuclei occupy the lowest energy level. The only factor which opposes this arrangement is the value of thermal energy (T) relative to that of energy difference (ΔE) between the two energy levels.

(a) If $\Delta E > kT$, then the lowest energy level is densely populated. Although, the upper level do contain some nuclei but this number decreases rapidly as ΔE becomes larger and larger.

(b) If $\Delta E < kT$, the upper levels are equally populated. Nevertheless, the population in the lowest level is still larger than those in the upper levels.

(c) If $\Delta E/kT = 0$, indicates that the two levels will have the same population. This is possible only when the temperature of the system is infinite, which is an impossible condition.

Thus, it may be concluded that the lowest level is always more populated than the upper level.

PROBLEMS BASED ON RELATIVE POPULATION OF ENERGY LEVELS

Q. 1. Calculate the relative population of protons in the two available energy levels at 300 K under field strength of 1 Tesla (or 10,000 gauss).

Ans. For proton, value of $m_I = +1/2$ and $-1/2$

$E = -B\mu_N g m_I$ (using equation 7.10)

$E_{+1/2} = -(5.047 \times 10^{-27}\ JT^{-1}) \times (5.584) \times (1/2) \times B$

$= -(1.410 \times 10^{-26}\ JT^{-1})\ B$

$E_{-1/2} = -(5.047 \times 10^{-27}\ JT^{-1}) \times (5.584) \times (-1/2) \times B$

$= +(1.410 \times 10^{-26}\ JT^{-1}) \times B$

$\Delta E = E_{-1/2} - E_{+1/2}$

$\Delta E = (1.410 \times 10^{-26})B - (-1.410 \times 10^{-26})B$

$\Delta E = (2.820 \times 10^{-26}\ JT^{-1})B$

Here, B = 1 Tesla (given)

$\Delta E = (2.820 \times 10^{-26}\ JT^{-1}) \times (1\ T)$

$\Delta E = 2.820 \times 10^{-26}\ J$

Also, $kT = (1.38 \times 10^{-23}\ JK^{-1}) \times 300K$

$kT = 4.14 \times 10^{-21}\ J.$

Thus, $n_2/n_1 = \exp(-\Delta E/kT)$

$n_2/n_1 = \exp(-2.820 \times 10^{-26}\ J/4.14 \times 10^{-21}\ J)$

$n_2/n_1 = \exp(-6.81 \times 10^{-6})$

$n_2/n_1 = 0.99.$

Q. 2. A sample containing protons is placed in a magnetic field of strength 2 Tesla. Calculate the ratio of the number of proton spins in the lower state to the higher state at 298 K. Given the nuclear g-factor, g_N for ^1H is 5.585.

Ans. $\mu_N = 5.047 \times 10^{-27}\ JT^{-1}$

$g = 5.585$

$B = 2\ T$

$\Delta E = B\mu_N g$ (using equation 7.12)

$\Delta E = (5.047 \times 10^{-12}\ JT^{-1}) \times (5.585) \times (2\ T)$

$\Delta E = 5.6 \times 10^{-26}$ Joules

$kT = (1.38 \times 10^{-23}\ JK^{-1}) \times (298\ K)$

$= 4.1 \times 10^{-21}$ J

Now, $n_2/n_1 = \exp(-\Delta E/kT)$

$n_2/n_1 = \exp(-5.6 \times 10^{-26}\ J/4.1 \times 10^{-21}\ J)$

$n_2/n_1 = \exp(-1.29 \times 10^{-5})$

$n_2/n_1 = 0.999987$

7.13 Intensity of NMR Signal

With the absorption of radiation, the nucleus can be promoted from the lower energy level to the upper one (upward transition). Similarly, the nuclei can come from the upper level to the lower level (downward transition) with the emission of radiation. Also the rate at which transitions from a level takes place depends on the population of the level. Since the lowest level is always more populated than the upper level, the rate of upper transitions are more than the downward transitions and hence there occurs net absorption.

The intensity of net absorption depends on the difference between the population of particles in the two levels. In case of NMR, the difference between the energy levels is small and thus the intensity of absorption is also small. However, the intensity of absorption can be increased either by carrying NMR at low temperature (so that $\Delta E > KT$) or by employing high magnetic field. Both these factors help in increasing the population difference between the two levels.

PROBLEM BASED ON INTENSITY OF NMR SIGNAL

Q. 1. Explain why the intensity of NMR signal is low at room temperature? How it can be enhanced?

7.14 Nuclear Magnetic Resonance (NMR) Spectrum

In the NMR technique, the protons in an organic molecule are exposed to the powerful magnetic field. On exposure of applied external magnetic field, the different set of protons will precess at different frequencies. Then we irradiate these precessing protons with a radiofrequency source. It is more convenient of keep the radiofrequency constant and the strength of the magnetic field is constantly varied. Protons being in different electronic environments in a molecule cannot resonate at a particular value of applied magnetic field. Thus, as the field strength increases, the precessional frequency of each proton increases and at some value of field strength, the energy required to flip the protons matches the energy of radiation. As a proton (or a set of equivalent protons) comes to the resonance with the radiofrequency source, absorption occurs and a signal is observed. Such a spectrum obtained is called **NMR spectrum.**

All the protons in a molecule do not absorb at the same applied external magnetic field. The effective field strength is different for different sets of protons, since one set of protons will have slightly different environment from other set of protons in a molecule. The NMR spectrum consists of series of peaks that corresponds to different applied field strengths. In this, we measure applied field strengths for each set of protons and the absorption peaks are plotted. The number of signals at different applied field strengths is equal to the different set of equivalent protons, i.e. each peak means a set of protons.

7.15 Instrumentation

Nuclear magnetic resonance spectrophotometer makes use of a magnet, a radiofrequency detector and an amplifier. The detection system is used to note that energy is being transferred from the radiofrequency beam to the nucleus of the sample.

The sample under investigation is taken in a glass tube which is placed between the poles of a magnet. A radiofrequency source (ν = 60 mega cycles sec^{-1}) is made to fall on the sample, which can be done by feeding energy (radiofrequency source) into a coil wound round the sample tube. A signal is detected if the nuclei in the sample resonates with the source, i.e. ΔE energy required to flip the proton is the same as that of the source. Energy is transferred from the source via nuclei to the detector coil after amplication is recorded on the chart paper.

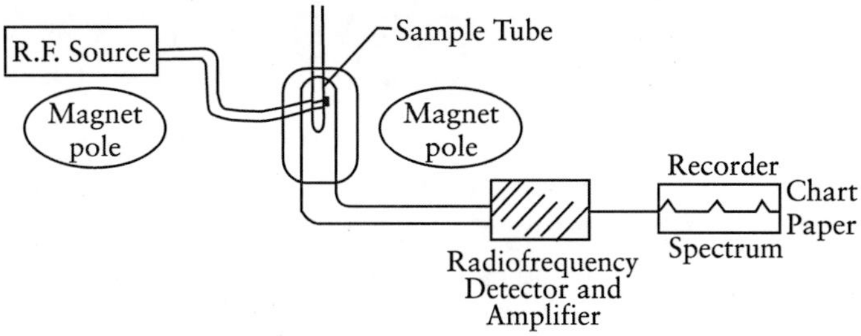

Fig. 7.12: Nuclear magnetic resonance spectrophotometer

As protons are in different electronic environments in a molecule, thus cannot resonate exactly at 60 mega cycles sec^{-1}. For practical purposes, radiofrequency source is held steady at the said frequency and field strength is varied by placing small electromagnet to the poles of the magnet. By increasing the current flowing through these electromagnets, the total field strength is increased.

As the field strength increases, the precessional frequency of each proton increases until resonance with the radiofrequency source takes place. As proton (or a set of equivalent protons) comes to resonance, the signal from the detector produces a peak on the chart paper. The NMR spectrum consists of series of peaks that correspond to different applied field strengths. Each peak includes a set of protons.

7.16 Measurement of Spectra

1. **Sample Handling:** About 0.4 mL of neat liquid or 20-50 mg of a solid is dissolved in a deuterated solvent (which does not give signals in the NMR spectrum). Choice of solvent depends upon the solubility of the compound. The commonly used solvents are D_2O, C_6D_6, CCl_4, CD_3Cl and DMSO. The solution should not be viscous and should be free from insoluble impurities.
2. **Reference Compound:** An internal reference compound TMS [tetramethylsilane $(CH_3)_4Si$] is added. The exact frequency value cannot be calibrated with an accuracy of ± 1 Hz, hence absolute positions of absorptions cannot be directly obtained from the NMR instrument. However, relative frequencies can be obtained with an accuracy

of ± 1 Hz. Therefore, the positions of absorption signals are expressed relative to a reference compound (normally TMS) as **chemical shift**.

TMS is chosen as internal standard as it is volatile, inert, economical. It has 12 equivalent protons and gives one signal at the extreme of frequencies for most of the organic compounds because all the hydrogens in TMS are equivalent. Also its boiling point (27°C) is low, so it can be easily removed from a sample of valuable organic compound to be recovered after spectral determination. It is miscible with almost all organic substances. It does not take part in intermolecular associations with the sample.

PROBLEMS BASED ON REFERENCE COMPOUND, TMS

Q. 1. What is the role of Trimethylsilane (TMS) in scanning the NMR spectrum of molecule?

Q. 2. Give two points in favour of TMS as a reference compound in NMR studies.

Q. 3. What are the characteristics of TMS which make it useful as a reference in NMR spectrum?

Ans. (a) It gives a single peak, since all the 12 hydrogen nuclei are equivalent.

(b) The hydrogen nuclei in TMS are highly shielded as compared to those in the organic compounds and hence exhibit resonance at a very high-field strength which can be easily recognized (1.4092 T for 60 MHz radiation).

(c) It is volatile and chemically inert.

(d) It has a low boiling point (27°C) and hence can be easily removed after the spectrum has been recorded.

7.17 Shielding and Deshielding Effects

Hydrogen nuclei (proton) in a molecule are surrounded by negatively charged electrons. These charged electrons generate a secondary magnetic field called **induced magnetic field** which shields the nucleus from the influence of applied magnetic field, B. Greater the electron density around the proton, greater will be the induced secondary magnetic field. Thus, the magnitude of induced magnetic field is directly proportional to the electron density and the applied magnetic field B. The induced magnetic field opposes the applied magnetic field (B) in the region of nucleus (Fig. 7.13). As a result nucleus will experience a weaker magnetic field (B_{eff}) and the effect is known as **shielding effect** and the nuclei is said to be shielded, $B_{eff} = B - \sigma B$, where σ is the **shielding constant**.

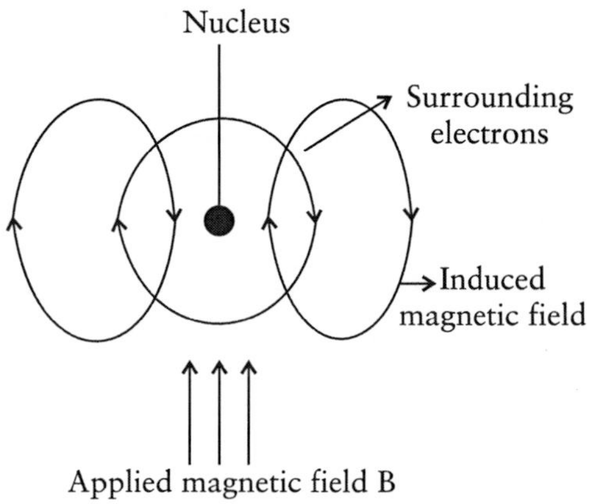

Fig. 7.13: Shielding of nucleus by circulating electrons

For a bare proton, $\sigma = 0$, thus $B_{eff} = B$, but in a molecule, σ is positive and hence $B_{eff} < B$, i.e. if a proton is attached to an electron donating group i.e., the nucleus is surrounded by increased electron density as a result magnitude of induced magnetic field around the nucleus would be more and thus nucleus will experience a decrease in B_{eff} or nucleus is shielded more from the applied field. In other words, greater the electron density, higher is the shielding effect. Thus, to overcome the shielding effect and to bring the protons to resonance, greater applied magnetic field is required.

In contrast, when a proton is attached to an electron withdrawing group, i.e. decrease in electron density around the nucleus, as a result magnitude of induced magnetic field around the nucleus is less and thus, nucleus will experience an increase in B_{eff} and nucleus is said to be deshielded. For the deshielded nucleus, smaller value of applied field will be needed to bring the proton to resonance.

7.18 Chemical Shift

The positions of the signals in the NMR spectrum help us to know the nature of protons i.e. aliphatic, aromatic, acetylenic, vinylic, etc. Each of these type of protons will have different electronic environments and thus they absorb at different applied magnetic field strengths. Thus, the number of signals in NMR spectrum tells the number of the sets of equivalent protons in a molecule. Thus, in NMR spectra the spread of absorption frequencies is caused by different chemical environments and the signals are described as **chemical shift** from some standard frequency. For measuring chemical shifts of various protons in a molecule, the signal for tetramethyl silane (TMS) is taken as a reference and chemical shifts for different kinds of protons are measured relative to it. **The difference in the absorption position of the proton with respect to TMS signal is called chemical shift** (**δ-value**).

In majority of organic compounds, protons resonate at a lower field (high frequency) than the protons of tetramethyl silane (TMS). Generally, chemical shift is measured as the difference of the frequency of the peak 'ν_s' and frequency of internal standard 'ν_{TMS}' (both measured in Hz) and divided by the operating frequency in MHz

$$\delta(\text{ppm}) = \frac{\nu_S(\text{Hz}) - \nu_{TMS}(\text{Hz})}{\text{operating frequency (MHz)}} \times 10^6$$

$$\delta(\text{ppm}) = \frac{\Delta\nu}{\text{operating frequency}} \times 10^6 \quad \ldots(7.18)$$

where, $\Delta\nu$ is the **frequency shift**

The value of 'δ' is expressed in parts per million (ppm). Most chemical shifts have δ values between 0 and 10.

- Sometimes in the formula of chemical shift, instead of operating frequency (ν), the applied magnetic field is given. In that case, the applied magnetic field is to be converted into frequency by using the relationship given below:

$$\text{i.e.} \quad \nu = \frac{\Delta E}{h} = \frac{\mu_N g B}{h}$$

$$\nu = \frac{(5.047 \times 10^{-27}\ \text{JT}^{-1}) \times (5.585)}{(6.626 \times 10^{-34}\ \text{Js})} \times B$$

$$\nu = (42.54 \times 10^6\ \text{T}^{-1}\ \text{s}^{-1}) \times B$$

$$\therefore \quad \delta = \frac{\nu_s(\text{Hz}) - \nu_{\text{TMS}}(\text{Hz})}{\text{operating frequency (MHz)}} \times 10^6$$

$$= \frac{\nu_S(\text{Hz}) - \nu_{\text{TMS}}(\text{Hz})}{(42.54 \times 10^6) \times (\text{B/T})\text{s}^{-1}} \times 10^6$$

$$\delta = \frac{\nu_S(\text{Hz}) - \nu_{\text{TMS}}(\text{Hz})}{42.54 \times (\text{B/T})\,\text{MHz}} \times 10^6 \qquad \text{...(7.19)}$$

- Chemical shift can also be expressed in terms of external magnetic field.

$$\delta(\text{ppm}) = \frac{B_{\text{sample}} - B_{\text{TMS}}}{B_0} \times 10^6 \qquad \text{...(7.20)}$$

where, B_{ref} (or B_{TMS}) and B_{sample} are the values of the external magnetic field at which resonances occur of the protons of the reference compound and that of the given compound respectively. B_0 is the applied magnetic field. The NMR signal is plotted with magnetic field strength increasing to the right. The TMS peak is set at the extreme right of the spectrum with δ = 0 ppm; as its protons absorbs at higher filed than that of most of the organic compounds.

For most of the organic compounds, signal appears downfield to the left of TMS signal. The δ-value increase in the low applied magnetic field and decrease in the high applied magnetic field (shown in Fig. 7.14).

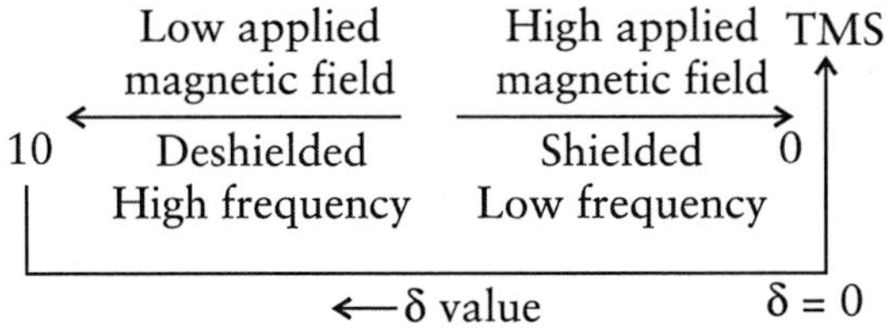

Fig. 7.14: Chemical shift in δ scale

Another commonly used standard scale is 'τ'. In the τ scale, signal for the standard reference, TMS is taken as 10 ppm (shown in Fig. 7.15). The two scales are related as:

$$\tau = 10 - \delta$$

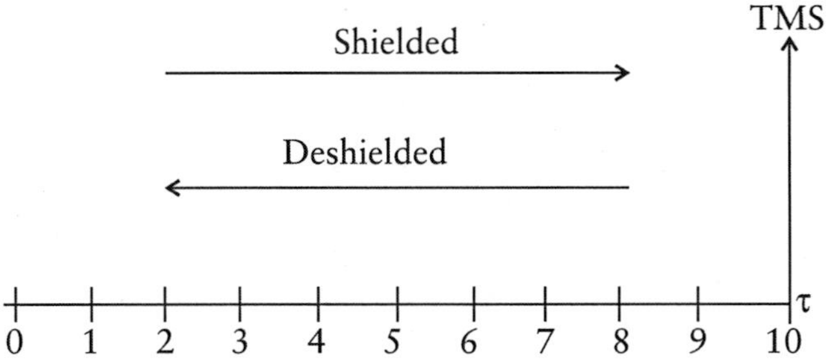

Fig. 7.15: Chemical shift in τ scale

7.19 Factors Effecting Chemical Shift

1. Inductive Effect: If a proton is attached with an electronegative group or atom, then the proton gets deshielded. Greater the electronegativity of the atom, greater is the deshielding of the proton and proton signal appears at higher δ value, i.e. proton resonates at higher frequency than expected and small value of applied field will be needed to bring the proton to resonance. Also, greater the distance of the electronegative atom from a particular proton, smaller the deshielding effect on it. For example, the chemical shifts of protons of methyl halides are in accordance with the electronegativity of the attached halogen.

	CH_3I	<	CH_3Br	<	CH_3Cl	<	CH_3F
δ:	2.16		2.68		3.05		4.26

2. Anisotropic Effect: In organic compounds having π bonds, e.g. alkenes, alkynes, aldehydes and aromatic compounds, the chemical shifts cannot be explained on the basis of electronegativity. In these cases, the chemical shifts are explained on the basis of anisotropic effect generated by π electrons in addition to the induced magnetic field generated by sigma electrons, under the influence of applied magnetic field. The values of δ in these cases can be justified by explaining the way in which the π electrons circulate under the influence of applied magnetic field.

(a) **Alkynes (Acetylenic Proton):** When the axis of acetylene molecule is aligned parallel to the applied magnetic field, the π electrons are induced to circulate around the axis in such a way that these π electrons generate an induced magnetic field in a direction opposite to the applied magnetic field. Thus, the protons of acetylene gets shielded by the induced magnetic field of π electrons. Thus, the protons resonate at lower frequency (or low δ value) and higher value of applied field will be needed to bring the proton to resonance (Fig. 7.16).

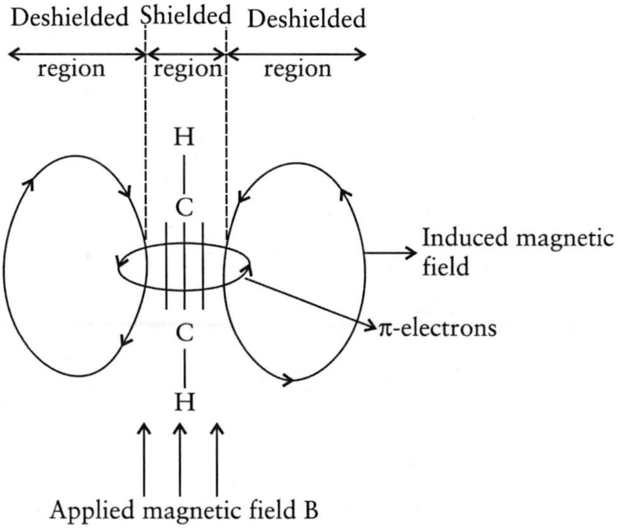

Fig. 7.16: Shielding of acetylenic protons

(b) **Alkene and Aldehydic Protons:** Alkene and aldehyde molecules orient perpendicular to the applied magnetic field B, and their π electrons generate induced magnetic field in the direction of the applied magnetic field. These protons experience a magnetic field greater in magnitude than the applied filed and small value of applied field will be needed to bring the protons to resonance. Hence, B_{eff} is increased in the region of protons (Fig. 7.17) and the protons are deshielded consequently. Alkene and aldehydic protons resonate at higher frequency than expected and small value of applied field will be needed to bring the protons to resonance.

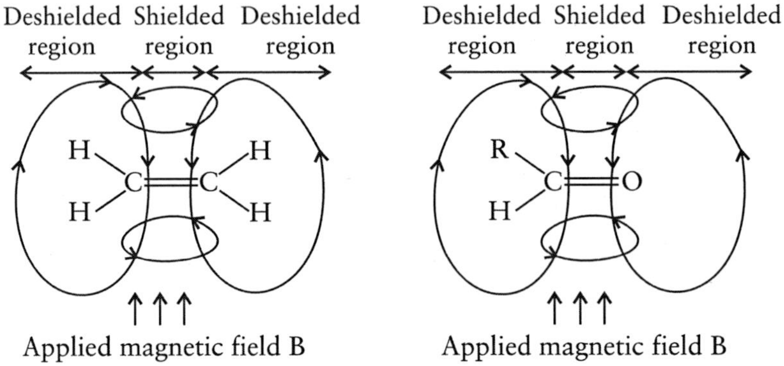

Fig. 7.17: Deshielding of alkene and aldehydic protons

(c) **Aromatic Protons (Benzene):** In benzene, in the presence of applied magnetic field, the π electrons are induced to circulate producing the ring current. The induced current produced by π electrons inside the ring is in a direction opposite to the applied magnetic field and the induced current outside the ring is in a direction parallel to the applied field (Fig. 7.18). In benzene, the protons are outside the ring and thus aromatic protons experience a magnetic field greater in magnitude than the applied field and smaller value of applied filed will be needed to bring the aromatic protons to resonance.

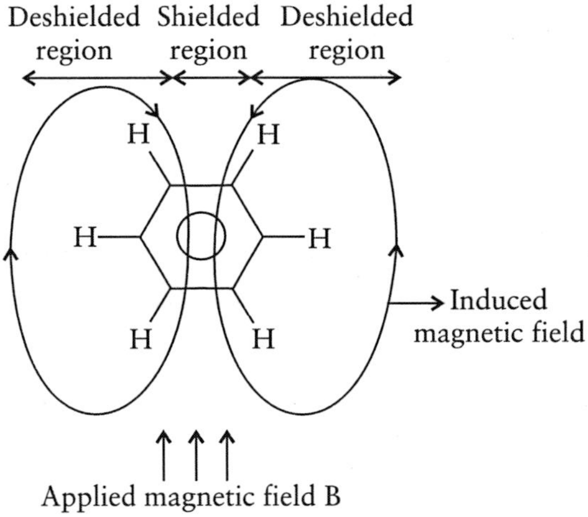

Fig. 7.18: Ring current effects (deshielding of aromatic protons)

3. Hydrogen Bonding: Due to high electronegativity of the atom to which the proton is hydrogen bonded, the proton is highly deshielded and smaller value of applied magnetic field will be needed to bring these hydrogen bonded protons than non-hydrogen bonded protons.

In case of intermolecular hydrogen bonding, the chemical shift of the hydroxyl proton is dependent upon temperature, solvent and concentration.

Table 7.1: Chemical shift values for various types of protons

Types of Protons		Chemical shift (δ) in ppm
Alkane	R—CH_3	0.9
Alkene	C=C–H	4.6–5.8
Alkyne	C≡C–H	2.0–3.5
Aromatic	Ar–H	6.0–9.0
Fluoride	F—C—H	4.0–4.5
Chloride	CI—C—H	3.0–4.0
Alcohol	HO—C—H	3.4–4.0
	H—C—OH	1.0–5.5
Phenol	Ar—OH	4.0–12.0
Ester	ROOC—H	2.0–2.2
Carboxylic Acid	HOOC—H	2.0–2.5
	H—C—COOH	10.5–12.0
Aldehyde	RCO—H	9.0–10.0
Ether	RO—C—H	3.3–4.0

4. Effect of variation of temperature on chemical shift:

Coalescence Temperature is defined as the temperature at which the appearance of the spectrum changes from that of two separate peaks to that of a single, flat-topped peak.

At low temperature, two separate methyl signals are seen, but as the temperature is raised, the rate of rotation about the bond increases. The lines broaden and then coalesce and thus lineshaped spectra is seen.

Like:

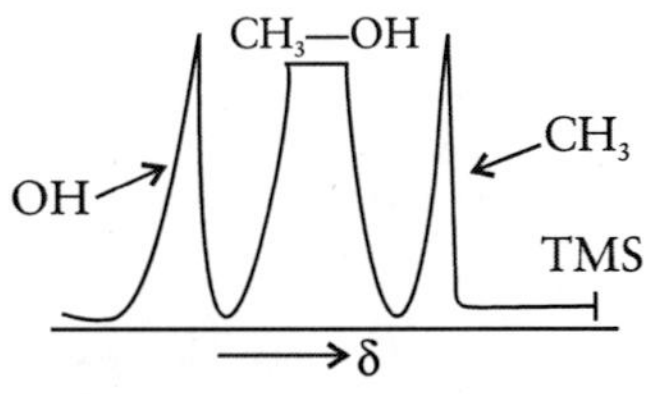

Fig. 7.19: Effect of temperature on chemical shift of CH_3–OH

PROBLEMS BASED ON CHEMICAL SHIFT

Q. 1. How chemical shift varies with magnetic field?

Q. 2. Explain the term chemical shift by giving appropriate example. What is the effect of external magnetic field on it?

Q. 3. The chemical shifts of protons in CH_3F and CH_3I are respectively 4.26 and 2.16 in the delta scale. Account for this in a qualitative manner.

Q. 4. What do you understand by the term chemical shift? Explain taking benzene and acetylene as examples.

Q. 5. The PMR peak for acetylene is found upfield of the expected value. Explain.

Q. 6. Why proton in acetylene molecule absorb at high magnetic field, while the proton in benzene absorb at low magnetic field for the same operating frequency in NMR spectrum?

Q. 7. What is the effect of variation of temperature on chemical shift?

Q. 8. At a magnetic field of 1.41 T, the frequency separation between protons in benzene and protons in TMS is 436.2 Hz. What is the chemical shift?

Ans. $$\delta = \frac{\nu_S(\text{Hz}) - \nu_{TMS}(\text{Hz})}{42.54 \times (\text{B/T})\,\text{MHz}} \times 10^6 \text{ (using equation 7.19)}$$

$$\delta = \frac{\Delta\nu(\text{Hz})}{42.54 \times (\text{B/T})\,\text{MHz}} \times 10^6$$

$$\delta = \frac{436.2}{(42.54) \times (1.41) \times 10^6\,\text{s}^{-1}} \times 10^6$$

$$\delta = \frac{436.2}{60.0096 \times \cancel{10^6}} \times \cancel{10^6}$$

$\delta = 7.27$ ppm

Q. 9. A compound shows a proton NMR peak at 240 Hz downfield from the TMS peak in a spectrometer operating at 60 MHz. What are the values of chemical shifts, δ and τ in ppm relative to TMS?

Ans. $$\delta = \frac{\nu_{\text{sample}} - \nu_{\text{TMS}}}{\nu_0} \times 10^6 \text{ (ppm)}$$

$$= \frac{240 \quad \text{s}^{-1}}{60 \times 10^6 \ \text{s}^{-1}} \times 10^6$$

$\delta = 6$ ppm

Now, $\tau = 10 - \delta = 10 - 6 = 4$ ppm.

7.20 Equivalent and Non-Equivalent Protons (Number of NMR Signals)

Number of signals in NMR spectrum indicates how many different kinds of equivalent protons are present in a molecule. Each signal corresponds to a set of equivalent protons. The protons with same chemical environment are said to be **chemically equivalent** and absorb at the same field strength or same chemical shift (spin-spin splitting is not observed for protons which are chemically equivalent). Whereas the protons with different chemical environments are non-equivalent protons and they have different chemical shifts.

For example:

(a) Methane ($\overset{a}{CH_4}$): All the four hydrogens of methane have identical environment and are chemically equivalent. Thus, these four hydrogens shows one peak in NMR spectra.

(b) Ethane ($\overset{a}{CH_3}—\overset{a}{CH_3}$): All the six hydrogens of ethane have identical environment and are chemically equivalent. Thus, these six hydrogens shows one peak in NMR spectra.

(c) Propane ($\overset{a}{CH_3}—\overset{b}{CH_2}—\overset{a}{CH_3}$): Propane has six equivalent primary protons (a), which shows one signal and two equivalent secondary protons (b), which shows another signal in NMR spectra.

(d) Benzene ($C_6\overset{a}{H}_6$): All the six hydrogens of benzene have identical environment and are chemically equivalent. Thus, these six hydrogens shows one peak in NMR spectra.

(e) Ethyl chloride ($\overset{b}{C}H_3{-}\overset{a}{C}H_2{-}Cl$): Shows two peaks in NMR spectra.

(f) Propyl chloride ($\overset{c}{C}H_3{-}\overset{b}{C}H_2{-}\overset{a}{C}H_2{-}Cl$): Shows three peaks in NMR spectra.

(g) Isopropyl Chloride ($\overset{a}{C}H_3{-}\underset{\displaystyle Cl}{\overset{b}{C}H}{-}\overset{a}{C}H_3$): Shows two peaks in NMR spectra.

(h) Acetone $\left(\begin{matrix}\overset{a}{C}H_3\\ \overset{a}{C}H_3\end{matrix}\!\!>C=O\right)$: Shows one peak in NMR spectra.

(i) Cyclobutane $\left(\begin{matrix}CH_2-CH_2\\ | \quad\quad |\\ CH_2-CH_2\end{matrix}\right)$: Shows one peak in NMR spectra.

(j) Methyl alcohol ($\overset{b}{C}H_3{-}\overset{a}{O}H$): Shows two peaks in NMR spectra.

(k) Ethyl amine ($\overset{c}{C}H_3{-}\overset{b}{C}H_2{-}\overset{a}{N}H_2$): Shows three peaks in NMR spectra.

(l) Isopropyl alcohol ($\overset{b}{C}H_3{-}\underset{\displaystyle \overset{a}{O}H}{\overset{c}{C}H}{-}\overset{b}{C}H_3$): Shows three peaks in NMR spectra.

7.21 Counting the Number of Protons from Peak Area

In NMR spectrum, different kinds of protons gives rise to different peaks. The intensity of the peak (as area of the peak) is directly proportional to the number of protons responsible for the peak. NMR instruments are equipped with electronic integrators which trace a series of steps at signals and gives the height of each step (the height is proportional to the area of respective peak). From the ratio of these set of numbers, the number of protons causing each signal is determined.

7.22 Low-Resolution Spectrum

The spectrum which shows only the effects of chemical shift, is known as the low-resolution spectrum. The low-resolution NMR spectrum shows a single peak for a hydrogen atom of particular molecular environment. To explain this, lets take few example:

(1) NMR spectrum of methyl alcohol: The extent of shielding varies in a molecule because of different chemical environment. Like in methyl alcohol (CH_3OH), it has three identical C—H bonds and one O—H bond. The chemical environment of hydrogen atoms in C—H bonds and O—H bonds are different. Since oxygen is more electronegative than carbon, the electron density on H in O—H bond will be less than that on H in the C—H bond. We can also say that the hydrogen atom in C—H bond is more shielded than the hydrogen atom in the O—H bond. Thus, the hydroxyl proton will show resonance earlier than the methyl protons as the external magnetic field is steadily increased. So the NMR signal of hydroxyl proton (O—H) will be observed downfield and that of methyl protons (C—H) at the upfield. Also the area under the signal of methyl protons will be three times larger than that under the hydroxyl proton, since methyl group has three identical protons. It must be noted that neither oxygen nor carbon will show any absorption as both of these are magnetically inactive nuclei.

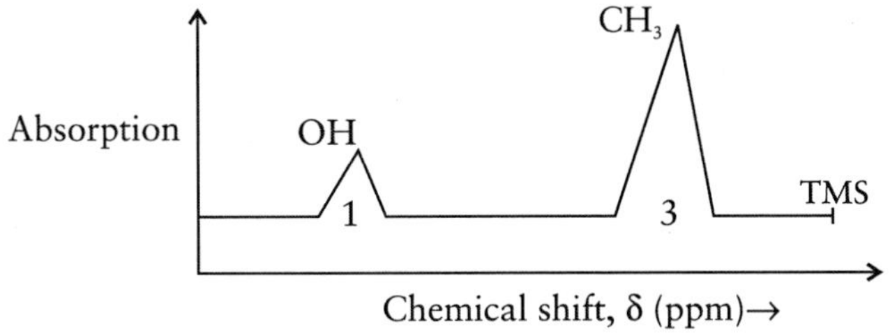

Fig. 7.20: Low resolution spectrum of methyl alcohol (CH_3OH)

(2) **NMR spectrum of ethyl alcohol (CH_3CH_2OH):** In ethyl alcohol, there are three different types of protons—hydroxyl proton (O—H), methylene protons (CH_2) and methyl protons (CH_3) with different chemical environments. So the find three different peaks corresponding to three different types of hydrogens which are in different environments.

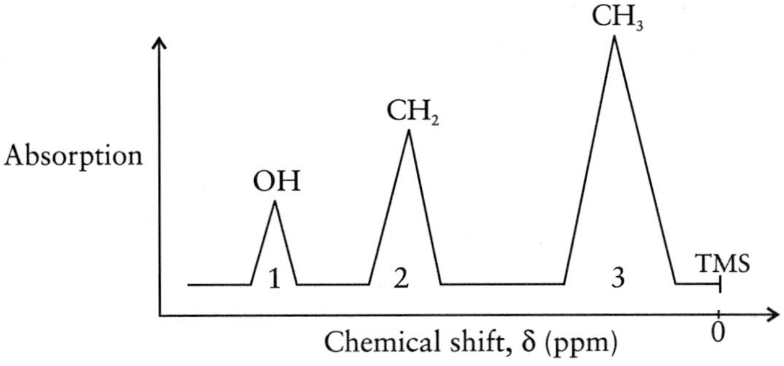

Fig. 7.21: Low resolution spectrum of ethanol (CH_3CH_2OH)

Out of these, the hydroxyl proton is expected to be the least shielded as it is attached to the more electronegative oxygen atom. Thus, the hydroxyl proton will come into resonance at a lower field than those of methylene and methyl protons. Of the methylene and methyl groups, the protons attached to the methylene group are expected to be less shielded as the CH_2 group is directly attached to the oxygen atom. Thus, the methylene protons will come into resonance earlier than those of the methyl protons. The spectrum of ethyl alcohol (Fig. 7.20) has three absorption signals. The area under these peaks will follow the ratio 1 : 2 : 3, as they represent one proton of the OH group, two protons of the CH_2 group and three protons of the CH_3 group, respectively.

7.23 Spin-Spin Coupling/Spin-Spin Splitting

In NMR spectrum, we expect a single peak for a single proton or group of chemically equivalent protons. But generally we observe a group of peaks (multiplets) because of spin-spin coupling or spin-spin splitting. The signal may split into two peaks (a **doublet**), three peaks (a **triplet**), four peaks (a **quartet**) and so on. A single peak for a given proton or group of chemically equivalent protons is observed only if there is no proton on the adjacent atom. Thus, splitting occurs only between nuclei with different chemical shifts, i.e. between chemically non-equivalent protons.

For example:

(a) 1, 2-dichloroethane $(Cl—\overset{a}{C}H_2—\overset{a}{C}H_2—Cl)$: All the four protons are chemically equivalent, thus these do not split each others signal and give a singlet in spectra.

Spin-spin coupling is the indirect coupling of proton spins through the intervening bonding electrons. A nuclear magnetic moment induces magnetic polarization in the bonding electron which passes it to another bond finally to the next proton and the spin orientation of one proton is transmitted to another.

The spin-spin interaction decreases very rapidly as the distance between the protons is increased and this interaction is transmitted through the bonds. In molecules, it operates only between the next nearest neighbours. Also, the interaction is always mutual, i.e. if the splitting of A occurs due to the interaction of the neighbouring proton B, then there will also occur the splitting of the signal of B. Thus, it is an intramolecular behaviour, i.e. the NMR signals of one molecule will not be affected by the protons of the other molecule.

7.23.1 Effect of Magnetic Field on Spin-Spin Coupling

Although chemical shift depends on external magnetic field but the spin-spin separation between the component lines are intramolecular interaction, thus found to be independent of the external magnetic field. Thus, recording a spectrum at different magnetic field merely changes the chemical shifts and not the spin-spin couplings. It is also noted that the spin-spin couplings remain unaltered, whereas the chemical shifts are magnified in proportional to the change in frequency of the radiation.

7.23.2 Effect of Temperature on Spin-Spin Coupling

Same as discussed in effect of variation of temperature on chemical shift (Refer section 7.19)

7.23.3 Multiplicity

In general, the number of peaks or multiplicity of a signal = n + 1, where n is the number of protons causing the splitting.

For example:

(a) **Nitroethane** $(\overset{b}{C}H_3\overset{a}{C}H_2NO_2)$: The three methyl protons (b) are equivalent but these are non-equivalent to methylene protons (a) and vice versa. Signal due to methyl protons splits due to spin-spin coupling by methylene protons into (2 + 1) = 3 peaks, i.e. a triplet. Similarly, spin of methyl protons couple with spin of methylene proton and split their signal into 3 + 1 = 4 peaks, i.e. signal of methylene protons will appear as a quartet.

(b) H_2, CH_4, $CH_3—CH_3$, C_6H_6, etc. molecules have chemically equivalent protons. They do not exhibit spin-spin splitting in the NMR spectrum in spite of the fact that there exist spin-spin interactions among them. Thus, only single proton resonance line is observed for these molecules.

If the spin-spin splitting is because of different non-equivalent protons then the number of peaks will be equal to (n + 1) (n′ + 1) (n″ + 1), where n, n′ and n″ are the number of different kinds of protons.

For example:

(a) 1, 1-dibromo-3, 3-difluoropropane $\left(\begin{smallmatrix}F\\F\end{smallmatrix}\!>\!\overset{a}{C}H\!-\!\overset{b}{C}H_2\!-\!\overset{c}{C}H\!<\!\begin{smallmatrix}Br\\Br\end{smallmatrix}\right)$: There are three different kinds of protons. The methylene protons (b) have different kinds of vicinal protons to undergo spin-spin coupling. Hence, the signal for methylene protons (b) will appear as

(n + 1) (n′ + 1) = (1 + 1) (1 + 1) = 4 lines or quartet

as n = 1, number of (a) protons

n = 1, number of (b) protons

7.23.4 Relative Intensities of Peaks (Pascal's Triangle)

In ^{1}H NMR spectra, for n non-equivalent protons coupling to the proton under consideration, (n + 1) lines are obtained and the relative intensities of these peaks is given by Pascal's triangle or by coefficients of the terms in the expression of $(x + 1)^n$.

Table 7.2: Relative intensities of peaks in NMR spectra

Number of vicinal protons causing splitting (*n*)	(*n* + 1) (multiplet)	Pascal triangle (Relative Intensities)
0	1 (singlet), s	1
1	2 (doublet), d	1 1
2	3 (triplet), t	1 2 1
3	4 (quartet), q	1 3 3 1
4	5 (quintet), m	1 4 6 4 1
5	6 (sextet), m	1 5 10 10 5 1
6	7 (septet)	1 6 15 20 15 6 1 and so on

If n = 1, then $(x + 1)^1 = x + 1$ hence doublet with ratio of intensities = 1 : 1.

If n = 2, then $(x + 1)^2 = x^2 + 2x + 1$ and the ratio for the triplet = 1 : 2 : 1 and so on.

7.23.5 Coupling Constant (J)

The spacing between the centres of two adjacent peaks is equal and this spacing is called **coupling constant J.** The coupling constant J denotes the effectiveness of spin-spin coupling. Value of J is expressed in Hz or cps and it does not vary with change in applied magnetic field but varies with the environment of the protons and their geometric relationship to each other.

Coupling constant, J_{ab} means the coupling constant for H_a split by H_b or for H_b split by H_a. For a pair of coupled protons the J value is the same in each of the two doublets.

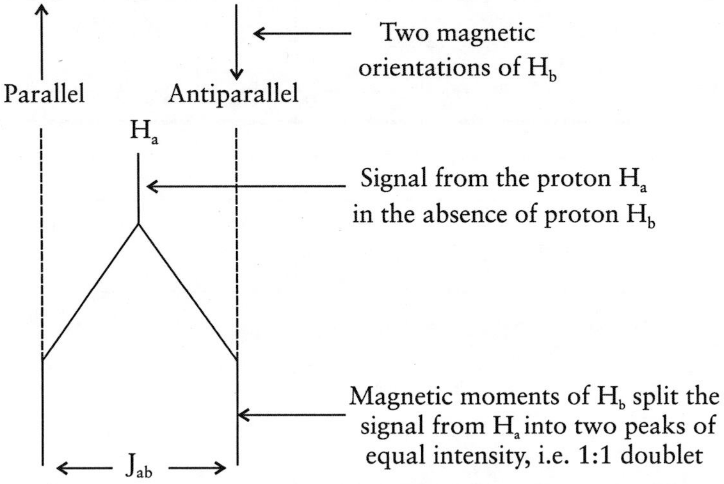

Fig. 7.22: Coupling constant, J_{ab}

Value of coupling constant depends upon the structural relationships between the couple protons. Various types of coupling are:

(a) **Geminal coupling:** In case of geminal protons (protons attached to the same carbon having different chemical environment) the value of J depends upon the bond angle.

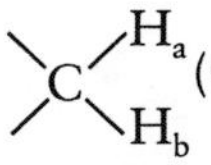

(Geminal protons)

(b) **Vicinal coupling:** In case of vicinal protons (protons attached to adjacent carbon atoms) the value of J depends upon the dihedral angle.

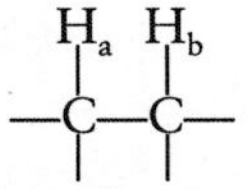

(Vicinal protons)

Three bond coupling (^3J) is referred to as vicinal coupling and is the most frequently observed type of coupling in ^1H NMR spectra, e.g. in nitroethane ($CH_3CH_2NO_2$), vicinal coupling between three protons of methyl group with two protons of the adjacent methylene group. Two bond or geminal coupling (^2J) has largest coupling constant but is less common than vicinal coupling.

Table 7.3: Coupling constant values due to geminal and vicinal coupling

Vicinal Protons	Type	Coupling constant (J_{ab}) in Hz
Alkenes	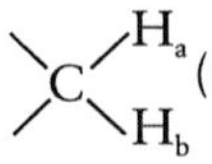 (Geminal protons)	0.0–25.0
	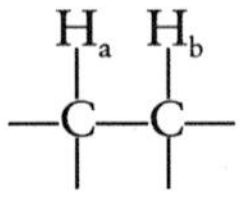 (Vicinal protons)	2.0–9.0
Alkenes	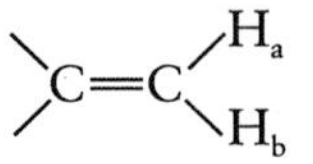	0.5–3.0
	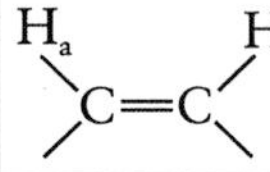 (Cis)	7.0–12.0
	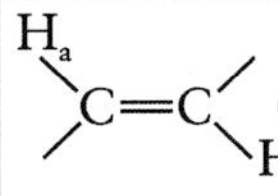 (Trans)	13.0–18.0
Alkynes	$H_a—C≡C—H_b$	9.1
Benzene	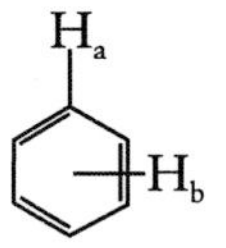	(i) ortho coupling = 6.9–9.0 (ii) meta coupling = 1.0–3.0 (iii) para coupling = 0.0–1.0

Calculation of value of Coupling Constant (J)

For the simple case of a doublet, the coupling constant is the difference between two peaks. The thing is that 'J' is measured in Hz, not in ppm. The first thing to do is to convert the peaks from ppm into Hz. Suppose we have one peak at 4.260 ppm and another at 4.247 ppm. To get in Hz, just multiply these values by the field strength in MHz. If we use a 500 MHz machine, our peaks are at 2130 Hz and 2123.5 Hz respectively. The 'J' value is just the difference. In this case it is 2130–2123.5 = 6.5 Hz.

PROBLEMS BASED ON COUPLING CONSTANT (J)

Q. 1. In the 400 MHz ^1H NMR spectrum of organic compound exhibited a doublet. The two lines of doublet are δ = 2.35 and δ = 2.38 ppm. Calculate the value of coupling constant.

Ans.

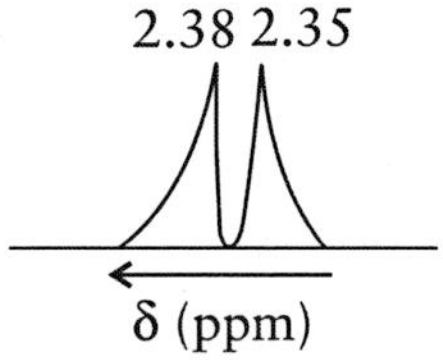

$J = (\delta_2 - \delta_1)\, \nu_{spectrometer}$

$J = (2.38 - 2.35) \times 400$ MHz

$J = 12$ Hz

Q. 2. In ^{1}H NMR spectrum of an organic compound recorded on 500 MHz spectrometer showed a quartet with line positions at 1759, 1753, 1747, 1741 Hz. Calculate the chemical shift(δ) and coupling constant (Hz) of the quartet.

Ans. J = (1753 – 1747) Hz = 6 Hz

$$\delta = \frac{1753+1747}{2} / 500 = \frac{3500}{2} / 500$$

$$\delta = 1750/500$$

$$\delta = 3.5 \text{ ppm}$$

Q. 3. The NMR spectrum of AX_3 type system exhibited lines at δ = 2.1 and 2.3 ppm (for X proton) and δ = 4.1, 4.3, 4.5, 4.7 ppm (for A type proton), measured at 100 MHz instrument. Calculate the chemical shift in ppm of A and X type proton and value of coupling constant (in Hz)

Ans.

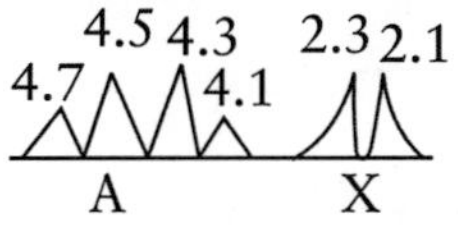

For A-type protons:

J = (4.5 – 4.3) × 100 MHz

J = 20 Hz

$$\delta\ (A) = \frac{4.5+4.3}{2} = 4.4 \text{ ppm}$$

For X-type protons:

J = (2.3 – 2.1) × 100 MHz

J = 20 Hz

$$\delta\ (X) = \frac{2.3+2.1}{2} = 2.2 \text{ ppm}$$

7.24 High Resolution Spectrum

The spin-spin splitting separation between the component lines usually has a small value as compared to the chemical shift of the involved protons and can be observed only if the spectrometer of high resolution are employed. It is for this reason, the spectrum recorded with spin-spin splitting is known as the **high-resolution spectrum.**

Peaks that appear as a single peak in low-resolution NMR, often appear as a cluster of peaks in high-resolution NMR because the magnetic fields generated by spinning of electrons interfere with those of neighbouring nuclei.

For example:

(1) **High Resolution NMR spectrum of methyl alcohol (CH_3OH):** In methyl alcohol, the hydroxyl proton (O—H) interacts with the three identical protons of methyl group (CH_3). The three hydrogen atoms of methyl group have the same chemical environment

and thus the chemical shifts of all the three hydrogen atoms are identical. Therefore the interaction of methyl protons with the hydroxyl protons are all equal. In general, if a proton is coupled with the neighbouring 'n' equivalent protons, the number of absorptions observed is (n + 1). Thus the hydroxyl proton will split into four peaks of intensities 1:3:3:1, due to coupling with methyl proton. Also in high resolution spectrum, the absorption peak of methyl proton will split into two peaks with relative intensities as 1:1, since it is attached to the hydroxyl group.

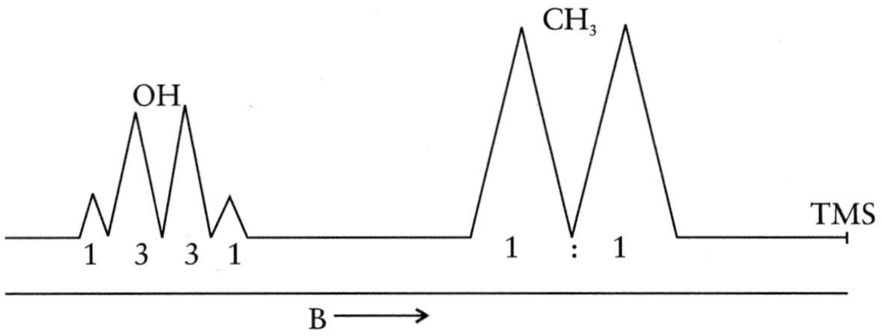

Fig. 7.23: High Resolution Spectrum of CH_3OH

(2) **High Resolution NMR Spectrum of Ethanol (CH_3CH_2OH):** In high resolution spectrum, the absorption peak of methyl protons (CH_3) will split into three peaks with relative intensities 1:2:1, since it is attached to the methylene group. The absorption peak of methylene protons (CH_2) will primarily split into four peaks with intensities 1:3:3:1, due to the interaction with the methyl protons. Each of these four peaks will further split into two peaks of equal intensity due to the coupling with the hydroxyl proton. Thus, in a very high resolution spectrum, the methylene protons will exhibit an octet with relative intensities 1:1:3:3:3:3:1:1. If the resolution is not very large, one often finds quintet with intensities 1:4:6:4:1, as if it has coupled with four protons (three from methyl and one from hydroxyl). The hydroxyl proton will split into three peaks of intensities 1:2:1, due to the coupling with methylene protons.

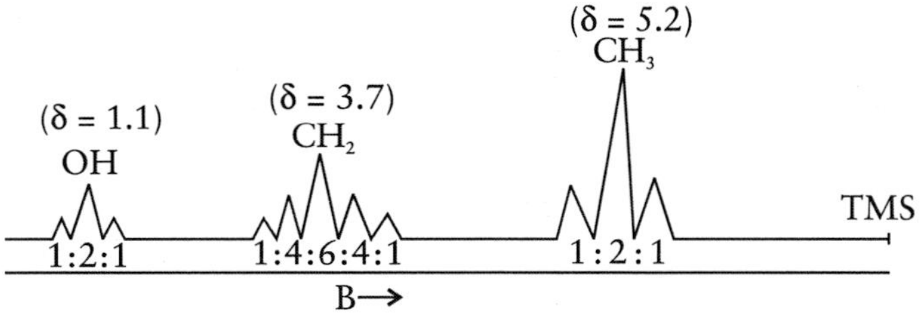

Fig. 7.24: High Resolution Spectrum of CH_3CH_2OH

7.25 High Resolution NMR Spectrum of CH_3CH_2OH in the Presence of Acid or Alkali

In pure ethyl alcohol, the OH proton does not undergo any exchange between the OH groups of different ethanol molecules. The time spent by the proton on any ethyl alcohol molecule is quite sufficient to allow the interaction with the neighbouring methylene protons and thus causing the spin-spin splitting. But in the presence of acid or alkali, the H^+ of the acid catalyzes the exchange of protons between the OH groups of different ethanol molecules forming

$CH_3—CH_2—\overset{\oplus}{O}H_2$. Thus, the rate of transfer of hydroxyl proton from one molecule to the other is very much enhanced and thus the time spend by the hydroxyl proton with any particular molecule is too small to allow the interaction with the methylene protons. In other words, there is no spin-spin interaction between the OH proton and the other protons. So, the splitting of hydroxyl and methylene protons due to each other is completely prevented. Hence, the high resolution NMR spectrum of acidified/alkaline ethanol consists of single peak of the hydroxyl proton. Likewise, the methylene protons show splitting due to neighbouring methyl protons but not due to hydroxyl proton. Thus the methylene protons show a quartet (1:3:3:1). The methyl group shows a triplet (1:2:1), since it is attached to the methylene group.

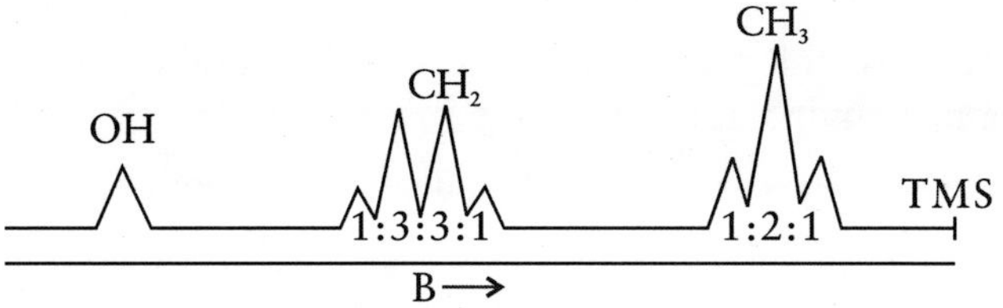

Fig. 7.25: High Resolution spectrum of acidified/alkaline CH_3CH_2OH

7.26 High Resolution NMR Spectrum of CH_3CH_2OH in the Presence of D_2O

On adding D_2O in ethyl alcohol, the hydroxyl proton is replaced by deuterium nucleus. Under such condition, the absorption peak of hydroxyl proton will disappear. Although deuterium nucleus is also magnetically active (I = 1) but no absorption peak corresponding to deuterium is observed in the magnetic field range over which the spectrum is scanned. This is due to the fact that the NMR frequency of deuterium nucleus occurs at a very small frequency as compared to the hydrogen nucleus. The NMR frequency of deuterium is 9.21 MHz as compared to 60 MHz for a proton, in a magnetic field of strength 1.4 T. In order to observe the absorption peak for deuterium nucleus, one has to employ a very large magnetic field of 9.1 Tesla, if the frequency of radiation is kept at 60 MHz. Also the magnetic moment of deuterium is much smaller than proton, the spin-spin interactions of deuterium with neighbouring protons may be neglected.

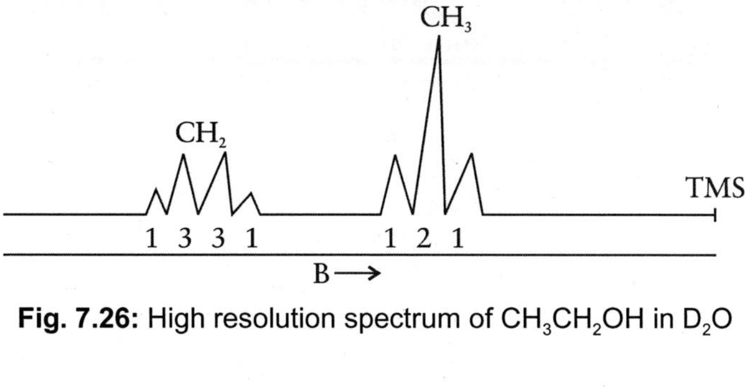

Fig. 7.26: High resolution spectrum of CH_3CH_2OH in D_2O

PROBLEMS BASED ON SPIN-SPIN COUPLING, LOW-RESOLUTION NMR AND HIGH RESOLUTION NMR SPECTRUM

Q. 1. How does NMR spectra helps in structure determination?

or

What information does proton-NMR spectra of a molecule gives?

Ans. (a) The **number of peaks** tells us about the kinds of protons present in a molecule.

(b) The **positions of the peaks** tell us about the electronic environment of each kind of proton.

(c) The **intensities of the peaks** tells us about the number of protons of each kind or same chemical environment that are present.

(d) The splitting of a peak into several peaks tells us about the environment of a proton with respect to other nearby protons in the molecule.

Q. 2. Predict the low-resolution NMR spectrum and the high-resolution NMR spectrum of acetone.

Ans. The low resolution NMR spectrum shows only chemical shift of NMR lines of different protons, whereas the high resolution NMR spectrum shows both the chemical shift and the spin-spin coupling.

In case of acetone, $\left(\begin{matrix} CH_3 \\ CH_3 \end{matrix} \rangle C = O \right)$, since all the protons are equivalent, thus **only one line** is observed in the low-resolution as well in the high-resolution NMR spectrum.

Q. 3. An organic compound C_3H_6O contains a carbonyl group $\left(\rangle C{=}O \right)$. How will you decide whether the compound is an aldehyde or a ketone using NMR technique?

(or)

How do you distinguish between an aldehyde and a ketone on the basis of their NMR spectra, if both have the same molecular formula, C_3H_6O?

Ans. The two possible structures for C_3H_6O are:

$$\overset{a}{C}H_3\overset{b}{C}H_2\overset{c}{C}H = O \quad \text{and} \quad \begin{matrix} \overset{a}{C}H_3 \\ \overset{a}{C}H_3 \end{matrix} \rangle C = O$$

(I) (II)

For aldehyde (structure I), NMR spectrum will consists of a triplet for H^a, a multiplet for H^b [(3+1) (1+1) = 4 × 2 = 8 lines, octet for H^b] and a triplet for H^c.

For ketone (structure II), since all the six protons are chemically equivalent, we will get only a singlet.

Thus, the NMR spectrum would decide whether the compound is an aldehyde or a ketone using NMR technique.

Q. 4. What do you understand by spin-spin coupling in NMR? Sketch the NMR spectrum of ethanol showing the splitting and intensity pattern.

Ans. Spin-spin coupling is the interaction between the spin magnetic moment of different nuclei. In NMR spectroscopy, the actual field observed by a proton, besides depending on the surrounding electron density, is also influenced by the neighbouring magnetic active nuclei. This is called **spin-spin coupling.**

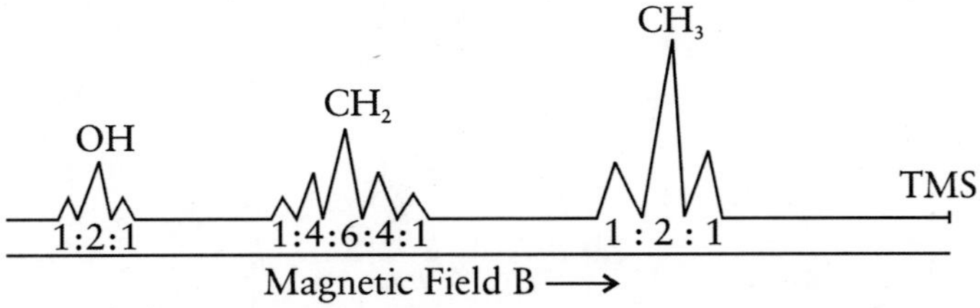

High resolution NMR spectra of CH_3CH_2OH

Q. 5. What is the effect of variation of temperature and external magnetic field on spin-spin coupling?

Q. 6. An isomer of $C_3H_6Cl_2$ gave a triplet and a pentet in its NMR spectrum. Account for its structure giving the relevant explanations. Also give the ratio of areas of split peaks.

(or)

An isomer of $C_3H_6Cl_2$ shows two peaks in its low resolution proton NMR. It shows a pentet and a triplet in its high resolution NMR spectrum. Arrive at its structure giving the basis of your answer.

(or)

Write down all the possible structural formula having molecular formula $C_3H_6Cl_2$. Identify them on the basis of low and high PMR spectra?

Ans. The possible isomers of $C_3H_6Cl_2$ are:

(1) $\overset{c}{C}H_3\overset{b}{C}H_2\overset{a}{C}HCl_2$

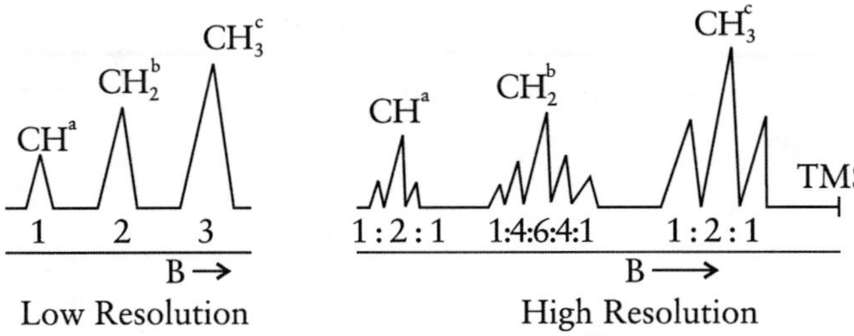

Although in the high resolution NMR spectrum of $CH_3CH_2CHCl_2$, the peak for H^b protons must be octet, since it is attached to methyl group whose protons (H^c) will split it into quartet. Then each of these four peaks will further split into two peaks of equal intensity due to the coupling with H^a protons [$(3+1) \times (1+1) = 4 \times 2 = 8$ lines].

Thus in a very high resolution spectrum, the methylene protons will exhibit an **octet** with relative intensities 1:1:3:3:3:3:1:1. If the resolution is not very large, one often finds quintet with intensities 1:4:6:4:1, as if it is coupled with four protons [three from methyl and one from hydroxyl, i.e., (3+1) + 1 = 5 lines].

(2) 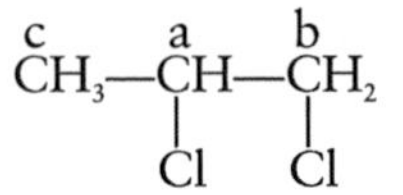

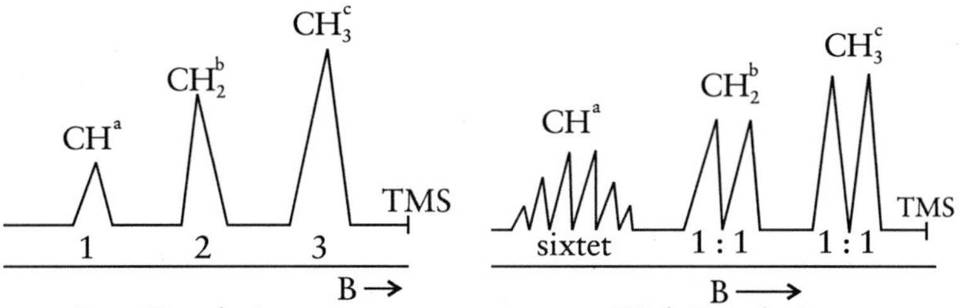

Low Resolution High Resolution

In the low resolution NMR spectrum of $CH_3CHCl\ CH_2Cl$, the peak for H^a protons lies lowfield than H^b protons because the secondary protons shows absorption at lower field than the primary protons.

Also in the high resolution spectrum of $CH_3CHCl\ CH_2Cl$, the peak for H^a protons must be having 12 lines, since it is attached to methyl group whose protons (H^c) will split it into quartet. Then each of these four peaks will further split into three peaks of equal intensity due to the coupling with H^b protons [(3+1) × (2+1) = 4 × 3 = 12 lines].

Thus in a very high resolution spectrum, the H^a protons will exhibit 12 lines. If the resolution is not very large, one often finds sextet with intensities 1:5:10:10:5:1, as if it is coupled with five protons [three from methyl and two from methylene, i.e. [(3+2) + 1 = 5 + 1 = 6 lines].

(3) 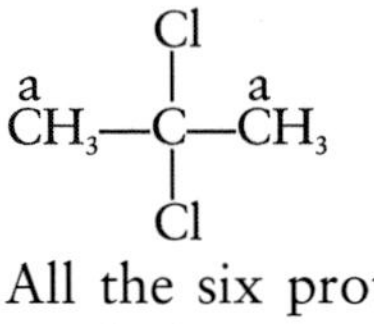

All the six protons are chemically equivalent, so singlet seen in both low and high resolution spectra.

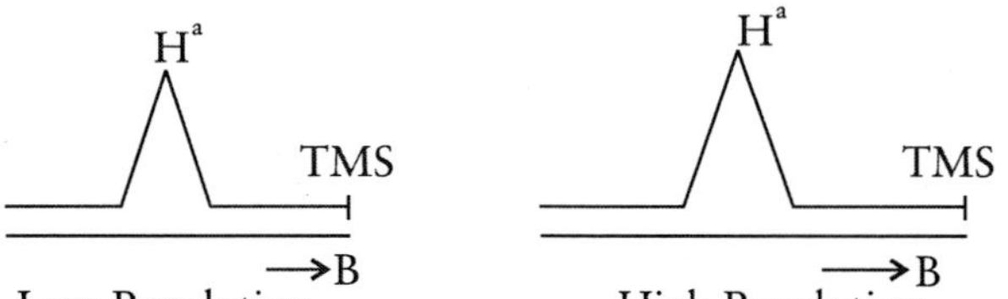

Low Resolution High Resolution

(4) $\overset{a}{Cl—CH_2}—\overset{b}{CH_2}—\overset{a}{CH_2}—Cl$

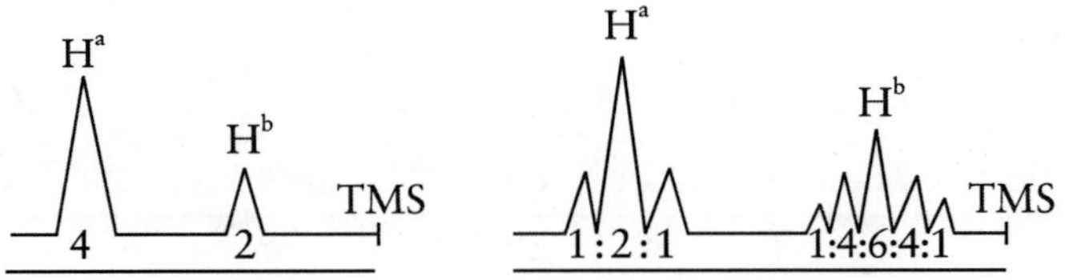

Low Resolution High Resolution

In the high resolution NMR spectrum, the peak of H^b protons must be having nine lines, since it is attached to two methylene groups. The first methylene group protons (H^a) will split it into triplet. Then each of these three peaks will further split into three peaks of equal intensity due to coupling with H^a protons [(2+1)(2+1) = 3 × 3 = 9 lines]. If the resolution is not very large, one often finds quintet with intensities 1:4:6:4:1, as if it is coupled with four protons [two from each methylene groups, i.e. (2+2) + 1 = 5 lines].

Q. 7. Predict the proton-NMR spectrum of ethyl acetate.

Ans. $\overset{a}{CH_3}—CO—O—\overset{b}{CH_2}—\overset{c}{CH_3}$

NMR spectrum will consist of:

(i) Single peak for H^a.

(ii) CH_3 protons (H^c) split the peak of CH_2 into four lines.
∴ Quartet for H^b.

(iii) CH_2 protons (H^b) split the peak of CH_3 into three lines.
∴ Triplet for H^c.

Q. 8. Predict the NMR spectrum of the $CH_3COCH_2C \equiv CCH_3$.

Ans. $\overset{a}{CH_3}\ CO\ \overset{b}{CH_2}\ C \equiv C\ \overset{c}{CH_3}$

The three group of protons (H^a, H^b and H^c) are far apart, there is no spin-spin coupling observed. Thus, H^a, H^b and H^c each will give rise to a singlet with relative intensities of 3:2:3 respectively.

Q. 9. How can the NMR spectrum used to distinguish between the isomers p-xylene and ethyl benzene?

Ans. 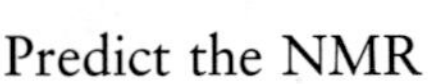 i.e. p–$\overset{a}{CH_3}\overset{b}{C_6H_4}\overset{a}{CH_3}$

(p–xylene)

It has two kinds of protons (H^a and H^b) and consists of two singlets.

Whereas

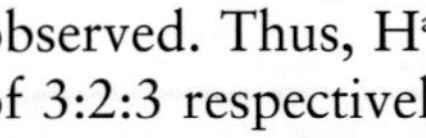 i.e. $\overset{a}{C_6H_4}\overset{b}{CH_2}\overset{c}{CH_3}$

(ethyl benzene)

It has three kinds of protons (H^a, H^b and H^c) and consists of three peaks in its NMR spectrum, i.e.

(i) singlet for H^a
(ii) quartet for H^b
(iii) triplet for H^c

Q. 10. Predict the high resolution PMR spectra of CH_3CH_2Cl and CH_3CHDCl.

Ans. Deuterium (D) does not have signal in the PMR spectrum nor does it split the nearby protons. Thus, we can ignore its presence in the molecule, i.e. ignore the interaction of D atoms with H.

(1) The PMR spectrum of $\overset{a}{C}H_3\overset{b}{C}H_2Cl$ consists of

Triplet for H^a and quartet for H^b.

(2) The PMR spectrum of $\overset{a}{C}H_3\overset{b}{C}HDCl$ consists of

Doublet for H^a and quartet for H^b.

Q. 11. How many kinds of protons and number of peaks are there in the following compounds:

Organic Compound	Kind of Protons	Number of Peaks
1. $\overset{a}{C}H_3—\overset{a}{C}H_3$	One kind; H^a	One peak
2. $\overset{a}{C}H_3—\overset{b}{C}H_2—\overset{a}{C}H_3$	Two kinds; H^a and H^b	Two peaks
3. 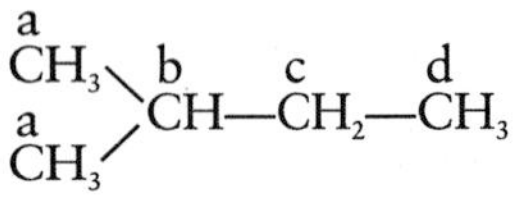	Four kinds; H^a, H^b, H^c and H^d	Four peaks
4. 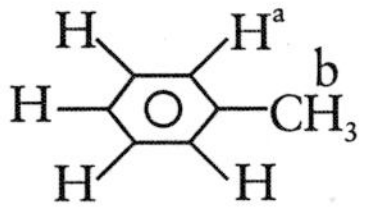	Two kinds; H^a and H^b	Two peaks

Theoretically, there are three kinds of aromatic protons, i.e. ortho, meta and para. However, the ring protons are little affected by alkyl groups and are thus nearly equivalent.

5. 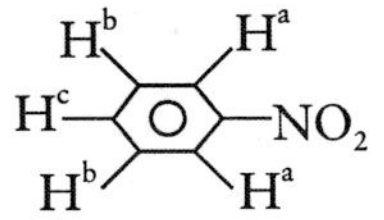	Three kinds; H^a, H^b and H^c	Three peaks (2 ortho, 2 meta and 1 para)
6. $\overset{a}{C}H_2=\overset{a}{C}H_2$	One kind; H^a	One peak
7. $CH_3—CH=CH_2$ 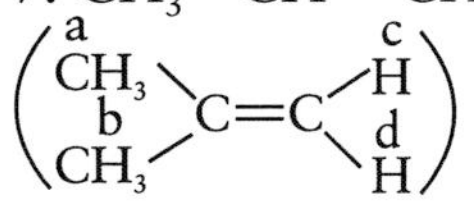	Four kind; H^a, H^b, H^c and H^d	Four peaks
8. 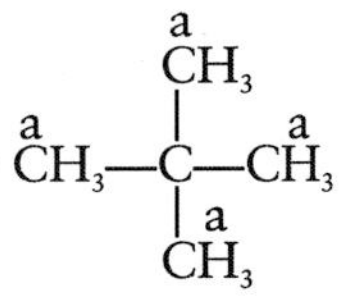	One kind; H^a	One peak

Organic Compound	Kind of Protons	Number of Peaks
9. 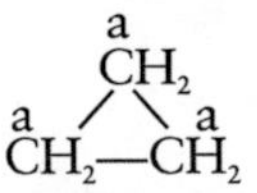	One kind; H[a]	One peak
10. $\overset{a}{CH_2}=C=\overset{a}{CH_2}$ (Allene, C_3H_4)	One kind; H[a]	One peak
11. $\overset{a}{CH_3}-C\equiv C-\overset{a}{CH_3}$	One kind; H[a]	One peak
12. $(\overset{a}{CH_3})_3C-C(\overset{a}{CH_3})_3$	One kind; H[a]	One peak
13. $\overset{a}{CH_3}-O-\overset{a}{CH_3}$	One kind; H[a]	One peak
14. $Br\ \overset{a}{CH_2}-\overset{a}{CH_2}-Br$	One kind; H[a]	One peak
15. 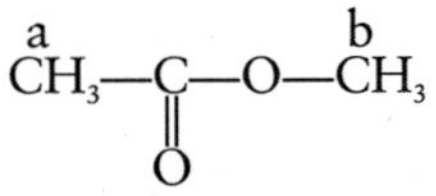	Two kinds; H[a] and H[b]	Two peaks
16. $\overset{a}{CH_3}-O-\overset{b}{CH_2}-O-\overset{a}{CH_3}$	Two kinds; H[a] and H[b]	Two peaks
17. $Cl-\overset{a}{CH_2}-O-\overset{b}{CH_3}$	Two kinds; H[a] and H[b]	Two peaks
18. 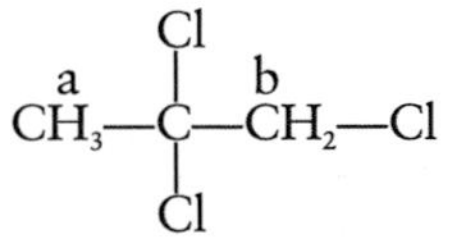	Two kinds; H[a] and H[b]	Two peaks
19. 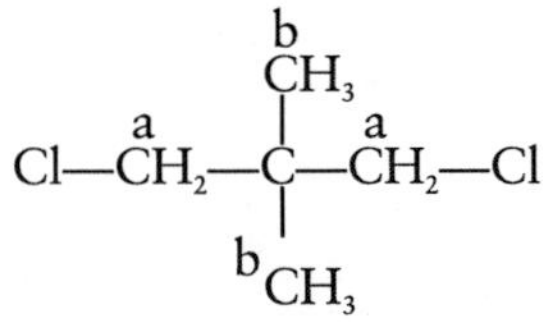	Two kinds; H[a] and H[b]	Two peaks
20. $Cl-\overset{a}{CH_2}-\overset{b}{CH_2}-Cl$	One kind; H[a]	One peak
21. 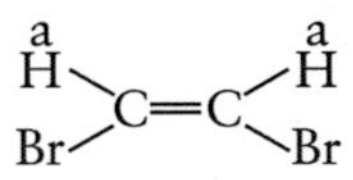	One kind; H[a]	One peak
22. 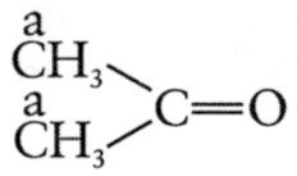	One kind; H[a]	One peak
23. 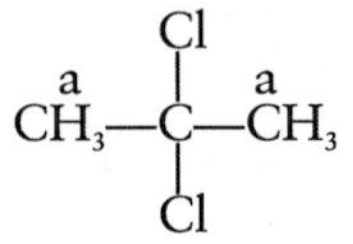	One kind; H[a]	One peak
24. $Cl-\overset{a}{CH_2}-\overset{b}{CH_2}-I$	Two kinds; H[a] and H[b]	Two peaks

Organic Compound	Kind of Protons	Number of Peaks
25. 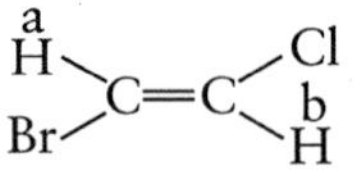	Two kinds; H^a and H^b	Two peaks
26. 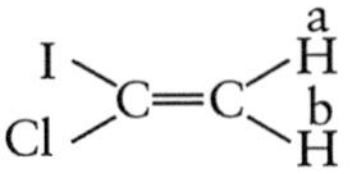	Two kinds; H^a and H^b	Two peaks
27. $\overset{b}{C}H_3—\overset{a}{C}H=O$	Two kinds; H^a and H^b	Two peaks

Q. 12. High resolution NMR spectrum of ethyl alcohol is shown in the figure.

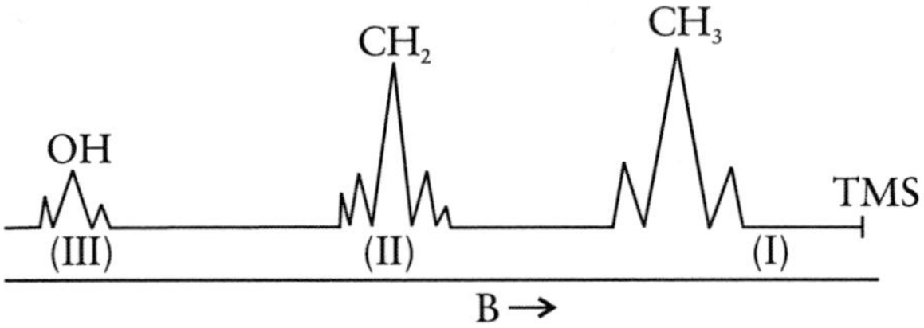

(i) What are the relative intensities the peaks in the spectrum.

Ans. I → 1:2:1
II → 1:4:6:4:1
III → 1:2:1

(ii) What is the effect on the spectrum in the presence of alkali.

Ans. In the presence of alkali, the peak III will appear as a singlet and peak II will appear as a quartet with relative intensity of 1:3:3:1. The peak I, will remain as such with relative intensity of 1:2:1.

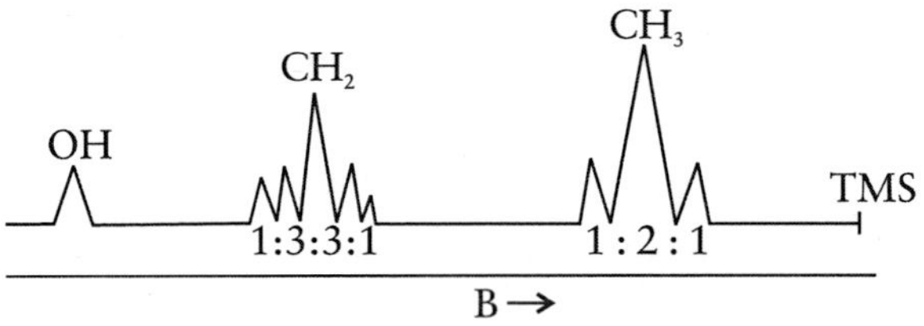

High Resolution Spectrum of CH_3CH_2OH in Alkali

(iii) What is the effect on addition of D_2O.

Ans. On addition of D_2O, the peak III due to hydroxyl group will disappear. The peak II will appear as a quartet with relative intensity of 1:3:3:1. The peak I, will remain as such with relative intensity of 1:2:1.

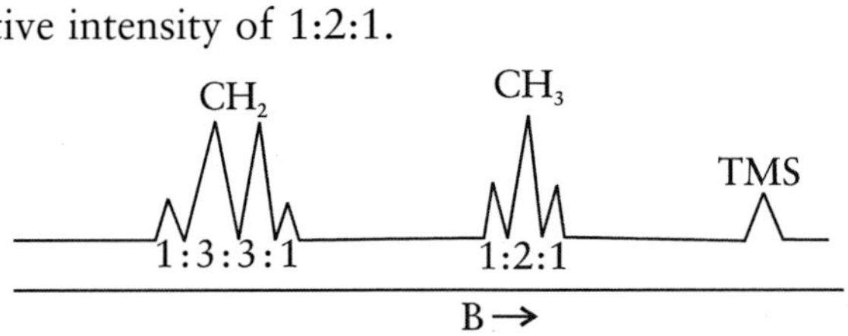

High Resolution Spectrum of CH_3CH_2OH in D_2O

(iv) Will you see peak due to deuterium, D? Give reason.

Ans. No, the peak due to deuterium is not observed. Although deuterium nucleus is magnetically active (I = 1), but no absorption peak corresponding to deuterium is observed in the magnetic field range over which the spectrum is scanned. This is due to the fact that the NMR-frequency of deuterium nucleus occurs at a very small frequency as compared to the hydrogen nucleus. The NMR frequency of deuterium is 9.21 MHz as compared to 60 MHz for a proton in a magnetic field of strength 1.4, T. In order to observe the absorption peak for deuterium nucleus, one has to employ a very large magnetic field of 9.1 T, if the frequency of radiation is kept at 60 MHz. Also the magnetic moment of deuterium is much smaller than proton, the spin-spin interactions of deuterium with neighbouring protons may be neglected.

Q. 13. The low resolution NMR spectra of anhydrous ethyl alcohol show three peaks at, δ = 5.2, 3.7 and 1.1 for —OH, —CH_2 and —CH_3 groups with intensity ratio of 1:2:3.

(a) Sketch the high resolution NMR spectrum of C_2H_5OH and explain the splitting.

(b) Passing HCl gas through anhydrous alcohol, may lead to the formation of $CH_3—CH_2—\overset{\oplus}{O}H_2$. Sketch the high resolution NMR spectra of this compound.

Q. 14. In anhydrous alcohol, there is hydrogen bonding as follows:

$CH_3—CH_2—O—H \cdots\cdots \underset{\large H}{\underset{|}{O}}—CH_2—CH_3$ on passing HCl gas:

(a) This H—bond breaks

(b) In anhydrous alcohol, slower intermolecular proton exchange takes place, but passage of HCl makes this proton exchange very rapid.

How will these observations effect the high resolution NMR spectra? Explain with the help of sketch.

Q. 15. Sketch the NMR spectrum of ethanol with respect to TMS under:

(i) low resolution

(ii) high resolution

Indicate the relative positions of the peaks with respect to the TMS peak and also give the ratio of the peak areas observed in low resolution spectrum and the relative intensities of the split peaks obtained in high resolution spectrum.

Q. 16. Sketch the low and high resolution of proton-NMR spectra for:

(i) $\overset{b}{C}H_3COO\overset{a}{C}H_3$

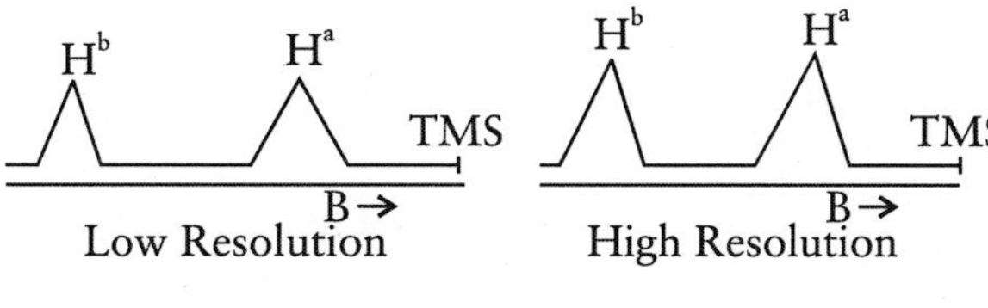

(ii) 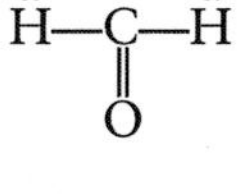

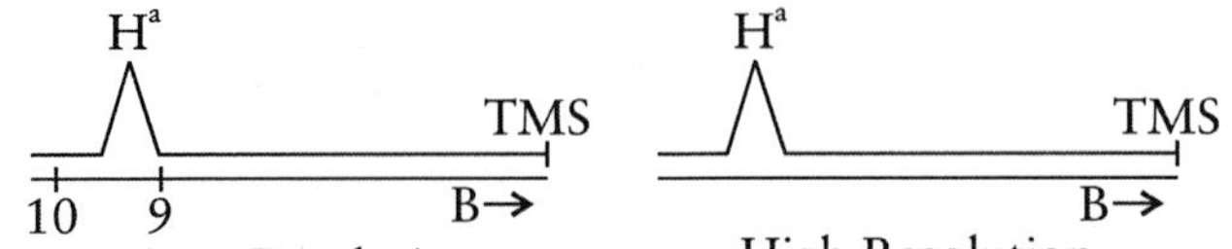

Low Resolution High Resolution

(iii) $\overset{a}{CH_3}$—C(=O)—$\overset{b}{H}$

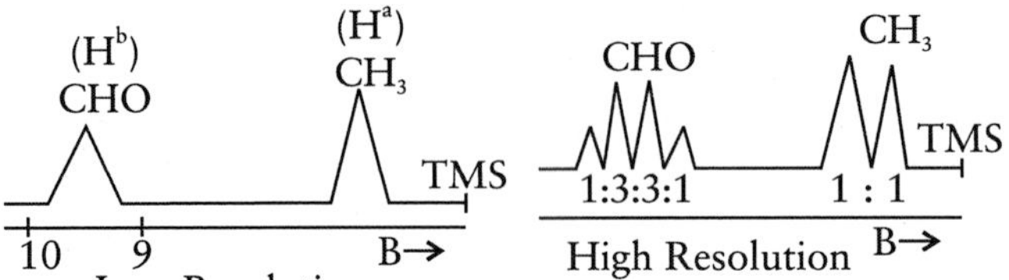

Low Resolution High Resolution

Q. 17. Draw and explain the low and high resolution NMR spectra of CH_3CHO. Show the position of bands relative to TMS.

Ans. (i) Low Resolution Spectrum of CH_3CHO:

CH_3—C(=O)—H

The three methyl protons are chemically equivalent and give one peak. Also one peak is seen for aldehydic proton.

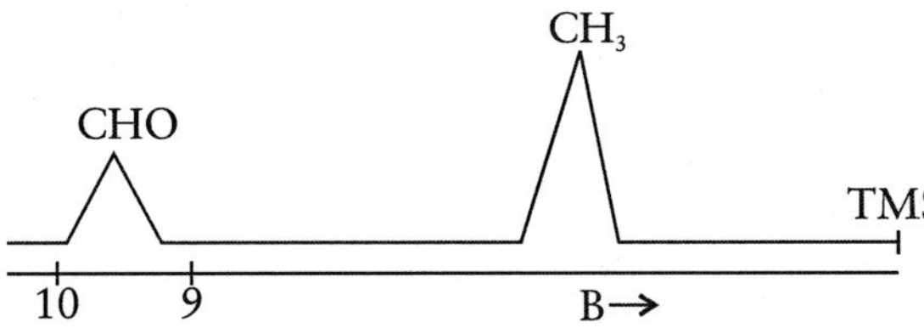

(ii) High Resolution Spectrum of CH_3CHO:

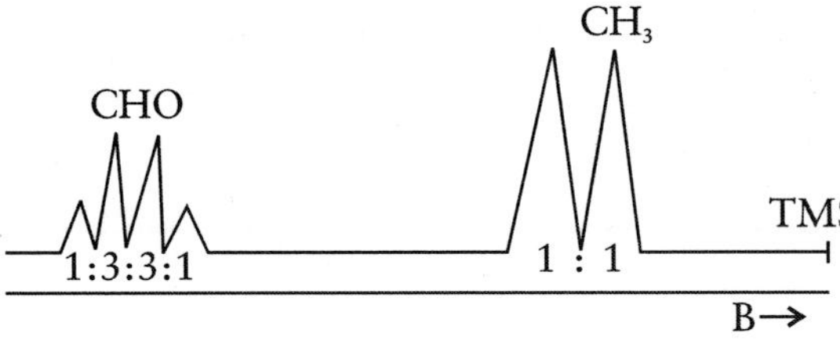

In high resolution NMR, the methyl and aldehydic group are in different chemical environment. Hence, the peak for methyl protons splits into doublet by aldehydic proton and the peak for aldehydic proton splits into quartet by methyl protons.

Q. 18. How can you distinguish between the following two molecules using low and/or high resolution NMR spectra? Explain:

CH_3—C(=O)—O—CH_2CH_3 and CH_3—O—C(=O)—CH_2CH_3

Ans. Low Resolution NMR Spectra:

The order of shielding constants of protons in the given compounds are

(i) $\overset{a}{CH_3}-\underset{\underset{O}{\|}}{C}-O-\overset{b}{CH_2}\overset{c}{CH_3}$ (ii) $\overset{a}{CH_3}-O-\underset{\underset{O}{\|}}{C}-\overset{b}{CH_2}\overset{c}{CH_3}$

Order of shielding: $\sigma_b < \sigma_a < \sigma_c$ $\sigma_a < \sigma_b < \sigma_c$

This is because the atom which is attached to electronegative atom, gets deshielded maximum.

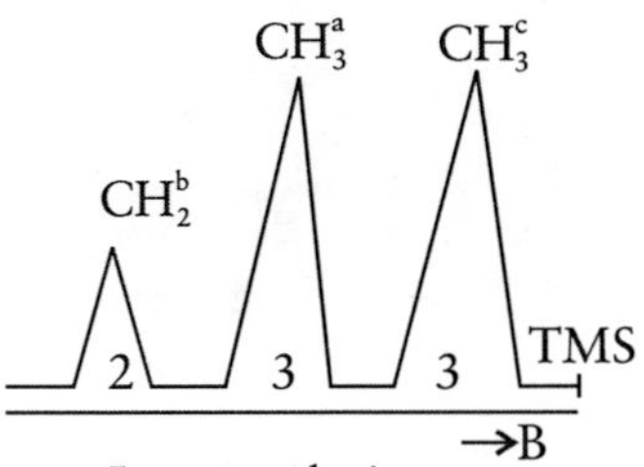

Low resolution spectra of compound (i)

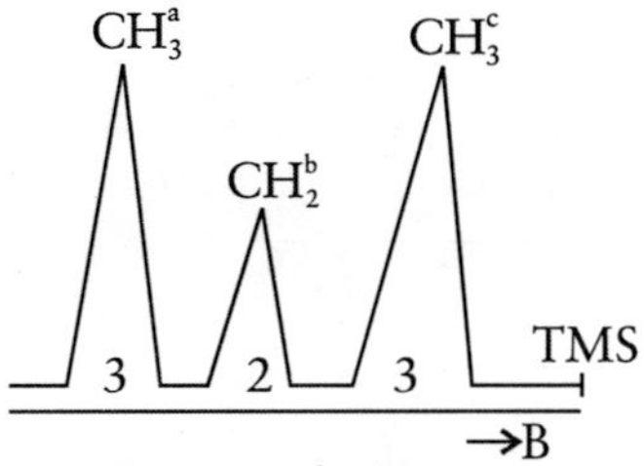

Low resolution spectra of compound (ii)

Since the two low resolution spectra are not identical, the two compounds can be distinguished from their low resolution spectra.

High Resolution PMR Spectra: The peak due to H^a protons will remain unaffected, i.e. will not split, as there is no neighbouring proton.

The peak due to H^b protons, in both cases will split into quartet, as it is attached to the methyl group.

Similarly, the peak due to H^c in both cases will split into triplet, as it is attached to the methylene group.

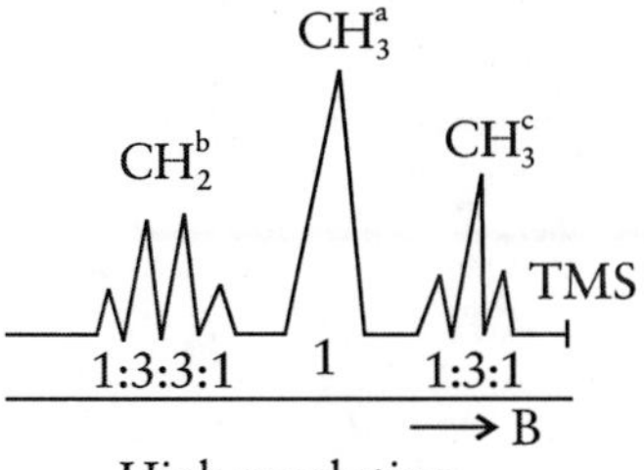

High resolution spectra of compound (i)

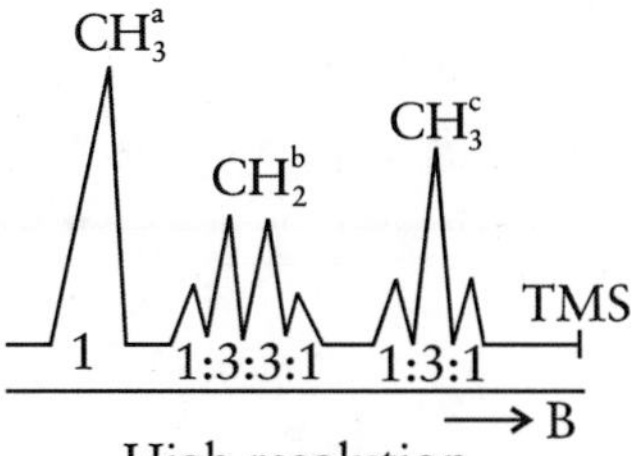

High resolution spectra of compound (ii)

Q. 19. Show the low and high resolution NMR spectrum of acidified ethanol.

Q. 20. A cyclic compound with a molecular formula $C_6H_8O_2$ gave one peak in its low resolution and high resolution NMR spectrum. Draw its structure.

Ans. 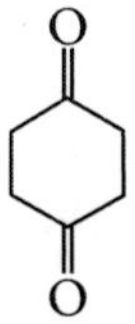

Q. 21. The low resolution NMR spectrum of $C_5H_{10}O_2$ exhibits three absorptions at δ values of 3.2, 2.1 and 1.2 with relative intensities 1:3:6 respectively. Predict the structure of the molecule giving reasons. Draw and explain the NMR spectrum expected under high resolution.

Ans. The expected low resolution spectrum would be

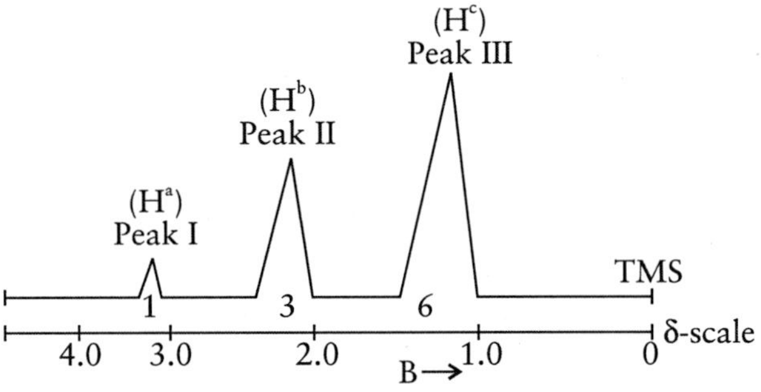

Analysis of the spectrum:

(i) Peak I at δ = 3.2, must be due to CH group. The position of the peak suggests that it must be attached with a strong electron withdrawing group.

(ii) Peak II at δ = 2.1, must be due to methyl (CH_3) group.

(iii) Peak III at δ = 1.1, must be due to two chemically equivalent methyl groups. The position of this peak indicates that these methyl group protons are more shielded than protons of other methyl groups.

If we subtract CH and three CH_3 groups from $C_5H_{10}O_2$, we get CO_2 as the remaining fragment. Thus, the given molecule must be a derivative of carboxylic acid.

The two identical methyl groups and one CH group may arise from the isopropyl group. Since CH shows peak at high δ-value, thus it must be attached to the electron withdrawing carboxylate group.

The probable structure must be

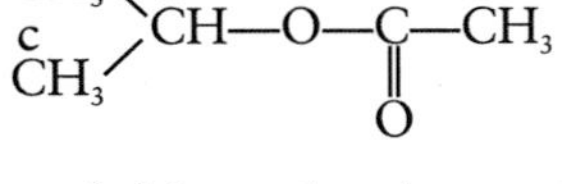

The high resolution spectra of this molecule would be:

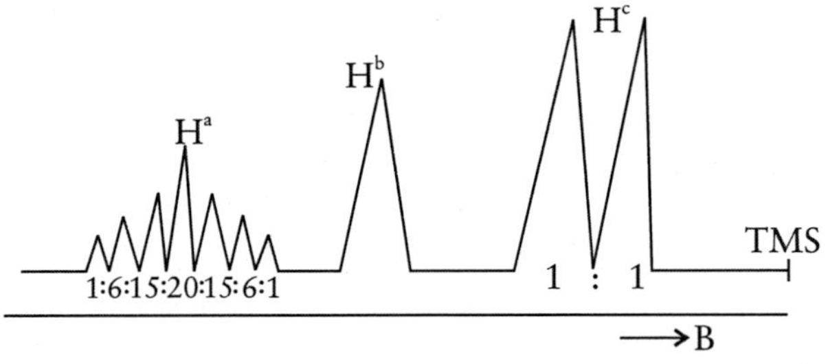

In the high resolution NMR spectra, peak I will split into septet due to six neighbouring protons (two methyl groups) and the peak III will split into doublet due to one neighbouring CH proton. Peak II will remain as such, it is not affected as there is no neighbouring proton and thus no splitting.

Q. 22. The low resolution spectrum of $C_6H_{10}O_2$. Predict the structure of the molecule giving reasons.

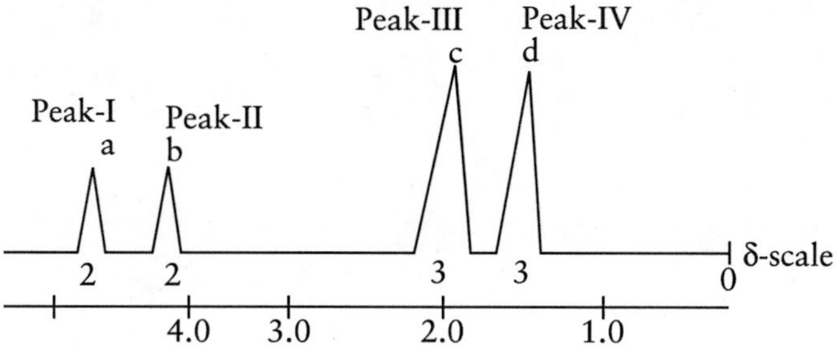

Ans. Analysis of Spectrum:

(i) The relative areas of peak-I and peak-II suggest the presence of two methylene groups.

(ii) The relative areas of peak-III and peak-IV suggest the presence of two methyl groups.

(iii) If we subtract two methylene and two methyl groups from $C_5H_{10}O_2$, we get C_2O_2. The latter may represent two CO groups.

The probable structure on the above facts is:

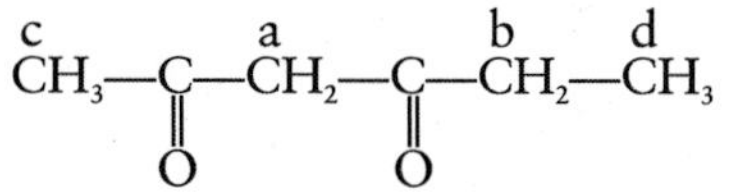

8

Electron Spin Resonance (ESR) or Electron Paramagnetic Resonance (EPR) Spectroscopy

8.1 Introduction

ESR was first observed by the Russian physicist E. Zavoisky in 1945, a year before the discovery of NMR by the American physicists Bloch and Purcelle. NMR and ESR are similar in their theoretical principles. However, whereas NMR spectra are experimentally observed in the radiofrequency region, the ESR spectra are observed in the microwave region of the electromagnetic spectrum. Both NMR and ESR are collectively referred to as **magnetic resonance phenomena.**

ESR is a branch of absorption spectroscopy which is observed primarily in systems containing unpaired electrons. Thus, the systems which can be investigated by ESR spectroscopy are species having odd number of electrons, organic or inorganic free radicals and ions of transition metals which contain unpaired d or f electrons. Since such substances are paramagnetic, thus this type of spectroscopy is often referred to as **electron paramagnetic resonance (EPR)**.

Substances with unpaired electrons may either arise naturally or to be produced artificially. Some of the examples are: NO, O_2, NO_2 (odd electron species), Fe^{3+}, $[Fe(CN)_6]^{3-}$, etc.

ESR has limited applications because it is observed primarily in systems containing unpaired electrons. Paired electrons cannot be detected since the spins of an electron pair within an orbital must remain antiparallel.

One of the most important uses of ESR technique lies in the detection of extremely short-lived free radical intermediates in chemical reactions, provided the lifetimes of such radicals are greater than about 10^{-6} sec. ESR spectra can be recorded for radical concentration of about 10^{-14} M irrespective of the number of non-radical species present. The magnetic moment of an unpaired electron is about 700 times that of the proton, thus the sensitivity of ESR detection is very much higher than in NMR. In NMR, we deal with the interaction of a positively charged proton with the external magnetic field, whereas in ESR, the interaction is between a negatively charged electron and the external magnetic field. The intensity of an ESR signal is directly proportional to the number of free radicals present. Hence, using ESR we can measure relative concentrations of free radicals produced under different conditions.

8.2 Angular Momentum

Just as linear momentum represents the tendency of an object to move in a straight line, the angular momentum represents the tendency of an object to move in angular motion or rotate. Since momentum has both a magnitude and a direction, it is a vector quantity. Linear momentum is represented by a vector in the direction of motion ($\vec{p}$). Whereas the angular momentum is represented through use of the Right Hand Rule (RHR): when the fingers of the right hand are curled in the direction of circular motion, the thumb points in the direction of the angular momentum vector. Mathematically, angular momentum (L) is the cross product of the position (r) and linear momentum (p) vectors.

$$\vec{L} = \vec{r} \times \vec{p}$$

There are two kinds of classical angular momentum: **orbital angular momentum** and **spin angular momentum** as shown in Fig. 8.1.

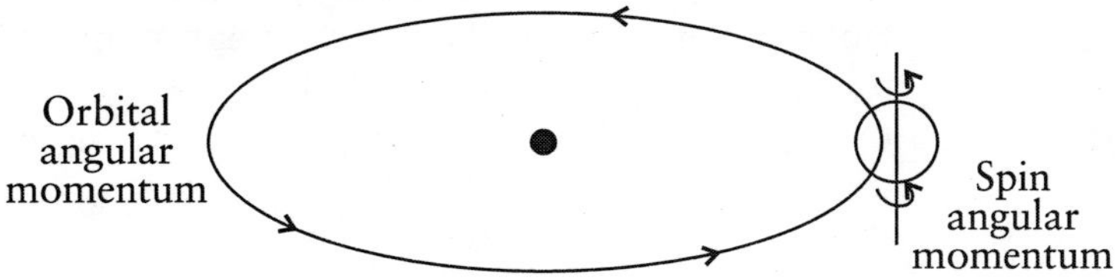

Fig. 8.1: Orbital and Spin angular momentum

(a) **Orbital angular momentum** is the circular motion of an object about a point, like the Earth orbiting about the Sun once a year.

(b) **Spin angular momentum** is the spinning motion of an object about its own axis, like the Earth spinning about its own axis once a day.

8.3 Quantisation of Angular Momentum

Just as energy is known to be quantized at the atomic level, angular momentum is also quantized. **Goudsmit** and **Uhlenbeck** in 1925 proposed that instead of orbital motion around the nucleus, the electron also possess spinning motion about its own axis. Thus electron has an intrinsic angular momentum due to spinning about its own axis, i.e. spin angular momentum. The value of the spin angular momentum is independent of the orbital angular momentum.

8.3.1 Orbital Angular Momentum of the Electron

The orbital motion and orbital angular momentum is described by two orbital quantum numbers: 'ℓ' and 'm_ℓ'.

Like, an electron in an atom may have **orbital angular momentum quantum numbers,** ℓ = 0, 1, 2, ... (which are more commonly denoted as s, p, d, ...) and **orbital magnetic quantum number,** $m_\ell = (-\ell, -\ell + 1, ..., \ell$ (which lead to p_x, p_y and p_z, for $\ell = 1$ and to the five d-orbitals for $\ell = 2$).

8.3.2 Spin Angular Momentum of the Electron

The spinning motion and thus spin angular momentum can be described by two spin quantum number: 's' and 'm_s'.

(a) **Spin angular momentum quantum number (s):** The spin angular momentum quantum number 's' represents the magnitude of the total spin angular momentum 'S'.

$$S = \sqrt{s(s+1)}\left(\frac{h}{2\pi}\right) \quad \text{...(8.1)}$$

(b) **Spin magnetic quantum number (m_s):** The spin magnetic quantum number 'm_s' represents, the z-component of the total spin angular momentum 'M_s', such that

$$M_S = m_s\left(\frac{h}{2\pi}\right) \quad \text{...(8.2)}$$

8.4 Orientations of Total Spin Angular Momentum Vector

Let z-axis is direction of the magnetic field. If θ is the angle between the total spin angular momentum and the z-axis, the z-component of the total spin angular momentum of the electron is given by

$$M_S = S \cos \theta$$

Using equation (8.1)

$$M_S = \sqrt{s(s+1)}\left(\frac{h}{2\pi}\right)\cos\theta \quad \text{...(8.3)}$$

Equating equation (8.2) and equation (8.3), we have

$$m_s\left(\frac{h}{2\pi}\right) = \sqrt{s(s+1)}\left(\frac{h}{2\pi}\right)\cos\theta$$

$$m_s = \sqrt{s(s+1)}\cos\theta \quad \text{...(8.4)}$$

The allowed values of m_s are:

$-s, -(s-1), ... (s-1), s$

But it is found that the spin angular momentum quantum number 's', has only one value of 1/2.

Thus, the only permitted value of m_s are +1/2 and −1/2.

Hence, the z-component of the total spin angular momentum is given by equation (8.2) becomes:

$$M_S = m_s\left(\frac{h}{2\pi}\right)$$

$$M_S = \pm\frac{1}{2}\left(\frac{h}{2\pi}\right)$$

i.e. **only two orientations of total spin angular momentum are possible.** Thus, the total spin angular momentum vector can never be aligned along the magnetic axis and hence it will always be oriented with respect to z-axis as shown in Fig. 8.2.

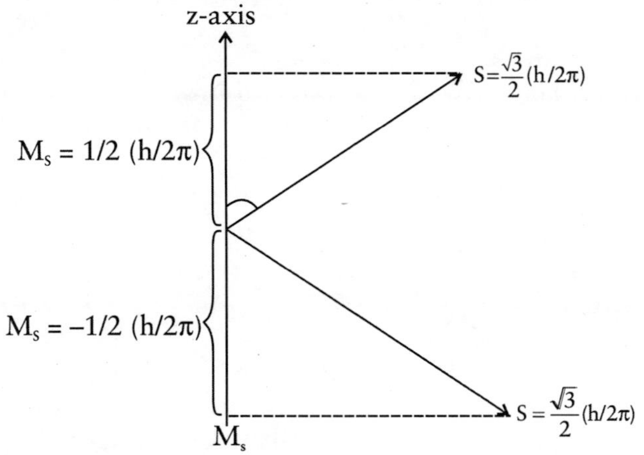

Fig. 8.2: Various projections of total spin angular momentum on z-axis

The magnitude of the spin angular momentum is

$$S = \sqrt{s(s+1)}\frac{h}{2\pi} = \sqrt{\frac{1}{2}(\frac{1}{2}+1)}\left(\frac{h}{2\pi}\right) = \frac{\sqrt{3}}{2}\left(\frac{h}{2\pi}\right) = 0.866\left(\frac{h}{2\pi}\right)$$

8.5 Principle Involved in ESR

Just as the ^1H nucleus has spin and therefore a magnetic moment (from NMR), similarly the electron with its spin is paramagnetic and also possess a magnetic moment. Like the proton, the electron will precess in an applied magnetic field with a precise precessional frequency and will undergo transitions between spin states (spin orientations) if energy of the correct frequency is applied.

In the absence of any external magnetic field, the spins of unpaired electrons in a substance are randomly oriented. When a uniform magnetic field, 'B', is applied, however, they will each have a preferred direction. Since the spin angular momentum quantum number, s = 1/2, thus the number of orientations with respect to the direction of the applied magnetic field 'B' will be 2s + 1 = (2 × 1/2) + 1 = 2.

Thus, spin magnetic quantum number will have two value of $m_s = \pm 1/2$.

So each electron can be thought of as spinning either clockwise or anticlockwise about the field direction, i.e. these will align with the applied magnetic field or against the magnetic field.

In the absence of magnetic field, the two values of m_s, i.e. +1/2 and –1/2 will give rise to a doubly degenerate spin energy state. If a magnetic field is applied, this degeneracy is removed and this leads to a two non-degenerate energy levels.

The low energy state will have the spin magnetic moment aligned with the applied magnetic field and corresponds to the quantum number, $m_s = -1/2$. On the other hand, the higher energy state will have the spin magnetic moment opposed to the applied magnetic field and corresponds to the quantum number $m_s = +1/2$, as shown in Fig. 8.3.

The ESR spectroscopy essentially measures the energy required to reverse the spin of an unpaired electron. The intensity of an ESR absorption is proportional to the concentration of the free radical or paramagnetic material present. For the electron, the energy necessary to induce spin transitions in modern ESR instruments is in the microwave region of the electromagnetic spectrum, with very much higher frequencies than radiofrequency.

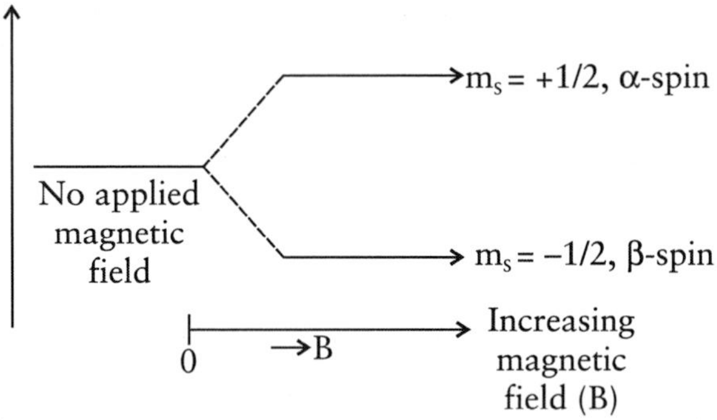

Fig. 8.3: Orientation of electron in an external magnetic field 'B'

8.6 Magnetic Moment Due to Spinning Motion of Electron

According to classical mechanics, like orbital motion, the spinning of the electron around its own axis generates a magnetic moment vector. Since the magnetic moment vector has a direction opposite to that of angular momentum vector, thus like angular momentum, the magnetic moment vector will also be directed along certain directions in space. The magnetic moment vector can never be aligned along the magnetic axis and hence it always be oriented with respect to z-axis, as shown in Fig. 8.4.

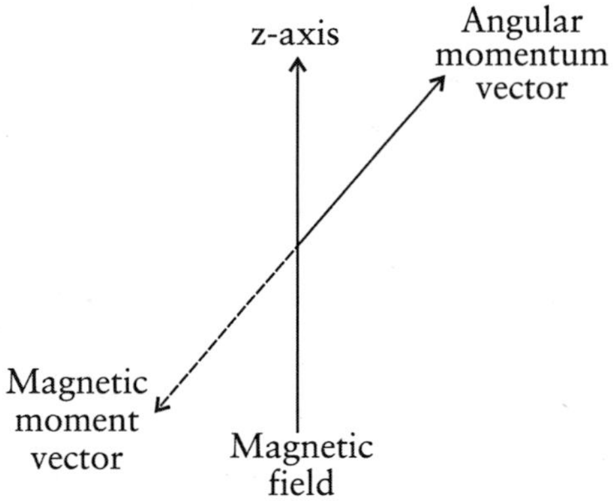

Fig. 8.4: Orientation of magnetic moment vector relative to angular momentum vector

The relation between the magnetic moment due to spinning motion (μ_m) and the total spin angular momentum (S) is:

$$\mu_m = -g\left(\frac{e}{2m_e}\right)S \qquad \text{[Refer section 7.6]}$$

Here, the negative sign indicates that the direction of the magnetic moment vector is opposite to that of angular momentum.

$$\mu_m = -g\left(\frac{e}{2m_e}\right)\sqrt{s(s+1)}\left(\frac{h}{2\pi}\right) \qquad \text{(using equation 8.1)}$$

$$\mu_m = -g\left(\frac{eh}{4\pi m_e}\right)\sqrt{s(s+1)}$$

$$\mu_m = -g\mu_B\sqrt{s(s+1)} \qquad \text{...(8.5)}$$

The factor 'g' is known as the **Lande splitting factor.** The numerical value of 'g' cannot be calculated theoretically and is determined from the experimental data. The value of 'g' for a free electron is found to be 2.0023.

Also, μ_B is the basic unit of magnetic moment for the electron, known as the **Bohr magneton.**

The value of **Bohr magneton** for the free electron can be calculated as:

$$\mu_B = \frac{eh}{4\pi m_e}$$

$$\mu_B = \frac{(1.602 \times 10^{-19} C) \times (6.626 \times 10^{-34} Js)}{4 \times (3.14) \times (9.11 \times 10^{-31} kg)}$$

$$\mu_B = 9.273 \times 10^{-24} \text{ J/T}$$

PROBLEMS BASED ON MAGNETIC MOMENT OF THE ELECTRON

Q. 1. What is Bohr magneton?

Ans. Magnetic moment of an electron is caused by its orbital or spin angular momentum. The physical quantity that represent this magnetic moment, which is caused by either its orbital or spin angular momentum is called Bohr magneton.

$$\mu_B = \frac{eh}{4\pi m_e}$$

$$\mu_B = \frac{(1.602 \times 10^{-19} C) \times (6.6 \times 10^{-34} Js)}{4 \times (3.14) \times (9.11 \times 10^{-31} kg)}$$

$$\mu_B = 9.273 \times 10^{-24} \text{ J/T}$$

where, e is the electronic charge and m_e is the mass of the electron.

Q. 2. Is Bohr magneton, greater than the nuclear magneton? Give reasons.

Ans. Bohr magneton is for electrons and nuclear magneton is for protons. Both these numbers are inversely proportional to their respective masses. Since a proton is 2000 times heavier than an electron, therefore, the nuclear magneton is 2000 times smaller than the Bohr magneton.

8.7 Potential Energy of an Electron in a Magnetic Field

Potential energy 'E' of interaction of a nucleus with external magnetic field 'B' is given by:

$$E = -B\mu_z \quad \text{...(8.6)}$$

where, μ_z is the z-component of the magnetic moment vector.

Let θ be the angle between the magnetic moment vector and z-axis, then

$$\mu_z = \mu_m \cos\theta \quad \text{...(8.7)}$$

Substituting equation (8.5) in equation (8.7), we get

$$\mu_z = (-g\mu_B\sqrt{s(s+1)}\cos\theta \quad \text{...(8.8)}$$

Thus, equation (8.6) becomes:

$$E = -B\mu_z$$

$$E = -B(-g\mu_B\sqrt{s(s+1)}\cos\theta$$

$$E = +Bg\mu_B\left(\sqrt{s(s+1)}\cos\theta\right)$$

Using equation (8.4)

$$E = Bg\mu_B m_S \qquad \text{...(8.9)}$$

For an electron, the value of spin angular momentum quantum number 's' has only one value of 1/2. Thus, the only permitted value of m_s are +1/2 and –1/2.

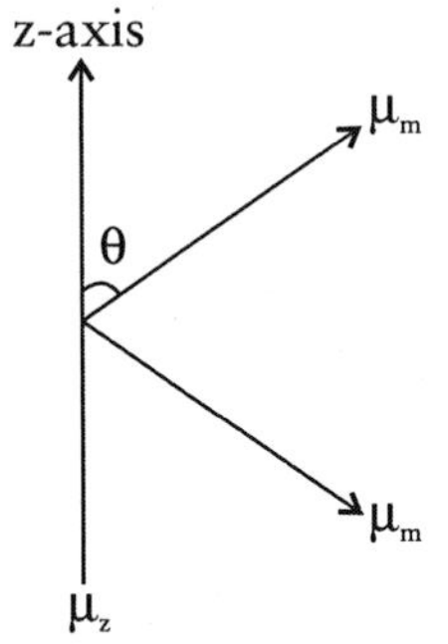

Fig. 8.5: Angle between magnetic moment and its z-component

$$E_{+1/2} = (1/2)g\mu_B\,B$$
$$= (1/2) \times (2.0023) \times (9.273 \times 10^{-24}\ JT^{-1}) \times B$$
$$= (9.273 \times 10^{-24})B \text{ Joules.}$$

$$E_{-1/2} = (-1/2)g\mu_B\,B$$
$$= (-1/2) \times (2.0023) \times (9.273 \times 10^{-24}\ JT^{-1}) \times B$$
$$= (-9.273 \times 10^{-24})B \text{ Joules.}$$

The energy difference between the two levels is given by

$$\Delta E = g\mu_B\,B \qquad \text{...(8.10)}$$

$$\Delta E = E_{+1/2} - E_{-1/2}$$

$$\Delta E = (9.273 \times 10^{-24})B - (-9.273 \times 10^{-24})B$$

$$\Delta E = (1.854 \times 10^{-24}\ JT^{-1})B$$

From the energy difference between the two levels of electron at a given magnetic field, the frequency required for flipping the electron can also be calculated.

$$\Delta E = h\nu$$

$$\nu = \frac{\Delta E}{h}$$

$$\nu = \frac{(1.854 \times 10^{-24}\ JT^{-1}) \times B(T)}{6.626 \times 10^{-34}\ Js}$$

$$\nu = (2.86 \times 10^{10}\ s^{-1}\,T^{-1})B$$

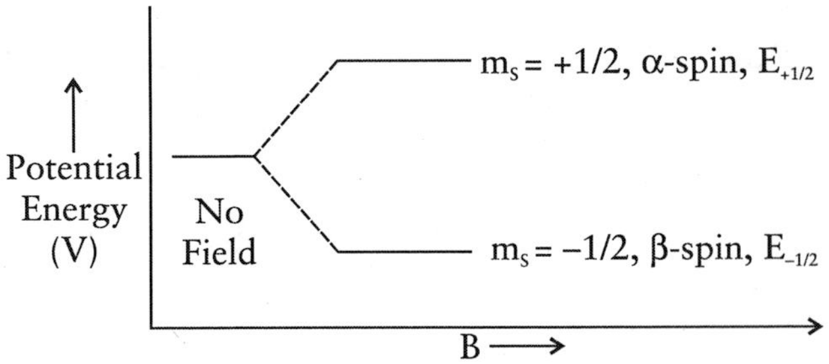

Fig. 8.6: Potential energy of an electron in α and β-states

Thus, an electron in the lower energy level (m_s = –1/2) can be promoted to the higher energy level (m_s = +1/2) in the presence of a magnetic field, B, by supplying the necessary energy in the form of electromagnetic radiation of frequency, ν, which is $(2.86 \times 10^{10}\ s^{-1}\ T^{-1})B$.

The magnitude of the applied magnetic field, B, normally used in most of ESR experiments is 0.3 Tesla. Thus for the field strength of 0.3 Tesla, the frequency of radiation required for the flipping of spin of electron is

$$\nu = (2.86 \times 10^{10}\ s^{-1}\ T^{-1}) \times (0.3\ T)$$

$$\nu = 8.58 \times 10^{9}\ s^{-1}$$

$$\nu = 8.58 \times 10^{9}\ Hz.$$

The above frequency lies in the microwave region of the electromagnetic spectrum $(3 \times 10^{6} - 3 \times 10^{10}$ Hz).

Thus, on comparing the energy difference between the two levels of an electron in ESR with the energy difference between the two levels of proton in NMR, we find that the former is larger than the latter. The reason behind this is the smaller mass of the electron relative to that of a proton.

PROBLEMS BASED ON PRECESSIONAL FREQUENCY AND MAGNETIC FIELD STRENGTH

Q. 1. ESR spectrum of atomic hydrogen is observed with a spectrometer operating at 9.5 GHz. If the Lande g-factor is 2.0026. Calculate the value of the external magnetic field required to record the spectrum ($\mu_B = 9.273 \times 10^{-24}\ JT^{-1}$).

Ans. $\nu = 9.5\ GHz = 9.5 \times 10^{9}\ s^{-1}$

$g = 2.0026.$

$\mu_B = 9.273 \times 10^{-24}\ JT^{-1}$

$\Delta E = h\nu$

$g\mu_B B = h\nu$ (using equation 8.10)

$$B = \frac{h\nu}{g\mu_B}$$

$$\frac{(6.626 \times 10^{-34}\ s) \times (9.5 \times 10^{9}\ s^{-1})}{(2.0026) \times (9.273 \times 10^{-24}\ T^{-1})}$$

$B = 0.339\ T$

Q. 2. The Lande g-factor of benzene radical anion is 2.0025. Calculate the value of strength of external magnetic field required for resonance in a spectrometer operating at 9.302 GHz.

Ans. $\nu = 9.302\ GHz = 9.302 \times 10^{9}\ s^{-1}$

$g = 2.0025$

$\mu_B = 9.273 \times 10^{-24}\ JT^{-1}$

$$B = \frac{h\nu}{g\mu_B}$$

$$B = \frac{(6.6 \times 10^{-34}\,Js) \times (9.302 \times 10^{9}\,s^{-1})}{(2.0025) \times (9.273 \times 10^{-24}\,JT^{-1})}$$

$B = 0.332$ T

Q. 3. Calculate the ESR frequency of an unpaired electron in a magnetic field strength of 0.33 T. If the Lande g-factor for the free electron is 2 ($\mu_B = 9.273 \times 10^{-24}$ JT^{-1}).

Ans. $$\nu = \frac{g\mu_B B}{h}$$

$$\nu = \frac{(2) \times (9.273 \times 10^{-24}\,JT^{-1}) \times (0.33T)}{6.626 \times 10^{-34}\,Js}$$

$\nu = 9000 \times 10^{6}\ s^{-1}$

$\nu = 9000 \times 10^{6}$ Hz

$\nu = 9 \times 10^{9}$ Hz

$\nu = 9$ GHz

Q. 4. Calculate the magnetic field at which the spectrometer is operating, if the ESR frequency of the free electron is 9.5 GHz.

Ans. $$B = \frac{h\nu}{g\mu_B}$$

$$B = \frac{(6.626 \times 10^{-34}\,Js) \times (9.5 \times 10^{9}\,s^{-1})}{(2) \times (9.273 \times 10^{-24}\,JT^{-1})}$$

$B = 0.339$ T

8.8 Relative Intensity of Energy Levels: Intensity of ESR Signal

The relative population of electrons in energy levels is given by the Boltzmann equation.

$$\frac{n_2}{n_1} = \exp\left(-\frac{\Delta E}{kT}\right)$$

where, n_1 and n_2 are the number of electrons in the lower and higher energy levels respectively. ΔE is the energy difference between the two levels and k is the Boltzmann constant ($k = R/N_A$). The relative population, i.e. the term n_2/n_1, depends on the ratio $\Delta E/kT$.

The excitation of an electron from the lower energy level to the higher energy level can occur only if the lower level is more populated than the higher one. Thus, the lower level is more populated than the upper level.

The population difference in case of ESR is larger than the population difference between the two energy levels in PMR because of the larger value of ΔE in ESR spectroscopy. Consequently, the ESR signals are expected to be more intense than those of PMR signals.

8.9 Presentation of ESR Spectrum

The ESR signals like PMR signals can be achieved by varying the field strength keeping the frequency of the oscillator constant. Thus, the ESR spectrum, similar to NMR spectrum, may be obtained by plotting intensity against the strength of a magnetic field.

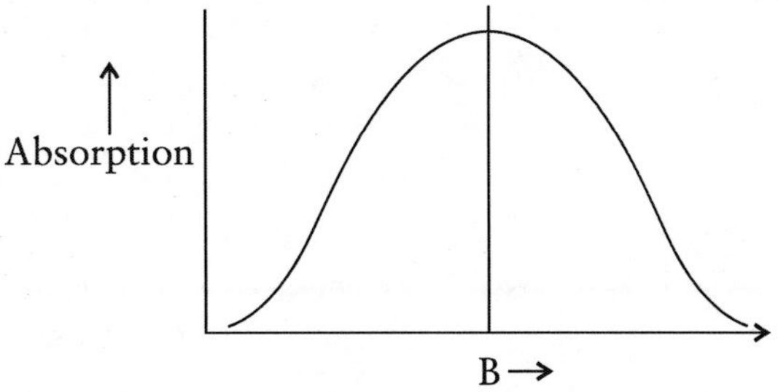

Fig. 8.7: Absorption peak in ESR

Since the ΔE between the two energy levels in case of ESR is large, thus it has larger relaxation time i.e. time required for the electron to revert from the excited level to the ground level would be large. As a result, ESR absorption peak have no fine structure (as shown in Fig. 8.7), i.e. wide signal is obtained in ESR.

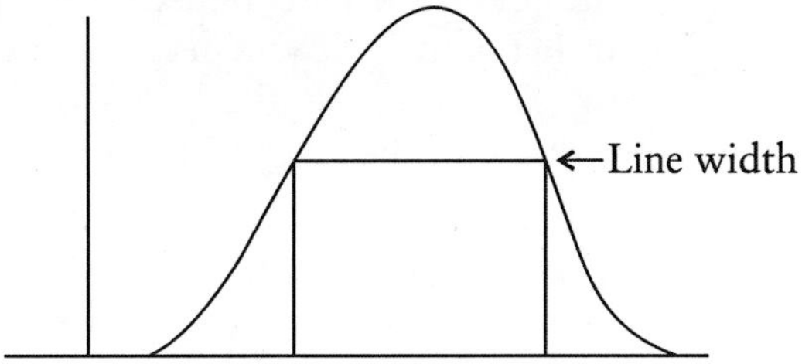

Fig. 8.8: Linewidth of a signal

Also finite linewidth is seen in case of ESR signals. The linewidth of the signal is the width at half height of the absorption signal. The linewidth is quite significant in case of ESR (as shown in Fig. 8.8).

However, in order to improve accuracy, the better way is to represent ESR spectrum as a first derivative of the absorption curve is plotted against the strength of the magnetic field (as shown in Fig. 8.9).

The results represented by derivative curves in ESR can be easily interpreted. Each negative slope in the derivative curve represents a peak or shoulder in the absorption spectrum. Every crossing of the derivative axis with a negative slope indicates a true maximum, whereas a crossing with positive slope indicates a minimum. It means that the number of peaks or shoulders in the absorption curve can be determined from the number of minima or maxima in the derivative curve.

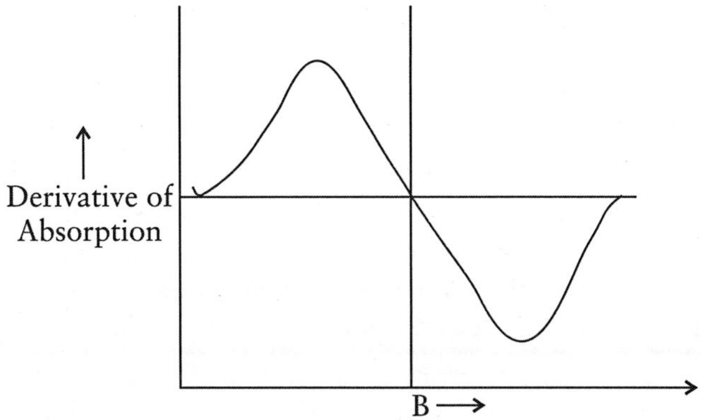

Fig. 8.9: Derivative curve of absorption peak in ESR

PROBLEMS BASED ON PRESENTATION OF ESR SPECTRUM

Q. 1. The ESR spectrum of a compound usually recorded as the derivative of the curve. Explain.

Ans. The ESR spectra of a compound is usually recorded as the first derivative of the curve of absorption against magnetic field because this arrangement gives a better signal-to-noise ratio. The result of ESR spectra interpret that each negative slope in the derivative curve represents a peak or shoulder in the absorption curve. Every crossing of the derivative axis with a negative slope indicates a true maxima, whereas a crossing with positive slope indicates a minimum. Thus, the number of peaks or shoulders in the absorption curve can be determined from the number of minima or maxima in the derivative curve.

Q. 2. Discuss the concept of natural linewidth of spectral line. It is quite significant in case of ESR compared to that of electronic transition. Explain.

Ans. The linewidth is also called as natural linewidth. The linewidth of the signal is the width at half height of the absorption signal.

The linewidth is related to the relaxation time (also called as natural lifetime). Since ΔE between the two energy levels in case of ESR is large, thus it has larger relaxation time, i.e. time, required for the electron to revert from the excited level to the ground level would be large. Therefore, wide signal with significant linewidth is seen in case of ESR.

Q. 3. Presentation of NMR and ESR spectra are different. Explain, why?

8.10 Hyperfine Splitting

The resonance lines in ESR spectroscopy are frequently split in a way similar to spin-spin splitting in NMR. This splitting in ESR is called **hyperfine splitting**. Actually, an unpaired electron can be associated with several atoms with the molecule in different degrees. The interaction of magnetic moment of an electron with those of surrounding magnetically active nuclei results in the splitting of ESR signal, i.e. a single absorption signal splits into more than one signal. This interaction energy is expressed as 'a', where 'a' is a constant term, known as the **hyperfine coupling constant** or the **hyperfine splitting constant**. It represents the extent of interaction between the electron and the interacting nucleus. The dimensions of the splitting are measured in '**Hertz**'.

The selection rule for hyperfine transition is $\Delta m_s = \pm 1$ and $\Delta m_\ell = 0$. This selection rule can be explained on the fact that the nuclear motion is much slower than the electronic motion. Thus, during the time, the electron changes its spin orientation, the nucleus being heavier, have no time to reorient its spin, so that $\Delta m_\ell = 0$.

The ESR spectrum exhibits hyperfine splitting which is caused by the interaction between the spinning electrons and the adjacent spinning magnetic nucleus. When a single electron interacts with one nucleus, the number of splitting will be equal to $(2I + 1)$, where I is the spin quantum number of the nucleus. In general, if a single electron interacts magnetically with 'n' equivalent nuclei, the electron signals will split up into $(2nI + 1)$ multiplet.

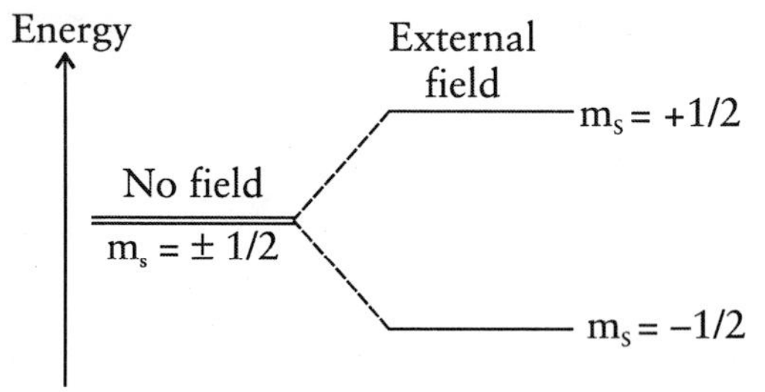

Fig. 8.10: Effect of external magnetic field on the energy states for an electron

Let us consider hyperfine interaction in the following cases:

(1) **Free Electron:** An unpair electron has $S = 1/2$. In the absence of magnetic field, the electron spin energy levels are degenerate and

thus the single electron with S = 1/2 give rise to doubly degenerate spin energy state. When the magnetic field is applied, the degeneracy is removed and two energy levels, one corresponding to $m_s = -1/2$ gets aligned with the field and other field corresponding to $m_s = +1/2$ gets aligned opposing the field (Fig. 8.10).

The ESR spectrum of a free electron would consist of a single peak corresponding to a transition between these levels (Fig. 8.11).

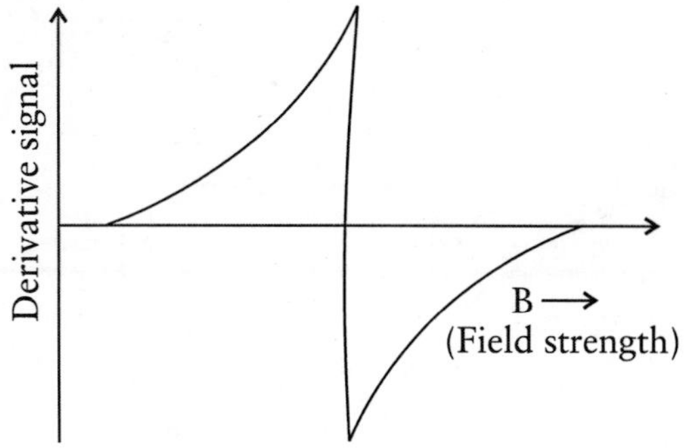

Fig. 8.11: ESR spectrum of an electron

(2) **Hydrogen Atom:** It contains one proton with nuclear spin, I = 1/2 and an unpaired electron with S = 1/2. In the absence of magnetic field, the electron spin energy levels are degenerate and thus the single electron with S = 1/2 gives rise to a doubly degenerate spin energy state. When the magnetic field is applied, the degeneracy is removed and two energy levels, one corresponding to $m_s = -1/2$ aligned with the field and other corresponding to $m_s = +1/2$ aligned opposing the field will be obtained. Each of these energy state of electron then interact with nuclear spin due to proton.

$$(2nI+1) = \left[\left(2\times1\times\frac{1}{2}\right)+1\right] = 2.$$

Thus, each energy state is further split up into two energy levels corresponding to $m_\ell = +1/2$ and $m_\ell = -1/2$. Thus, corresponding to two energy states, four different energy levels are obtained.

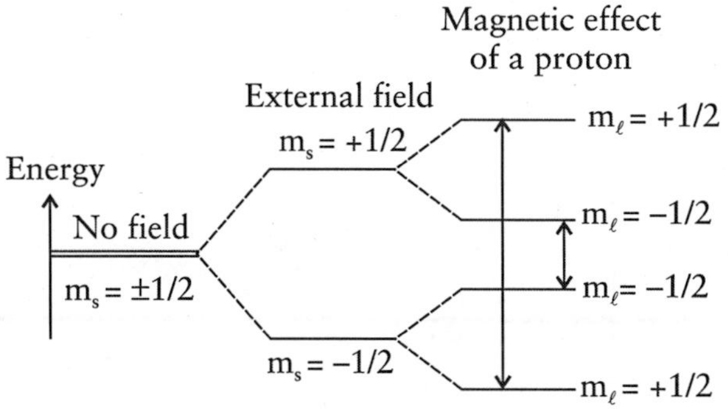

Fig. 8.12: Hyperfine splitting in hydrogen atom

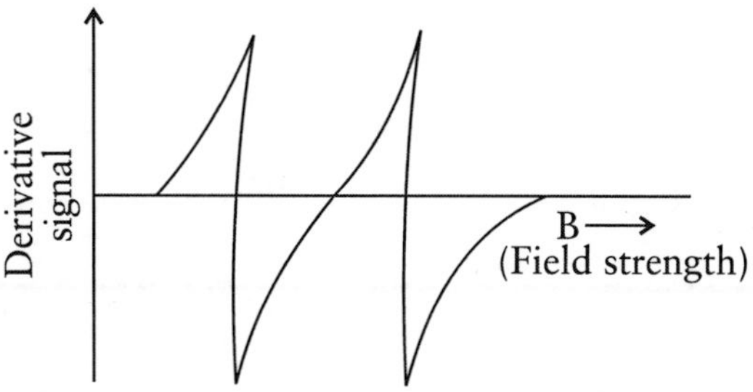

Fig. 8.13: ESR spectrum of hydrogen atom

(3) **Deuterium** (^2_1H): The deuterium is a simple example of a system with I = 1. Energy levels are computed in a similar manner as done in the case of hydrogen atom.

$(2nI + 1) = [(2 \times 1 \times 1) + 1] = [2 + 1] = 3$

Thus, each energy state is further split up into three energy levels corresponding to $m_\ell = +1$, 0 and −1.

Thus corresponding to two energy states, six different energy levels are obtained.

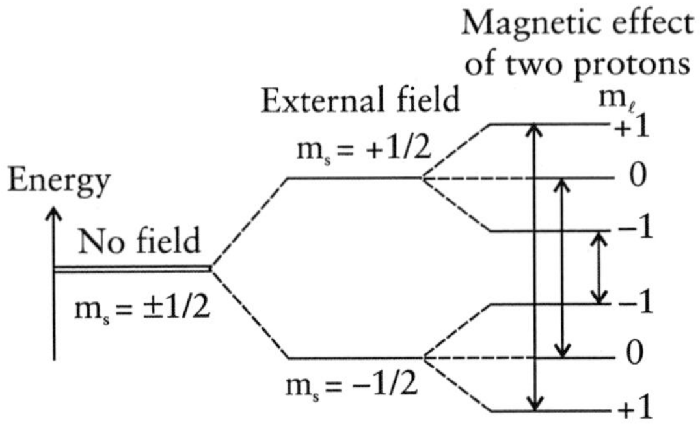

Fig. 8.14: Hyperfine splitting in deuterium

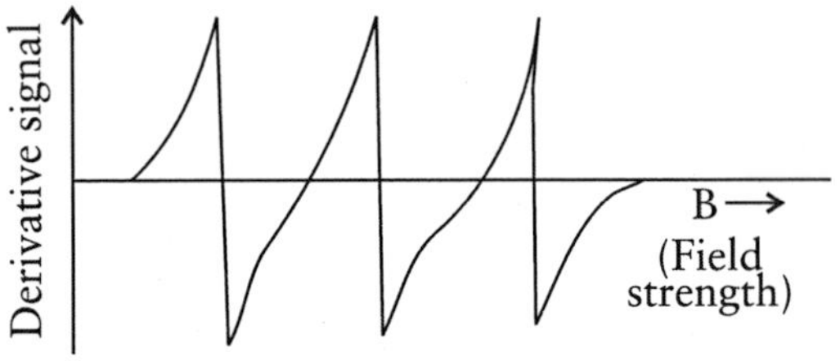

Fig. 8.15: ESR spectrum of deuterium

(4) **Methyl Radical (•CH_3):** It contains an unpaired electron with S = 1/2 and three equivalent protons. Since each proton has nuclear spin, I = 1/2, thus it is evident that for three equivalent protons, I = 3/2.

$$(2nI+1) = \left[\left(2\times3\times\frac{1}{2}\right)+1\right] = (3+1) = 4$$

Thus, when the interaction is considered between the single unpaired electron on the carbon atom and the three hydrogen nuclei, there are four values of m_ℓ, i.e. −3/2, −1/2, +1/2, +3/2 corresponding to each m_s value.

On applying the selection rule, $\Delta m_\ell = 0$ and $\Delta m_s = \pm 1$, four transitions are possible resulting in four lines in the ESR spectrum of methyl radical. The observed relative intensities for the four lines are in the ratio 1:3:3:1. The four transitions that occur in the ESR spectrum of methyl radical (as shown in Fig. 8.16). The corresponding ESR spectrum is shown in Fig. 8.17.

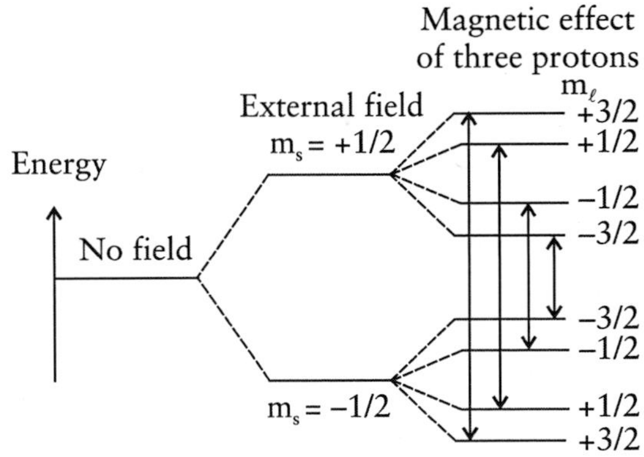

Fig. 8.16: Hyperfine splitting in methyl radical

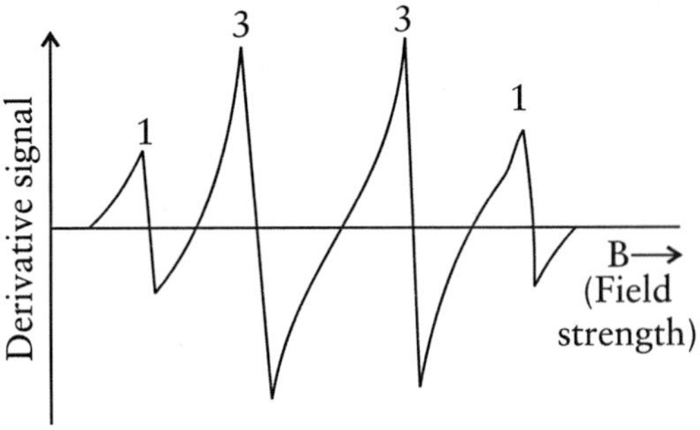

Fig. 8.17: ESR spectrum of methyl radical

(5) **Diethyl Ether Radical ($CH_3\dot{C}HOCH_2CH_3$):** The various hyperfine splittings correspond to coupling with H_3, H and H_2 in order of decreasing 'a_H' values. The γ—CH_3 group is too distant to contribute further splitting. Since for proton, I = 1/2, thus

$$(2nI+1) = \left[\left(2n\times\frac{1}{2}\right)+1\right] = (n+1)$$ i.e. the same (n + 1) rule holds here.

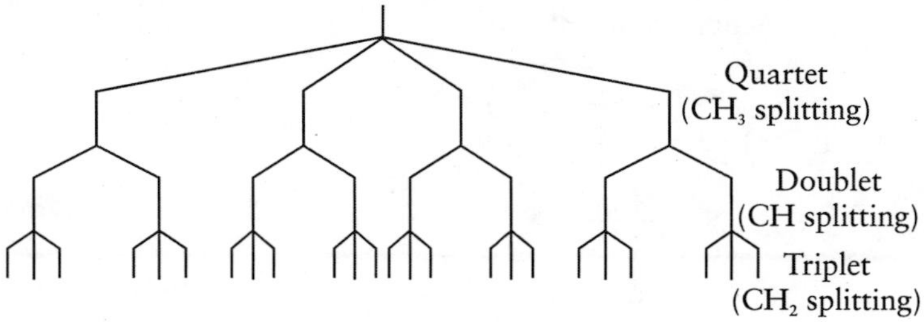

Fig. 8.18: Hyperfine splitting in diethyl ether radical

(6) **Benzene Radical Anion ($C_6H_6^-$):** It gives rise to seven line signal, since the six protons cause splitting, i.e. the same (n + 1) rule holds here, shown in Fig. 8.19.

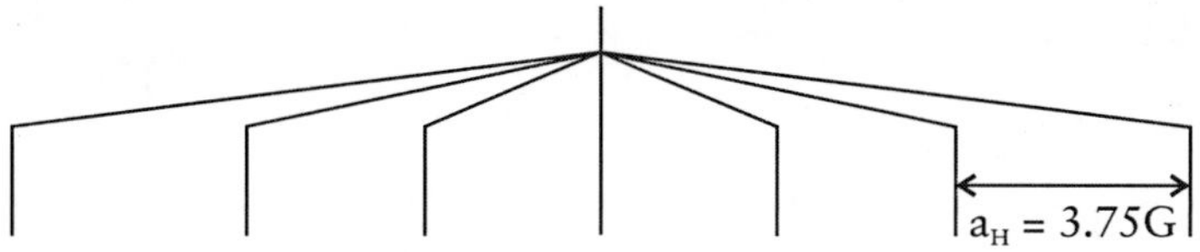

Fig. 8.19: Hyperfine splitting in benzene radical anion

The appearance of seven equally spaced lines in the ESR spectrum of benzene radical anion is proof that all six hydrogens are equivalent and therefore that the unpaired electron is coupling equally with all six hydrogens, rather than localized at one particular position. Since ^{12}C is non-magnetic, no further hyperfine splitting is produced.

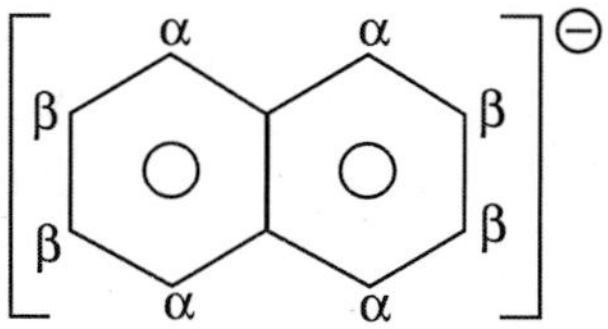

(7) **Naphthalene Anion:** It contains two sets (α and β) of four equivalent protons each. The ESR spectrum would thus show

$[(4 + 1) \times (4 + 1)] = 5 \times 5 = 25$ lines.

(8) **Anthracene Anion:** It contains three sets of equivalent protons: two sets (α and β) consists of four equivalent protons each and third set γ consists of two equivalent protons. The ESR spectrum consists of

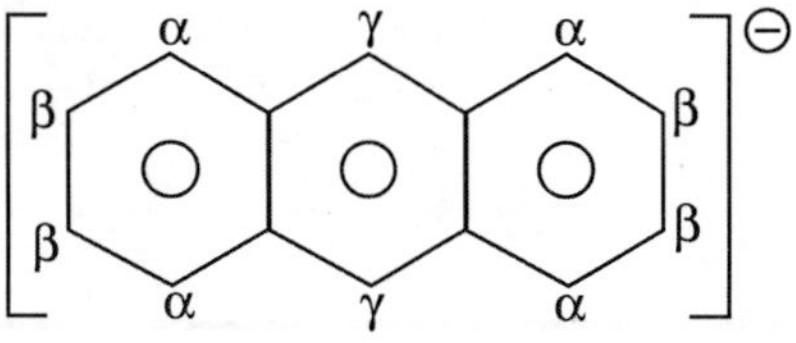

$[(4 + 1) \times (4 + 1) \times (2 + 1)] = 5 \times 5 \times 3 = 75$ lines.

(9) **$(SO_3)_2NO^-$ Radical Anion:** In this anion, the unpaired electron interacts with the nitrogen nucleus (I = 1). The ESR spectrum would thus show

$(2nI + 1) = [(2 \times 1 \times 1) + 1] = 2 + 1 = 3$

Thus, each energy state is further split up into three energy levels corresponding to $m_\ell = 1, 0, -1$.

Thus, ESR spectrum contains three equally spaced lines, i.e. triplet.

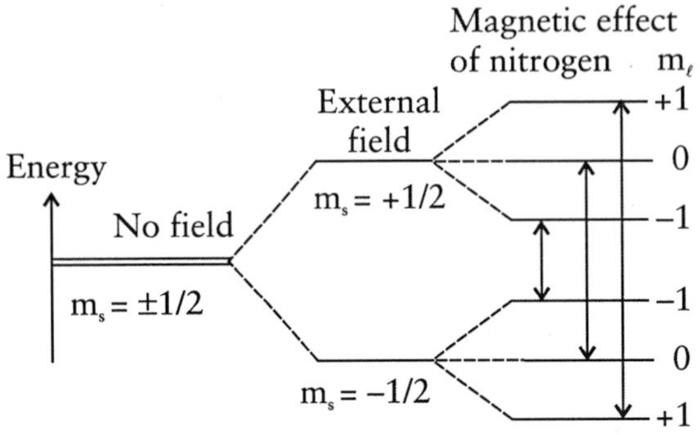

Fig: 8.20: Hyperfine splitting in $(SO_3)_2$ NO^- radical anion

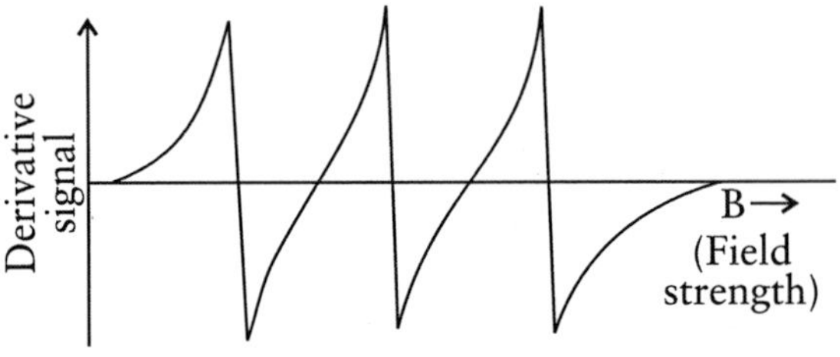

Fig: 8.21: ESR spectrum of $(SO_3)_2$ NO^- radical anion

(10) **p-Benzoquinone Radical Anion:** The anion has the structure 

It has one unpaired electron (S = 1/2), interacting with four equivalent protons. Since each proton has nuclear spin, I = 1/2, thus it is evident that for three equivalent protons, I = 3/2

$$(2nI+1) = \left[\left(2\times 4\times \frac{1}{2}\right)+1\right] = 4+1 = 5$$

Thus, there are five values of m_ℓ = + 2, + 1, 0, −1, −2. The ESR spectrum contains a quintet, i.e. five equally spaced lines. The observed relative intensities for the five lines are in the ratio 1:4:6:4:1.

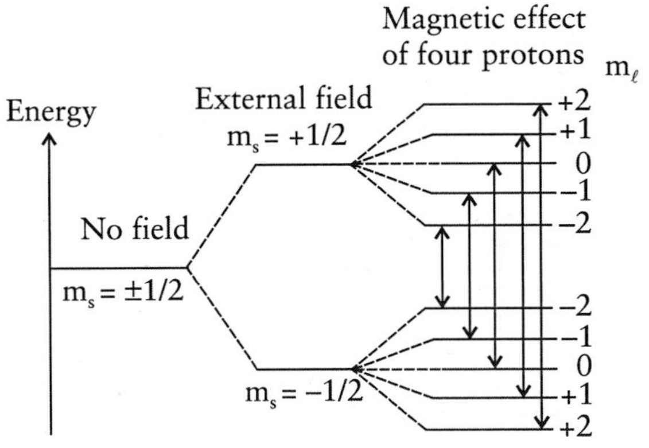

Fig: 8.22: Hyperfine splitting in p-benzoquinone radical anion

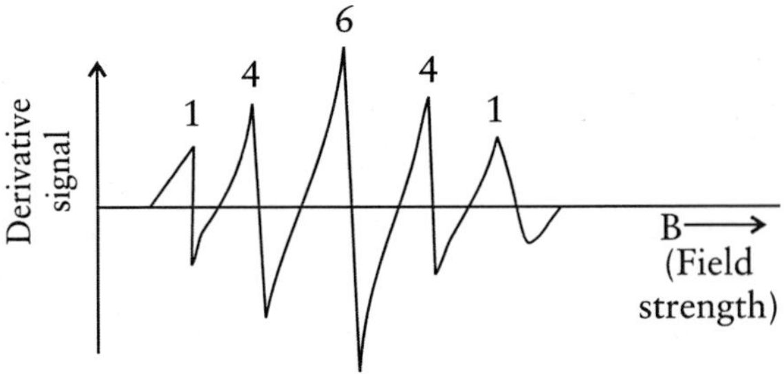

Fig: 8.23: ESR spectrum of p-benzoquinone radical anion

PROBLEMS BASED ON HYPERFINE SPLITTING

Q. 1. Sketch the hyperfine ESR spectrum of the $\dot{C}H_3$ (methyl radical) and explain by giving the intensities of the peaks.

Q. 2. Sketch the ESR spectrum of p-benzoquinone anion radical.

8.11 Difference between NMR and ESR/EPR

NMR	ESR
1. NMR deals with the interaction of positively charged proton with the external magnetic field.	1. ESR deals with the interaction of negatively charged electron with the external magnetic field.
2. NMR signals are plot of absorption against field strength. 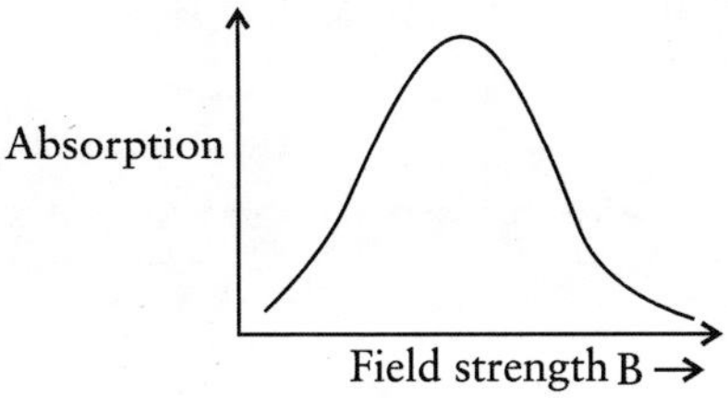	2. ESR spectra are plotted as first-derivative spectra, i.e. ESR signals are a plot of the rate of change of absorption against field strength. 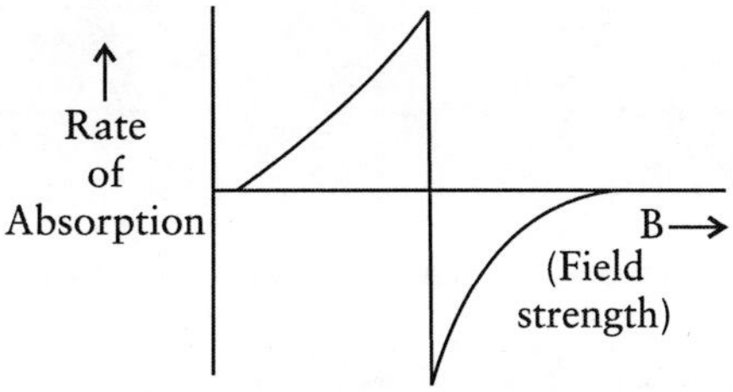
3. NMR absorptions are narrow.	3. ESR absorptions are broad compared to NMR.
4. A molecule may contain several ^1H nuclei and give rise to several ^1H NMR signals (usually split by coupling).	4. A radical contains only one unpaired electron and therefore give rise to only one signal (split by coupling).
5. In NMR, the chemical shift or resonance signals are expressed as 'δ' values which is a function of both field strength and frequency.	5. In ESR, the resonance positions are expressed as 'g' values, which is also a function of field strength and frequency. Actually, 'g' is a proportionality constant and has the value of 2.0023 for an unbound electron. When the unpaired electron is present in an organic substrate, magnetic interactions will shift 'g' from this value.

Index

E

F

M

N

O

P

Q

R

U

V

W

Z